Pseudocereals

Pseudocereals: Production, Processing, and Nutrition provides an overview of the chemistry, processing, and technology of pseudocereals which have become super grains. The cultivation of pseudocereals has spread to over 70 different countries due to their attractive nutritional properties and for food security. This book discusses necessary information on different pseudocereals as well as practical information on cultivation procedures, equipment, food processing using pseudocereals, and the use of by-products for bioactive compound extraction. It addresses concerns regarding globalization, food security, climate change, and the needs of underdeveloped or developing countries.

Key Features:

- Covers both common as well as less exploited pseudocereals
- Explains the grain structure and engineering properties of different pseudocereals
- Studies the effect of food processing on the bioactivity and nutritional value of pseudocereals and their products

Pseudocereals

Production, Processing, and Nutrition

Edited by
Sanju Bala Dhull, Aarti Bains, Prince Chawla, and Sawinder Kaur

CRC Press
Taylor & Francis Group
Boca Raton London New York

CRC Press is an imprint of the
Taylor & Francis Group, an **informa** business

Designed cover image: Shutterstock

First edition published 2024
by CRC Press
2385 NW Executive Center Drive, Suite 320, Boca Raton FL 33431

and by CRC Press
4 Park Square, Milton Park, Abingdon, Oxon, OX14 4RN

CRC Press is an imprint of Taylor & Francis Group, LLC

Library of Congress Cataloging-in-Publication Data

Names: Dhull, Sanju Bala, editor. | Bains, Aarti, editor. | Chawla, Prince, editor. | Kaur, Sawinder, editor.
Title: Pseudocereals : production, processing, and nutrition / edited by Sanju Bala Dhull, Aarti Bains, Prince Chawla, Sawinder Kaur
Description: First edition | Boca Raton, FL : CRC Press, 2024 | Includes bibliographical references and index
Identifiers: LCCN 2023039209 (print) | LCCN 2023039210 (ebook) | ISBN 9781032350967 (hardback) | ISBN 9781032350974 (paperback) | ISBN 9781003325277 (ebook)
Subjects: LCSH: Grain. | Quinoa. | Grain amaranths. | Buckwheat.
Classification: LCC SB189 .P777 2024 (print) | LCC SB189 (ebook) | DDC 633.1–dc23/eng/20240130
LC record available at https://lccn.loc.gov/2023039209
LC ebook record available at https://lccn.loc.gov/2023039210

ISBN: 978-1-032-35096-7 (hbk)
ISBN: 978-1-032-35097-4 (pbk)
ISBN: 978-1-003-32527-7 (ebk)

DOI: 10.1201/9781003325277

Typeset in Garamond
by Deanta Global Publishing Services, Chennai, India

Contents

v

Preface

Pseudocereals have attracted significant attention in recent years due to their nutritional advantages, adaptability to various agroclimatic conditions, and diverse culinary applications. Quinoa, amaranth, buckwheat, and chia have attracted the interest of farmers, researchers, and consumers worldwide. These pseudocereals possess exceptional nutritional profiles, containing varying concentrations of essential amino acids, dietary fiber, minerals, and vitamins. Moreover, they are widely embraced by individuals with specific dietary requirements, including those on gluten-free or plant-based diets. On the other hand, pseudocereals surpass cereals in terms of both protein quality and quantity. They claim elevated levels of lysine, an essential amino acid that is relatively limited in cereals. Amaranth and quinoa, in particular, hold significance in children's nutrition due to their high content of arginine and histidine. These two amino acids play a crucial role in the growth and development of infants and children. Therefore, incorporating amaranth and quinoa into the diets of young people can provide them with essential nutrients for optimal health. Traditionally, pseudocereals have not been afforded their due importance compared to some other crops such as soybean and staple crops, such as wheat, corn, and rice. However, in recent years, these pseudocereal food crops have gained increased importance due to their value according to their balanced nutritional profile and health benefits. Moreover, pseudocereal flours and ingredients are finding new uses in diverse food applications, especially gluten-free food products with enhanced nutritional and sensory properties. A number of factors affect the utilization of pseudocereals, such as type and cultivar selection, agronomic and storage conditions, handling infrastructure, processing methods, and final product preparation. Furthermore, nutrient content and bioavailability are also dramatically influenced by these different factors. In recent years, pseudocereals have been cited for imparting several health benefits, including obesity management, anti-diabetic, hypocholesterolemic response, and mitigation of colorectal cancer. Enhanced pseudocereal utilization, focused on improved dietary health, is an opportunity within both subsistent and developed populations.

This book provides a contemporary source of information that brings together current knowledge and practices in the value chain of pseudocereal production, processing, and nutrition. This work provides in-depth coverage of a wide variety of pertinent topics: production, composition, processing technologies, quality, nutritional profile, significance to human health, and food applications. An experienced team of over 30 contributors has written 15 chapters. These contributors come from a field of diverse disciplines, including horticulture, food science and technology, food biochemistry, food engineering, and nutritional sciences. Of the 15 chapters in the book, Chapters 1 and 2 discuss the cultivation strategy, production, and engineering aspects of different pseudocereals. Further, Chapters 3, 4, and 5 focus on the nutritional profile, processing, and utilization of the most commonly utilized pseudocereals, i.e., quinoa, amaranth, and buckwheat. Chapters 6 and 7 explore different ingredients, such as starches, gums, and proteins extracted from different pseudocereals, while Chapter 8 discusses different minerals and their bioavailability in pseudocereals. Chapter 9 explores the health benefits of pseudocereals, while Chapter 10 compiles information on the processing of the pseudocereals. The machines and equipment used for pseudocereal processing are discussed in Chapter 11. In addition, Chapters 12 and 13 discuss the traditional and recent trends in the utilization of pseudocereals. The role of pseudocereals in the management of different human diseases is explored in Chapter 14. Finally, Chapter 15 covers the quality management system for pseudocereals. Overall, this value-chain approach to the presented topics is a distinctive feature of this book. The editors acknowledge many individuals for their support from conception through to the final development of this book. Foremost are our sincere thanks and gratitude to all authors for their contributions and for bearing with us during the review and finalization process of their chapters. We are grateful to our family members for their understanding and support, enabling us to complete this work. We welcome any suggestions and comments from readers for future improvements of the book in subsequent editions.

Sanju Bala Dhull
Aarti Bains
Prince Chawla
Sawinder Kaur

About the Editors

Dr. Sanju Bala Dhull is an associate professor in the Department of Food Science and Technology, Chaudhary Devi Lal University, Sirsa, India, having more than 15 years of teaching and research experience. Her areas of interest include characterization and modification of biomolecules such as starches and gums, preparation and application of edible films, hydrogels, nanoparticles, nanoemulsions, and new product development. She has published five books, more than 50 research papers in different journals, and 25 chapters in books of national and international repute. She has presented more than 25 research papers at various national and international conferences. She is a life member of the Association of Food Scientists and Technologists (India) and the Association of Microbiologists of India. She is Associate Editor of the *Journal of Food Processing and Preservation* (Hindawi) and a member of the editorial board of *Legume Science* (Wiley). She is a reviewer of several national and international journals.

Dr. Aarti Bains is an assistant professor in the Department of Microbiology, Lovely Professional University, Phagwara, Punjab, India. She holds an M.Sc. in Microbiology from Himachal Pradesh University and completed her MPhil and Ph.D. in Microbiology at Shoolini University, Solan, India. Her area of interest includes the extraction and identification of bioactive compounds from wild edible mushrooms and their biological activity. She has published ten research papers and four international book chapters. She has eight years of research and teaching experience and has guided seven M.Sc. students. She is a reviewer of several national and international journals.

Dr. Prince Chawla recently joined Lovely Professional University, Phagwara, Punjab, India as an assistant professor in food technology and nutrition (School of Agriculture). Dr. Chawla is an alumnus of Chaudhary Devi Lal University, Sirsa, and Shoolini University, Solan. Dr. Chawla has a chief interest in mineral fortification, functional foods, protein modification, and the detection of adulterants in foods, using nanotechnology. He has worked on research projects funded by the Department of Biotechnology and the Department of Science and Technology and has eight years of research experience. He is the inventor on three patents, and the author/co-author of two books, 30 international research papers, and seven chapters in international books. Dr. Chawla is a recognized reviewer for more than 30 international journals.

Dr. Sawinder Kaur is an associate professor and Head of Department, Food Technology and Nutrition (School of Agriculture), Lovely Professional University, Phagwara, Punjab, India. She has published more than 50 research articles and chapters in journals and books of national and international repute. Her research areas are mainly focused on new product development, extraction of bioactive compounds, and natural food colorants. She has seven patent applications to her credit. She has participated in various national and international research projects. She is a referee and recognized reviewer for several international and national journals.

List of Contributors

Rohit Biswas
Department of Agricultural and Food
  Engineering
Indian Institute of Technology
Kharagpur, West Bengal, India

Subhankar Das
Biotechnology Unit
Mangalore University
Karnataka, India

Arpan Dubey
Department of Agricultural and Food
  Engineering
Indian Institute of Technology
Kharagpur, West Bengal, India

Vidhi Gupta
Department of Agricultural and Food
  Engineering
Indian Institute of Technology
Kharagpur, West Bengal,
  India

Manjula Ishwara Kalyani
Department of Microbiology
Mangalore University Janna
  Kaveri
Kodagu-Mangalore, Karnataka,
  India

Ashwani Kumar
Department of Postharvest Technology
College of Horticulture and Forestry
Rani Lakshmi Bai Central Agricultural
  University
Jhansi, India

Mukul Kumar
Department of Food Technology and
  Nutrition
Lovely Professional University
Phagwara, Punjab, India

Prakash Kumar
Department of Agricultural and Food
  Engineering
Indian Institute of Technology
Kharagpur, West Bengal, India

Vikas Kumar
University Institute of Biotechnology
Chandigarh University
Gharuan, Mohali, Punjab, India

Smita Mall
Faculty of Applied Sciences and
  BiotechnologyShoolini University
  of Biotechnology and Management
  Sciences
Bajhol, PO Sultanpur, Solan District,
  Himachal Pradesh, India

Danish Shafi Mir
University Institute of Biotechnology
Chandigarh University
Gharuan, Mohali, Punjab, India

Atul Anand Mishra
Department of Processing and Food
    Engineering
Sam Higginbottom University of
    Agriculture Technology and
    Sciences
Prayagraj, Uttar Pradesh, India

Indumathi Mullaiselvan
Department of Food Technology
JCT College of Engineering and
    Technology
Coimbatore, India

Baneeprajnya Nayak
Department of Processing and Food
    Engineering
Sam Higginbottom University of
    Agriculture Technology and Sciences
Prayagraj, Uttar Pradesh, India

Praveen Nayak
Department of Processing and Food
    Engineering
Sam Higginbottom University of
    Agriculture Technology and Sciences
Prayagraj, Uttar Pradesh, India

Puja Nelluri
Department of Agricultural and Food
    Engineering
Indian Institute of Technology
Kharagpur, West Bengal, India

Baghya Nisha
Department of Food Technology
Manakulavinayagar College of
    Engineering and Technology
Pondicherry, India

Harshvardhan Patel
Department of Food Technology and
    Nutrition
Lovely Professional
    University
Phagwara, Punjab, India

Nikhil Patil
Department of Food Technology and
    Nutrition
Lovely Professional University
Phagwara, Punjab, India

Abhishek Pradhan
Department of Agricultural and Food
    Engineering
Indian Institute of Technology
Kharagpur, West Bengal, India

Alpana Prajapati
Department of Processing and Food
    Engineering
Sam Higginbottom University of
    Agriculture Technology and Sciences
Prayagraj, Uttar Pradesh, India

Debapam Saha
Department of Agricultural and Food
    Engineering
Indian Institute of Technology
Kharagpur, West Bengal, India

Ayan Sarkar
Department of Agricultural and Food
    Engineering
Indian Institute of Technology
Kharagpur, West Bengal,
    India

Madhu Sharma
Department of Food Technology and
    Nutrition
Lovely Professional University
Phagwara, Punjab, India

Nitin Sharma
Department of Biotechnology
Chandigarh College of Technology
CGC Landran, Mohali, Punjab,
    India

Rama Nath Shukla
Department of Processing and Food
    Engineering
Sam Higginbottom University of
    Agriculture Technology and
    Sciences
Prayagraj, Uttar Pradesh, India

Ajay Kumar Singh
Department of Processing and Food
    Engineering
Sam Higginbottom University of
    Agriculture Technology and Sciences
Prayagraj, Uttar Pradesh, India

Neha Singh
Rajendra Mishra School of Engineering
    Entrepreneurship
Indian Institute of Technology
Kharagpur, West Bengal, India

K. Sneha
Department of Food Technology and
    Nutrition
Lovely Professional University
Phagwara, Punjab, India

Punyadarshini Punam Tripathy
Department of Agricultural and Food
    Engineering
Indian Institute of Technology
Kharagpur, West Bengal, India

Chapter 1

Cultivation and Production of Pseudocereals

Harshvardhan Patel, Madhu Sharma,
Aarti Bains, Prince Chawla

1.1 Introduction

Pseudocereals have received a lot of interest in recent years because of their nutritional benefits, adaptability to different agroclimatic conditions, and various culinary uses. This chapter focuses on the growing and production practices of four popular pseudocereals, quinoa, amaranth, buckwheat, and chia (Angeli et al. 2020). These are some of the best-known pseudocereals that have piqued the interest of farmers, academics, and consumers all over the world. These crops have outstanding nutritional profiles, with different concentrations of essential amino acids, dietary fiber, minerals, and vitamins. Furthermore, they are frequently acceptable to people with specific dietary demands, such as those who follow gluten-free or plant-based diets. The chapter opens with a review of pseudocereal cultivation and production practices, emphasizing each crop's unique qualities and requirements. Readers will acquire insights into the subtleties of farming these crops from understanding the special agroclimatic conditions favorable for their growth to researching sowing strategies and growth stages. Quinoa, a native of the Andes, has achieved worldwide recognition due to its outstanding nutritional characteristics and tolerance of a wide range of temperatures. Its cultivation has spread beyond its traditional locations, with farmers worldwide exploring its possibilities (Perez et al. 2015). The chapter delves into quinoa production data around the world,

DOI: 10.1201/9781003325277-1

showing the top-producing countries and their contributions to global production. It also looks at quinoa cultivation in India, covering its growth, problems, and contribution to global output. Another notable pseudocereal is amaranth, which has cultural and historical significance in many countries (Malik et al. 2023). Its cultivation has grown in popularity because of its adaptability to a variety of soil types and climates (Filho et al. 2017). The chapter investigates amaranth cultivation and production practices, offering insights into its development requirements and the possibilities for its incorporation into various agricultural systems. Buckwheat, with its distinctive triangular seeds, has been grown for generations in both temperate and subtropical climates. The chapter delves into buckwheat cultivation procedures, such as soil requirements, planting practices, and post-harvest processing (Joshi 2023). It investigates the crop's flexibility, from its use as a staple food to its value as a cover crop and its potential in specialist markets. Chia, a crop recognized for its omega-3 fatty acid content and water absorption capabilities, has gained popularity as a superfood (Saini et al. 2021; Punia and Dhull, 2019). The chapter examines its cultivation and production methods, emphasizing the necessity for understanding its special growing requirements, notably in terms of soil moisture management and harvesting procedures. Quinoa, amaranth, buckwheat, and chia cultivation and production have attracted global attention for their nutritional benefits and diversity. While statistics on output differ by country, Peru and Bolivia have historically led global quinoa production, and Russia and China are major buckwheat growers, but country-specific production estimates for amaranth and chia are relatively limited (Kumar et al. 2021). India's share of the worldwide pseudocereal business is rapidly increasing. With attempts to boost cultivation and increase output, India has emerged as a major player in quinoa production, and efforts are being made to improve amaranth and buckwheat cultivation and production throughout the country (Ebert 2022). This chapter seeks to offer readers a full overview of the global pseudocereal market and India's role within it by examining global production data and India's contributions. The chapter emphasizes the relevance of pseudocereals in diversifying agricultural systems, ensuring food security, and satisfying global nutritional needs. Pseudocereals are under-utilized crops that are gluten-free, high in protein, and rich in important phytochemicals, such as saponins, which have a variety of agricultural, pharmaceutical, and industrial applications (Mir et al. 2018; Dhull et al. 2020a, 2020b, 2020c). Saponins have hemolytic and antilipemic properties, which can help decrease cholesterol levels in blood serum, but have no deleterious impact on protein digestibility.

1.2 Quinoa (*Chenopodium quinoa*)

Quinoa (*Chenopodium quinoa* Willd.), pronounced "KEEN-waa," is a dicotyledonous plant that originates from South America, making it known as an Andean grain. It has been a part of the diet of Andean cultures for seven millennia, dating back to pre-Columbian times (Folk 2020). Despite its grain-like seed shape, quinoa is a pseudocereal. It is classified into five ecotypes based on their geographic adaptation at the center of diversity. These ecotypes are Valley, Altiplano, Salares, Sea-level, and Subtropical or yungas, each adapted to different altitudes and environmental conditions in Colombia, Ecuador, Peru, Bolivia, and Chile (Roman 2021). Quinoa is highly resilient and offers superior nutritional quality compared with common cereals. It thrives in environments where other crops struggle to grow, showcasing its ability to withstand adverse conditions. The designation of 2013 as the International Year of Quinoa by the UN/FAO has contributed to increased global interest in this crop (Murphy and Matanguihan 2015). Quinoa's balanced nutritional profile and its potential as a sustainable alternative for feeding the growing world population, especially in marginal environments, have attracted attention. The natural selection process of quinoa cultivars occurred in the challenging conditions of the Andes, where limited rainfall, extreme aridity, and salt-affected soils prevailed. Consequently, quinoa has built-in tolerance to abiotic stresses such as aridity, salinity, high-altitude environments, and frost, making it well-suited to marginal lands. However, high-temperature stress, particularly during flowering, can negatively affect seed set and ultimately yield. Although quinoa can tolerate a wide temperature range, temperatures above 35°C pose a significant risk to its productivity (Sadok and Jagadish 2020). Quinoa is a pseudocereal known for its high protein content and tolerance to a variety of growth environments.

The timeline of quinoa's global growth varies, although commercial cultivation beyond its native regions garnered substantial attention in the late twentieth and early twenty-first centuries (Jan et al. 2023). During the 1980s and 1990s, countries such as the United States and Canada experimented with quinoa farming. However, quinoa production did not achieve significant global traction until the early 2000s. Quinoa farming is a relatively new occurrence in India but is rapidly spreading. It is grown mostly in areas with favorable weather conditions. Quinoa production in India began to increase in the early 2000s, with trials and research activities taking place across the country (Zvelebil and Pluciennik 2011). States like Uttarakhand, Himachal Pradesh, Jammu and Kashmir, and Sikkim are known in India for their quinoa cultivation. Quinoa cultivation conditions vary based on locale, but there are some fundamental requirements. Quinoa is a hardy crop that can thrive under a range of conditions, from harsh high-altitude Andean highlands to coastal places. It does well in temperate to subtropical climates, with ideal temperatures ranging from 15 to 25°C (59 to 77°F) (Hussain et al. 2020). Quinoa grows best in well-drained sandy or loamy soils with a pH of 6 to 8. It is tolerant of environments ranging from slightly acidic to mildly alkaline. The crop requires a modest quantity of rainfall each year, between 300 and 1000 millimeters (12 and 39 inches), whereas excessive wetness might inhibit growth (Peterson and Murphy 2015). Adequate drainage is critical in locations with heavy rainfall, to avoid waterlogging. Quinoa is also classified as a day-neutral plant, which means that its flowering is unaffected by daylength. This property allows it to be grown at different latitudes. Shorter daylengths, on the other hand, can affect crop growth and maturity (Angeli et al. 2020). Appropriate practices are necessary to cultivate quinoa effectively. Before seeding, the field should be plowed, harrowed, and leveled. The seeds are then sown 25–30 cm apart within the row at 1–2 cm depth. Plants with adequate spacing receive enough sunlight and airflow, lowering the danger of disease outbreaks (Asher et al. 2020). Quinoa cultivation has spread beyond its indigenous Andean regions in South America to other parts of the world, including North and South America, Europe, Asia, and Australia. Bolivia, Peru, Ecuador, and the United States have taken the lead in quinoa production in North and South America, owing to favorable agroecological conditions and traditional knowledge (Mustafa and Temel 2018). Historically, Bolivia and Peru have been the world's major quinoa growers and exporters. Quinoa farming has increased in nations such as France, the Netherlands, Italy, Spain, and the United Kingdom. Because of the high demand for quinoa in European markets, European farmers have conducted trials and tailored their practices to local conditions (Beccari et al. 2021). Quinoa production has also increased in Asia, with countries such as India, China, and Nepal actively farming the grain. India has emerged as a significant quinoa producer, capitalizing on its diverse agroclimatic conditions and the availability of huge dry and semi-arid quinoa production zones (Sharma et al. 2015). Due to growing consumer demand and increased cultivation efforts, quinoa output has been steadily rising on a global scale. With an expected 2020 output of over 134,400 metric tonnes, Peru continues to be the top producer of quinoa, followed by Bolivia with 88,500 metric tonnes, while the United States has seen an upsurge in quinoa production, generating roughly 6,000 metric tonnes, mainly in states like Colorado and California. Ecuador actively participates in the quinoa industry and produces an estimated 15,800 metric tonnes (Bazile et al. 2016). Quinoa output has also increased in Canada, reaching over 1,200 metric tonnes. With Peru and Bolivia being the top producers, both nations are essential to satisfying the world's quinoa demand (Zohry 2020). India offers considerable potential for quinoa cultivation because of its numerous agroclimatic zones. High-altitude and cold-weather regions like Uttarakhand, Himachal Pradesh, and Jammu and Kashmir provide favorable circumstances like quinoa's original Andean habitat (El-Hakim et al. 2022). Quinoa production has gradually increased in India in recent years, owing to farmers' recognition of its nutritional potential and market demand. In 2020, India produced over 800 metric tonnes. Although India's production remains minor in comparison to leading quinoa-producing countries, it reflects the country's expanding position in the global quinoa landscape. The increased growth of quinoa in India provides farmers with an opportunity to diversify crops, increase income generation, and solve nutritional issues (Lozano-Isla et al. 2023). Quinoa has transcended its origins in South America's Andean area and garnered a global reputation as a highly sought-after pseudocereal. Its high-quality nutritional profile, gluten-free nature, and flexibility have elevated it to superfood status in modern diets (Yazar and Kaya 2014). Quinoa cultivation and production are predominantly driven by countries such as Peru, Bolivia, Ecuador, the United States, and Canada, which have established themselves as key quinoa market participants. The arrival of India into the quinoa production landscape is a noteworthy milestone that emphasizes the crop's flexibility and potential for diversification into a range of agroclimatic settings (Matías et al. 2021). Although India's present quinoa production volume is tiny in comparison with other countries, it is likely to rise in the future. Quinoa production in India provides an excellent chance to solve nutritional concerns, enhance farmer livelihoods, and contribute to the global quinoa market. As quinoa demand grows, it is critical to focus on additional research, technical improvements, and sustainable practices to optimize growing techniques, increase productivity, and ensure the long-term sustainability of quinoa production globally (Rodriguez et al. 2020). Exploring and expanding quinoa planting in various places not only contributes to the optimization of global food security but also opens new options for agricultural development, which benefit both farmers and consumers.

1.3 Amaranth (*Amaranthus* spp.)

Amaranth (*Amaranthus* L.) is an ancient crop that has been domesticated and cultivated for nearly 8,000 years, dating back to the Mayan culture in South and Central America. The precise origin of *Amaranthus* is unknown. Around 1,400 years ago, it was frequently used as a staple meal with maize and beans in Mexico, but production dropped after the collapse of the Central American civilization (Stetter et al. 2016). *Amaranthus* spp. are high-protein, high-vitamin, and high-mineral pseudocereals (Malik et al. 2023). Amaranth has regained popularity and is being marketed as a good crop for food and nutritional security due to its nutritional content and capacity to survive a variety of climates (Rodríguez et al. 2020). Amaranth leaves are eaten as vegetables, whereas whole grains are utilized as a cereal. *Amaranthus* Sauer and *Blitopsis* Dumort are the two Sections in the amaranth family (Amaranthaceae). Based on its uses, the genus *Amaranthus* is further classified into grain amaranth, vegetable amaranth, ornamental, and weedy amaranths. Grain amaranth species include *Amaranthus hypochondriacus*, *Amaranthus cruentus*, *Amaranthus caudatus*, and *Amaranthus edulis*, whereas vegetable amaranth is classified in Section *Blitopsis* and includes significant species such as *Amaranthus tricolor* and *Amaranthus lividis* (Martinez-Lopez et al. 2020). Following acknowledgment of amaranth by the United States National Academy of Sciences for its excellent nutritional content and agronomic promise, food technologists are investigating the functional features of amaranth. In India, amaranth is primarily grown in temperate, tropical, and subtropical regions. It requires warm conditions for optimal growth, with a temperature range of 20–35°C (68–95°F). However, some cultivars can withstand higher or lower temperatures depending on their adaptability. Amaranth can tolerate high temperatures but is sensitive to frost. Adequate rainfall or irrigation is necessary during the growing season (Rodríguez et al. 2020). Moreover, amaranth grows well in a variety of soil types (Ramzan and Rehman 2021). It prefers well-drained soils but can tolerate a wide range of soil textures, including sandy, loamy, or clayey soils. Amaranth performs best in fertile soils with good organic matter content and a pH range of 6.0–7.5. It can grow in sandy soils with low fertility as well as in more fertile loamy or clayey soils. Well-drained soils are essential to prevent waterlogging, which can negatively impact plant growth. Soil pH levels ranging from slightly acidic to neutral (pH 6.0–7.5) are suitable for *Amaranthus* cultivation (Woods et al. 2020). Amaranth is a sun-loving crop and requires full sunlight exposure for at least 6–8 hours per day. Adequate sunlight ensures proper photosynthesis and promotes healthy growth and development of the plant. Amaranth has moderate water requirements. It requires regular watering, especially during dry periods or when grown in regions with limited rainfall. However, over-watering should be avoided as it can lead to waterlogging and subsequent damage to the crop. Proper irrigation management is crucial to maintaining optimal soil moisture levels (Chaudhary 2022).

The following points highlight some key aspects of amaranth production worldwide. China is one of the largest producers of amaranth globally, benefiting from its vast agricultural land area and favorable climatic conditions (Schnetzler 2018). India has a long history of amaranth cultivation and is a significant contributor to global production. Various states in India, including Uttar Pradesh, Bihar, Maharashtra, Gujarat, and Rajasthan, are known for their amaranth cultivation. Russia has been steadily increasing its amaranth production in recent years, leveraging its vast land resources and favorable climate for cultivation. Ukraine has emerged as a notable amaranth producer, utilizing its fertile soils and favorable climatic conditions. Peru is one of the leading amaranth-producing countries in South America; the crop has been cultivated in that region for centuries and remains an important part of local agriculture (Serrano-Arellano et al. 2015). The selection of appropriate amaranth varieties, based on regional conditions, and the utilization of local knowledge and resources are crucial for successful cultivation. Local agricultural extension services and research institutions play a vital role in providing guidance and support to farmers in selecting the most suitable varieties and implementing effective cultivation practices (Kulakow 2018). In addition to variety selection, farmers should also consider factors such as water availability, soil fertility, pest and disease management, and harvesting techniques. Proper irrigation management is particularly important for amaranth, as it requires adequate moisture during the growth stages. Collaborative efforts between farmers, researchers, and agricultural experts can lead to the development of region-specific best practices for amaranth cultivation. Sharing knowledge and experiences among farmers and conducting research trials can help refine cultivation techniques and improve overall productivity (Bhunia et al. 2021). By giving due attention to these factors and leveraging local expertise, farmers can maximize the potential of amaranth cultivation, contributing to increased production and meeting the growing demand for this versatile and nutritious crop. In the end, amaranth cultivation and production have grown in popularity both globally and in India. Its adaptability, nutritional value, and market demand have all contributed to its widespread cultivation. Amaranth is important in India as both a traditional crop and a commercial crop, with government support and initiatives helping to increase production (Stetter et al.2017).

1.4 Buckwheat (*Fagopyrum esculentum*)

Buckwheat (*Fagopyrum esculentum* Moench) is a pseudocereal crop that has gained significant attention due to its nutritional value, adaptability to diverse climatic conditions, and multiple uses. In this subsection, we will explore the cultivation and production of buckwheat worldwide, with a specific focus on India. The growth conditions, agricultural practices, and role of buckwheat in the agricultural landscapes of different regions will be examined (Jacquemart et al. 2012).

Buckwheat, a pseudocereal crop, is a member of the Polygonaceae family and the genus Fagopyrum. Buckwheat, despite its name, is not related to wheat and is not a member of the grass family (Farooq et al. 2016). It is more closely linked to rhubarb and sorrel. Buckwheat's scientific name is *Fagopyrum esculentum*. Buckwheat has a long history and has been grown for millennia in various parts of the world. It is thought to have originated in East Asia, specifically in what is now known as China and Tibet. Buckwheat farming extended from there to other parts of Asia, Europe, and, eventually, North America (Yao et al. 2023). Buckwheat is classified into two species: *Fagopyrum esculentum*, often known as common buckwheat, and *Fagopyrum tartaricum* Gaertn., also known as Tartary buckwheat. The more extensively farmed and consumed variety is common buckwheat (*F. esculentum*), while Tartary buckwheat is less widespread but important in some areas (Song et al. 2022). The grain-like seeds of common buckwheat are utilized in a variety of culinary applications. The triangular seeds, also known as buckwheat groats, have a characteristic nutty flavor, and are frequently used in porridge, pancakes, noodles, and as a rice or grain substitute (Lauranne Aubert et al. 2021). Buckwheat groat flour is gluten free and frequently used in gluten-free baking. Tartary buckwheat (*F. tartaricum*) is primarily grown in colder climates, such as China, Russia, and Eastern Europe (Ji et al. 2019). It has higher nutritional content than common buckwheat and is commonly utilized as a food staple in these areas. Tartary buckwheat seeds are smaller and darker in color, with a more bitter flavor. They are frequently roasted or turned into flour and used in classic meal items such as noodles, pancakes, or bread (Aubert and Quinet 2022). Tartary buckwheat and common buckwheat are both noted for their endurance and adaptability to a variety of environmental situations. Because of their ability to suppress weeds and promote soil health, they are frequently used as cover crops. Buckwheat plants are extremely appealing to pollinators, which makes them useful for bee populations (Siracusa et al. 2017). Buckwheat cultivation thrives in cool to moderate climes, making it a favored cool-season crop. It has a relatively short growing season, from 70 to 90 days, allowing it to be grown in places with shorter frost-free seasons. Temperature is critical for buckwheat growth and development. During the growth season, the crop thrives in temperatures ranging from 10°C to 25°C (Dar et al. 2018). Temperatures that are optimal stimulate robust vegetative development, flowering, and seed production. On the other hand, buckwheat is vulnerable to excessive heat and cold. Temperatures above 30°C can limit flowering and seed set, while frost or freezing temperatures can harm the crop. Buckwheat is a short-day plant, which means that daylengths shorter than the critical value are needed for the plant to produce floral hormones, flower and form seeds (Aubert and Quinet 2022). It is classified as a quantitative short-day plant because it may flower later or even fail to produce seeds in regions where the daylength exceeds the critical threshold (Luthar et al. 2021), and its response to photoperiod differs between varieties. Moving on to soil requirements, buckwheat grows best in well-drained soils. Loamy soils with high water-holding capacity are ideal because they retain moisture without causing waterlogging. Buckwheat grows well on a variety of soil types, including sandy loam, loam, or clay loam (Zhou et al. 2022). Heavy clay soils, on the other hand, should be avoided due to poor drainage. Another important aspect in buckwheat cultivation is soil pH. It grows best in slightly acidic to neutral soils with pH levels ranging from 5.0 to 7.0. Soil pH outside of this range can have an impact on nutrient availability and uptake (Singh, Malhotra, and Sharma 2020). If the soil is too acidic, lime may be required to alter the pH for optimal growth. Buckwheat cultivation benefits from elevated soil organic matter content. Organic matter improves soil structure, increases water retention, increases nutrient availability, and promotes beneficial microbial activity. Before planting buckwheat, it is recommended that well-decomposed organic matter, such as compost or well-rotted manure, be incorporated into the soil to improve soil fertility and overall crop performance (Thiyagarajan et al. 2016). Buckwheat cultivation involves careful attention to sowing date, seed rate and spacing, water management, and weed control when it comes to agricultural practices. Buckwheat is usually drilled directly in the field. Planting times vary according to location and can be altered to take advantage of favorable temperature and moisture conditions (Singh et al. 2020). The recommended seed rate for buckwheat ranges from 30 to 40 kilograms per hectare, and spacing between rows is typically maintained at 30 to 45 cm to allow sufficient room for plant growth and airflow. While buckwheat is primarily a rainfed crop, supplementary irrigation may be required in areas with inadequate rainfall during the growing season, particularly during important growth stages like flowering and seed production. Adequate moisture is required for excellent yields. Weed control is also essential for buckwheat cultivation. Weeds can compete for resources with the crop, reducing yields dramatically (Weijuan et al. 2017). Soil preparation before planting, inter-row cultivation, and the use of herbicides (where suitable and using approved recommendations) can assist in managing weed growth and crop health.

It is critical to select buckwheat varieties that are suitable for the local environment and growth circumstances (Amador et al. 2014). Varieties may differ in maturation period, plant height, disease resistance, and yield potential. It is best to choose varieties that are well adapted to the given region and have shown good performance there. Using certified seeds from reputable sources promotes genetic purity and improved crop production because they have been subjected to quality control techniques to ensure seed viability and vigor (Luthar et al. 2021).

Buckwheat, a versatile crop with numerous applications, was grown in several locations throughout the world in 2019 and buckwheat production in the world was roughly 4.7 million metric tonnes that year (Rodríguez et al. 2020). Russia was the biggest contributor among the major buckwheat-growing countries, regularly keeping the top spot in recent years. The broad agricultural practices and favorable environmental circumstances in the country permitted significant buckwheat cultivation, resulting in a significant proportion of global production (Graziano et al. 2022). China, which is known for its rich culinary traditions centered on buckwheat, has emerged as a prominent player in the worldwide market. China, which has a long history of buckwheat growing, contributes significantly to overall global production. Ukraine, with its favorable agroclimatic conditions, also plays an important role in buckwheat production. Buckwheat growing is made possible by the country's agricultural geography and favorable temperatures (Stringer 2013). Buckwheat agriculture has a long history in France, particularly in places such as Brittany and Normandy. Buckwheat production in the country has been impacted by the country's cultural heritage and culinary practices, contributing to global supply (Singh et al. 2021). Buckwheat farming in the United States is predominantly centered in regions such as North Dakota, Minnesota, and New York, where farmers take advantage of favorable growing conditions. While buckwheat production in India was tiny in comparison with the big producers, it was culturally and regionally significant, especially in the northern states. Buckwheat production in India is expected to be over 35,000 metric tonnes in 2019 (Pirzadah and Bilal 2021). The mountainous regions of northern states, such as Uttar Pradesh, Uttarakhand, and Himachal Pradesh, afford favorable climatic conditions for buckwheat production, adding to its regional significance. Buckwheat is also culturally significant in India, where it is widely utilized in traditional recipes during holidays like Navratri. However, given the magnitude of India's production and the global production of around 4.7 million metric tonnes, India's contribution to global buckwheat production in 2019 was rather minor (Agregán et al. 2022). Buckwheat cultivation requires specific climate and soil conditions to ensure successful growth and yields. Understanding the ideal conditions for buckwheat cultivation in India and around the world enables farmers to optimize their agricultural practices and increase crop productivity. Farmers can harness the full potential of buckwheat cultivation and contribute to its sustainable production by implementing appropriate techniques for seed selection, soil preparation, water management, pest control, and post-harvest practices (Pandey and Rao 2016).

1.5 Chia (*Salvia hispanica*)

Chia (*Salvia hispanica* L.) is an ancient crop from the Lamiaceae (mint) family. It is native to Central and South America, and indigenous civilizations, such as the Aztecs and Mayans, have long cultivated it (Bochicchio et al. 2015). For thousands of years, chia seeds have been an important part of their diet and culture. The chia plant is an annual plant that can grow to be one to two meters tall. It has square stems and opposite leaves, as well as small spikes of purple or white flowers (Baginsky et al. 2016). To thrive, the plant needs a tropical or subtropical climate, and it is typically grown in areas with mild winters and long growing seasons. Chia seeds, the plant's most valuable part, are small, oval-shaped seeds with a mottled appearance. They are extremely nutritious and have grown in popularity in recent years due to their health benefits and versatility in culinary applications (Chew et al. 2018; Punia and Dhull, 2019). The Lamiaceae is a large plant family that includes well-known herbs such as basil, rosemary, thyme, and oregano. Chia resembles other plants in this family in that it has square stems and aromatic leaves (Ikumi et al. 2019). Chia seeds are high in essential nutrients, making them a nutrient-dense food. They are high in omega-3 fatty acids, fiber, protein, antioxidants, vitamins, and minerals such as calcium, magnesium, and phosphorus. When mixed with water, chia seeds absorb liquid and form a gel-like consistency, making them a popular ingredient in puddings, smoothies, and other recipes (Ikumi et al. 2019). Chia seeds have been linked to a variety of health benefits in addition to their nutritional benefits (Punia and Dhull 2019). Chia seeds' high fiber content aids digestion and promotes a feeling of fullness, making them useful for weight management. Omega-3 fatty acids have been linked to reduced inflammation and improved heart health (Muñoz et al. 2013; Dhull and Punia 2020a, 2020c; Dhull et al. 2020d). Chia seeds are also valued for their wide range of uses. They can be eaten raw, added to baked goods, sprinkled on cereals or salads, or used in recipes as a thickening agent. Chia gel, made by soaking chia seeds in water, can be used in vegan baking as an egg substitute and fat replacer (Alagawany et al. 2022; Punia and Dhull 2019).

Chia seed cultivation is practiced in many different parts of the world, each with its own set of conditions and cultivation practices. It is primarily grown in Mexico's central and northern regions, particularly in the states of Chiapas, Sonora, and Sinaloa. The semi-arid to sub-humid climate in these areas, with moderate rainfall and temperatures ranging from 20°C to 30°C, is ideal for chia cultivation. Chia seeds are typically sown during the rainy season when soil moisture is sufficient (Grancieri et al. 2019). Bolivia, which is in the Andes, has also emerged as a significant chia-producing country. Bolivian chia cultivation is concentrated in areas with altitudes ranging from 1,500 to 3,000 meters above sea level (Basuny et al. 2021). Chia cultivation thrives in the cool, dry climate of these high-altitude areas, with temperatures ranging from 15°C to 25°C. The crop is well adapted to the Bolivian climate, and its cultivation helps many local farmers make a living (de Falco et al. 2017). Argentina, another prominent chia producer, focuses on the northwest and central regions of the country. The semi-arid climate in these areas, with temperatures ranging from 15°C to 35°C, provides suitable conditions for chia cultivation. Chia seeds are usually sown during the spring or summer seasons, taking advantage of the higher temperatures and longer daylight hours (Xingú López et al. 2017). Due to limited rainfall in these regions, water management practices, such as careful irrigation, are necessary to ensure proper plant growth and development. In Peru, chia cultivation is primarily found in the high-altitude regions of the Andes, such as the Puno and Ayacucho regions. The cool, dry climate in these areas, with temperatures ranging from 10°C to 25°C, is ideal for chia cultivation. Chia is often intercropped with other crops, such as quinoa or amaranth, to maximize land utilization and provide shade protection for the delicate chia plants (Bordin-Rodrigues et al. 2021). Australia has seen a significant increase in chia seed production, particularly in regions with temperate or Mediterranean climates, with Western Australia being a major chia-producing region. The crop requires well-drained soils and is often cultivated in rotation with other crops such as wheat or canola (Grimes 2021). Adequate rainfall or irrigation during the growing season is crucial for successful chia cultivation in Australia. In India, chia seed cultivation has attracted attention in recent years due to its nutritional benefits and adaptability to certain agroclimatic regions. The arid and semi-arid regions of the country, including states like Rajasthan, Gujarat, and Maharashtra, provide suitable conditions for chia cultivation (Felemban et al. 2021). The hot and dry climate, with temperatures ranging from 25°C to 35°C, is favorable for chia growth. Chia is typically grown as a summer crop in India, taking advantage of the high temperatures and longer days (Grimes et al. 2019). Across all these regions, appropriate agricultural practices play a vital role in successful chia seed cultivation. This includes selecting suitable chia varieties, preparing the soil, sowing the seeds at the right time, managing irrigation and water requirements, implementing weed control measures, and monitoring for crops and diseases. Farmers in these regions adapt their cultivation practices to the specific climatic and soil conditions, ensuring optimal chia seed production (Grimes et al. 2019). In 2019, global chia seed production was estimated to be around 39,500 metric tons, and it was expected to grow further in the following years. However, it's important to note that the production figures can vary from year to year, depending on several factors, such as weather conditions, market demand, and agricultural practices (Rasha et al. 2020). Chia seeds have grown in popularity due to their nutritional value and versatility in culinary applications. According to the United Nations Food and Agriculture Organization (FAO), global chia seed production has been steadily increasing, reaching approximately 170,000 metric tonnes in 2019 (Cai et al. 2021). Historically, Mexico has been the leading producer of chia seeds, followed by Bolivia, Paraguay, Argentina, and Peru. These countries have favorable agroclimatic conditions for chia cultivation, including semi-arid to arid climatic regions. Chia plants thrive in well-drained soils and can withstand high temperatures and low rainfall, making them ideal for these environments

(Ferreira et al. 2023). Because of the growing demand for superfoods and healthy food options, chia cultivation has gained traction in India in recent years. Chia seeds are primarily grown in the Indian states of Rajasthan, Gujarat, Tamil Nadu, and Karnataka. These regions have ideal climatic conditions for chia plant growth, with hot and dry summers (Coates 2011). Chia seed production data in India may vary from year to year and depend on a variety of factors such as acreage, yield, and market demand. Chia cultivation is gradually expanding in India, as more farmers recognize its potential as a profitable crop. Chia plants are typically sown in India during the summer season, from February to March (Komu et al. 2021). To ensure optimal growth and seed production, they require well-drained soils and proper irrigation management. Farmers in India frequently use organic farming practices for chia cultivation, in response to the market's increasing demand for organic and pesticide-free chia products (Domancar and Valverde 2017). Chia seed cultivation in India has several advantages. It diversifies farmers' income and contributes to the country's agricultural diversity. Chia seeds are in high demand around the world because of their high nutritional profile, which includes omega-3 fatty acids, fiber, protein, and a variety of vitamins and minerals (Bochicchio et al. 2015). Chia seed cultivation in India is expected to expand further, as public awareness of the health benefits of chia seed grows, offering both domestic and export opportunities. In conclusion, chia seed cultivation has attracted significant global attention due to its nutritional value and versatile applications. Major chia-producing countries include Mexico, Bolivia, Argentina, Ecuador, Australia, and Peru. India has also shown potential for chia cultivation, although specific production data are not readily available. The country has been gradually increasing its efforts in chia cultivation, supported by government initiatives. Challenges exist in terms of awareness, access to quality seeds, and market linkages in India. Ongoing research and development efforts are crucial for maximizing the potential of chia cultivation worldwide and in India. Promising opportunities lie in sustainable farming practices, pest and disease management, value addition, and market development. To obtain the most accurate and up-to-date information, it is recommended to consult recent research and industry-specific sources for the latest developments in chia seed cultivation and production globally, including in India.

1.6 Conclusion

In recent years, the cultivation and production of pseudocereals, including quinoa, amaranth, buckwheat, and chia, have garnered significant attention and recognition worldwide. These crops have emerged as promising alternatives in the agricultural landscape, thanks to their exceptional nutritional profiles, adaptability to diverse environments, and versatile applications in various industries. One of the key factors contributing to the popularity of pseudocereals is their nutritional value. These crops are rich in essential nutrients, including proteins, dietary fibers, vitamins, and minerals. Additionally, they are often gluten free, making them suitable for individuals with gluten sensitivities or celiac disease. The high protein content and balanced amino acid profiles of pseudocereals make them valuable ingredients for vegetarian and vegan diets, as well as for addressing protein deficiencies in populations worldwide. Another advantage of pseudocereals is their adaptability to different climates and growing conditions. They can be cultivated in a wide range of environments, from high-altitude regions to arid or marginal lands. For example, quinoa is native to the Andean region and thrives in harsh conditions with low rainfall and poor soil quality. Amaranth is resilient and can tolerate both high temperatures and drought, while buckwheat is well-suited to cool climates. Chia can be grown in a variety of regions and climates, making it highly adaptable. The versatility of pseudocereals is demonstrated through their numerous culinary applications. They can be used as whole grains, flour, flakes, or in processed forms to create a range of food products such as bread, pasta, snacks, cereals, and beverages. Additionally, pseudocereals can be incorporated into gluten-free products or used as functional ingredients in the food and pharmaceutical industries due to their unique properties. For example, chia seeds are known for their hydrophilic properties, which make them suitable for use in gels, emulsions, and as a thickening agent. However, the cultivation and production of pseudocereals are not without challenges. Farmers and agricultural communities face hurdles such as limited access to quality seeds, inadequate infrastructure, and market linkages. Furthermore, there is a need for sustainable farming practices that minimize environmental impacts, efficient processing technologies, and market development initiatives to fully harness the potential of pseudocereals.

References

Agregán, Rubén, Nihal Guzel, Mustafa Guzel, Sneh Punia Bangar, Gökhan Zengin, Manoj Kumar, and José Manuel Lorenzo. 2022. "The Effects of Processing Technologies on Nutritional and Anti-nutritional Properties of Pseudocereals and Minor Cereal." *Food and Bioprocess Technology*, November. https://doi.org/10.1007/s11947-022-02936-8.

Alagawany, Mahmoud, Shaaban S. Elnesr, R. Mayada, Karim Farag, Othman El-Sabrout, A. O. Mahmoud, Hidayatullah Dawood, and Sameh A. Soomro. 2022. "Nutritional Significance and Health Benefits of Omega-3,-6 and-9 Fatty Acids in Animals." *Animal Biotechnology* 33(7): 1678–90.

Amador, Moreno, Isabel María Comino María De Lourdes, and Carolina Montilla. 2014. *Alternative Grains as Potential Raw Material for Gluten-Free Food Development in the Diet of Celiac and Gluten-Sensitive Patients.* Austin Journal of Nutrition and Metabolism, 2(3): 1–9.

Angeli, Viktória, Pedro Miguel Silva, Danilo Crispim Massuela, Muhammad Waleed Khan, Alicia Hamar, Forough Khajehei, Simone Graeff-Hönninger, and Cinzia Piatti. 2020. "Quinoa (*Chenopodium quinoa* Willd.): An Overview of the Potentials of the 'Golden Grain' and Socio-economic and Environmental Aspects of Its Cultivation and Marketization." *Foods (Basel, Switzerland)* 9(2): 216. https://doi.org/10.3390/foods9020216.

Asher, Aviv, Shmuel Galili, Travis Whitney, and Lior Rubinovich. 2020. "The Potential of Quinoa (*Chenopodium quinoa*) Cultivation in Israel as a Dual-Purpose Crop for Grain Production and Livestock Feed." *Scientia Horticulturae* 272: 109534. https://doi.org/10.1016/j.scienta.2020.109534.

Aubert, Lauranne, Christian Decamps, Guillaume Jacquemin, and Muriel Quinet. 2021. "Comparison of Plant Morphology, Yield and Nutritional Quality of *Fagopyrum esculentum* and *Fagopyrum tataricum* Grown under Field Conditions in Belgium." *Plants* 10(2): 258. https://doi.org/10.3390/plants10020258.

Aubert, Lauranne, and Muriel Quinet. 2022. "*Comparison of Heat and Drought Stress Responses among Twelve Tartary Buckwheat (Fagopyrum tataricum) Varieties.*" *Plants* 11(11): 1517.

Baginsky, Cecilia, Jorge Arenas, Hugo Escobar, Marco Garrido, Natalia Valero, Diego Tello, Leslie Pizarro, Alfonso Valenzuela, Luis Morales, and Herman Silva. 2016. "Growth and Yield of Chia (*Salvia hispanica* L.) in the Mediterranean and Desert Climates of Chile." *Chilean Journal of Agricultural Research* 76(3): 255–64.

Basuny, A. M., S. M. Arafat, and D. M. Hikal. 2021. "Chia (*Salvia hispanica* L.) Seed Oil Rich in Omega-3 Fatty Acid: A Healthy Alternative for Milk Fat in Ice Milk." *Food and Nutrition Sciences* 12(6): 479–93.

Bazile, Didier, Cataldo Pulvento, Alexis Verniau, Mohammad S. Al-Nusairi, Djibi Ba, Joelle Breidy, Layth Hassan, et al. 2016. "Worldwide Evaluations of Quinoa: Preliminary Results from Post International Year of Quinoa FAO Projects in Nine Countries." *Frontiers in Plant Science* 7(June): 850. https://doi.org/10.3389/fpls.2016.00850.

Beccari, Giovanni, Mara Quaglia, Francesco Tini, Euro Pannacci, and Lorenzo Covarelli. 2021. "Phytopathological Threats Associated with Quinoa (*Chenopodium quinoa* Willd.) Cultivation and Seed Production in an Area of Central Italy." *Plants* 10(9): 1933. https://doi.org/10.3390/plants10091933.

Bhunia, Shantanu, Ankita Bhowmik, Rambilash Mallick, Anupam Debsarcar, and Joydeep Mukherjee. 2021. "Application of Recycled Slaughterhouse Wastes as an Organic Fertilizer for Successive Cultivations of Bell Pepper and Amaranth." *Scientia Horticulturae* 280: 109927. https://doi.org/10.1016/j.scienta.2021.109927.

Bochicchio, Rocco, Tim D. Philips, Stella Lovelli, Rosanna Labella, Fernanda Galgano, Antonio Di Marisco, Michele Perniola, and Mariana Amato. 2015. "Innovative Crop Productions for Healthy Food: The Case of Chia (*Salvia hispanica* L.)." In *The Sustainability of Agro-Food and Natural Resource Systems in the Mediterranean Basin*, 29–45. Springer International Publishing. https://doi.org/10.1007/978-3-319-163 57-4_3.

Bordin-Rodrigues, Jaqueline Calzavara, Tiago Roque Benetoli da Silva, Debora Fernandes Del Moura Soares, Juliana Stracieri, Rhaizza Lana Pereira Ducheski, and Gessica Daiane da Silva. 2021. "Bean and Chia Development in Accordance with Fertilization Management." *Heliyon* 7(6): e07316. https://doi.org/10.1016/j.heliyon.2021.e07316.

Cai, Junning, Alessandro Lovatelli, José Aguilar-Manjarrez, Lynn Cornish, Lionel Dabbadie, Anne Desrochers, and Simon Diffey. 2021. "Seaweeds and Microalgae: An Overview for Unlocking Their Potential in Global Aquaculture Development." *FAO Fisheries and Aquaculture Circular*: 1229.

Chaudhary, E. 2022. *Garden Up: Your One Stop Guide to Growing Plants at Home*. Penguin Random House India Private Limited.

Chew, Kit Wayne, Shir Reen Chia, Pau Loke Show, Yee Jiun Yap, Tau Chuan Ling, and Jo-Shu Chang. 2018. "Effects of Water Culture Medium, Cultivation Systems and Growth Modes for Microalgae Cultivation: A Review." *Journal of the Taiwan Institute of Chemical Engineers* 91(October): 332–44. https://doi.org/10.1016/j.jtice.2018.05.039.

Coates, Wayne. 2011. *Whole and Ground Chia (Salvia hispanica L.) Seeds, Chia Oil-Effects on Plasma Lipids and Fatty Acids.* Academic Press.

Dar, Fayaz A., Tanveer B. Pirzadah, Bisma Malik, Inayatullah Tahir, and Reiaz U. Rehman. 2018. "Molecular Genetics of Buckwheat and Its Role in Crop Improvement." In *Buckwheat Germplasm in the World*, 271–86. Elsevier. https://doi.org/10.1016/b978-0-12-811006-5.00026-4.

Dhull, S. B., and S. Punia. 2020a. "Essential Fatty Acids: Introduction." In *Essential Fatty Acids: Sources, Processing Effects, and Health Benefits*, edited by Dhull, S. B., Punia, S. P., & Sandhu, K. S., 1–18. CRC Press.

Dhull, S. B., and S. Punia. 2020b. "Sources: Plants, Animals and Microbial." In *Essential Fatty Acids: Sources, Processing Effects, and Health Benefits*, edited by Dhull, S. B., Punia, S., & Sandhu, K. S., 19–56. CRC Press.

Dhull, S. B., M. Kaur, and K. S. Sandhu. 2020a. "Antioxidant Characterization and In Vitro DNA Damage Protection Potential of Some Indian Fenugreek (*Trigonella foenum-graecum*) Cultivars: Effect of Solvents." *Journal of Food Science and Technology* 57: 1–10.

Dhull, S. B., S. Punia, M. K. Kidwai, M. Kaur, P. Chawla, S. S. Purewal, M. Sangwan, and S. Palthania. 2020b. "Solid-State Fermentation of Lentil (*Lens culinaris* L.) with *Aspergillus awamori*: Effect on Phenolic Compounds, Mineral Content, and Their Bioavailability." *Legume Science* 2(3): e37.

Dhull, S. B., S. Punia, R. Kumar, M. Kumar, K. B. Nain, K. Jangra, and C. Chudamani. 2020c. "Solid State Fermentation of Fenugreek (*Trigonella foenum-graecum*): Implications on Bioactive Compounds, Mineral Content and In Vitro Bioavailability." *Journal of Food Science and Technology* 58: 1927–1936.

Dhull, S. B., S. Punia, and K. S. Sandhu (Eds.). 2020d. *Essential Fatty Acids: Sources, Processing Effects, and Health Benefits.* CRC Press.

Domancar, María Elena, and Octavio Valverde. 2017. "Chia – The New Golden Seed for the 21st Century: Nutraceutical Properties and Technological Uses." In *Sustainable Protein Sources*, 265–81. Academic Press.

Ebert, Andreas W. 2022. "Sprouts and Microgreens – Novel Food Sources for Healthy Diets." *Plants* 11(4): 571. https://doi.org/10.3390/plants11040571.

El-Hakim, A., F. Ahmed, E. Mady, A. M. Abou Tahoun, M. S. Ghaly, and M. A. Eissa. 2022. "Seed Quality and Protein Classification of Some Quinoa Varieties." *Journal of Ecological Engineering* 23(1): 24–33. https://doi.org/10.12911/22998993/143866.

Falco, Bruna de, Mariana Amato, and Virginia Lanzotti. 2017. "Chia Seeds Products: An Overview." *Phytochemistry Reviews: Proceedings of the Phytochemical Society of Europe* 16(4): 745–60. https://doi.org/10.1007/s11101-017-9511-7.

Farooq, S., R. Ul Rehman, T. B. Pirzadah, B. Malik, F. Ahmad Dar, and I. Tahir. 2016. "Cultivation, Agronomic Practices, and Growth Performance of Buckwheat." In *Molecular Breeding and Nutritional Aspects of Buckwheat*, 299–319. Elsevier. https://doi.org/10.1016/b978-0-12-803692-1.00023-7.

Felemban, Loai F., Atef M. Al-Attar, and Isam M. Abu Zeid. 2021. "Medicinal and Nutraceutical Benefits of Chia Seed (*Salvia hispanica*)." *Journal of Pharmaceutical Research International*, January, 15–26. https://doi.org/10.9734/jpri/2020/v32i4131040.

Ferreira, Diana Melo, Maria Antónia Nunes, Liliana Espírito Santo, Susana Machado, Anabela S. G. Costa, Manuel Álvarez-Ortí, José E. Pardo, Maria Beatriz P. P. Oliveira, and Rita C. Alves. 2023. "Characterization of Chia Seeds, Cold-Pressed Oil, and Defatted Cake: An Ancient Grain for Modern Food Production." *Molecules (Basel, Switzerland)* 28(2). https://doi.org/10.3390/molecules28020723.

Filho, Antonio Manoel Maradini, Mônica Ribeiro Pirozi, João Tomaz Da Silva Borges, Helena Maria Pinheiro Sant'Ana, José Benício Paes Chaves, and Jane Sélia Dos Reis Coimbra. 2017. "Quinoa: Nutritional, Functional, and Antinutritional Aspects." *Critical Reviews in Food Science and Nutrition* 57(8): 1618–30. https://doi.org/10.1080/10408398.2014.1001811.

Folk, M. E. 2020. *Migration and Agricultural Practice: A Paleoethnobotanical Analysis of los Batanes in Southern Peru* (Doctoral Dissertation).

Grancieri, Mariana, Hercia Stampini Duarte Martino, and Elvira Gonzalez de Mejia. 2019. "Chia Seed (Salvia hispanica L.) as a Source of Proteins and Bioactive Peptides with Health Benefits: A Review." *Comprehensive Reviews in Food Science and Food Safety* 18(2): 480–99. https://doi.org/10.1111/1541-4337.12423.

Graziano, Sara, Caterina Agrimonti, Nelson Marmiroli, and Mariolina Gullì. 2022. "Utilisation and Limitations of Pseudocereals (Quinoa, Amaranth, and Buckwheat) in Food Production: A Review." *Trends in Food Science and Technology* 125(July): 154–65. https://doi.org/10.1016/j.tifs.2022.04.007.

Grimes, Samantha. 2021. *Screening and Cultivation of Chia (Salvia hispanica L.) under Central European Conditions: The Potential of a Re-Emerged Multipurpose Crop.*

Grimes, Samantha, Timothy Phillips, Filippo Capezzone, and Simone Graeff-Hönninger. 2019. "Impact of Row Spacing, Sowing Density and Nitrogen Fertilization on Yield and Quality Traits of Chia (*Salvia hispanica* L.) Cultivated in Southwestern Germany." *Agronomy (Basel, Switzerland)* 9(3): 136. https://doi.org/10.3390/agronomy9030136.

Hussain, Muhammad Iftikhar, Muhammad Farooq, Adele Muscolo, and Abdul Rehman. 2020. "Crop Diversification and Saline Water Irrigation as Potential Strategies to Save Freshwater Resources and Reclamation of Marginal Soils-a Review." *Environmental Science and Pollution Research International* 27(23): 28695–729. https://doi.org/10.1007/s11356-020-09111-6.

Ikumi, P., M. W. Mburu, and D. M. Njoroge. 2019. *Chia (Salvia hispanica L.) - A Potential Crop for Food and Nutrition Security in Africa. Journal of Food Research* 8(6): 104. https://doi.org/10.5539/jfr.v8n6p104

Ikumi, Pauline, Monica Mburu, and Daniel Njoroge. 2019. "Chia (*Salvia hispanica* L.) – A Potential Crop for Food and Nutrition Security in Africa." *Journal of Food Research* 8(6): 104. https://doi.org/10.5539/jfr.v8n6p104.

Jacquemart, A. L., V. Cawoy, J. M. Kinet, J. F. Ledent, and M. Quinet. 2012. "Is Buckwheat (*Fagopyrum esculentum* Moench) Still a Valuable Crop Today." *The European Journal of Plant Science and Biotechnology* 6(2): 1–10.

Jan, Nusrat, Syed Zameer Hussain, Bazila Naseer, and Tashooq A. Bhat. 2023. "Amaranth and Quinoa as Potential Nutraceuticals: A Review of Anti-nutritional Factors, Health Benefits and Their Applications in Food, Medicinal and Cosmetic Sectors." *Food Chemistry: X*, no. 100687 (April): 100687. https://doi.org/10.1016/j.fochx.2023.100687.

Ji, Xiaolong, Lin Han, Fang Liu, Sheng Yin, Qiang Peng, and Min Wang. 2019. "A Mini-Review of Isolation, Chemical Properties and Bioactivities of Polysaccharides from Buckwheat (*Fagopyrum* Mill)." *International Journal of Biological Macromolecules* 127(April): 204–9. https://doi.org/10.1016/j.ijbiomac.2019.01.043.

Joshi, B. K. 2023. "Buckwheat (*Fagopyrum esculentum* Moench and *F. tataricum* Gaertn)." In *Neglected and Underutilized Crops*, 151–200. Academic Press.

Komu, Clement, Monica Mburu, Daniel Njoroge, and Richard Koskei. 2021. "Physicochemical Profile of Essential Oils Obtained from Chia (*Salvia hispanica* L.) Seeds Grown in Different Agro-Ecological Zones of Kenya." *European Journal of Advanced Chemistry Research* 2(3): 21–6. https://doi.org/10.24018/ejchem.2021.2.3.56.

Kulakow, Peter A. 2018. "Genetic Characterization of Grain Amaranth." In *Amaranth Biology, Chemistry, and Technology*, 9–22. CRC Press. https://doi.org/10.1201/9781351069601-2.

Kumar, Ajay, Thattantavide Anju, Sushil Kumar, Sushil Satish Chhapekar, Sajana Sreedharan, Sonam Singh, Nirala Ramchiary, Su Ryun Choi, and Yong Pyo Lim. 2021. "Linking Omics and Gene Editing Tools for Rapid Improvement of Traditional Food Plants for Diversified Foods and Sustainable Food Security." *Preprints*. https://doi.org/10.20944/preprints202106.0363.v1.

Lozano-Isla, Flavio, José-David Apaza, Angel Mujica Sanchez, Raúl Blas Sevillano, Bettina I. G. Haussmann, and Karl Schmid. 2023. "Enhancing Quinoa Cultivation in the Andean Highlands of Peru: A Breeding Strategy for Improved Yield and Early Maturity Adaptation to Climate Change Using Traditional Cultivars." *Euphytica; Netherlands Journal of Plant Breeding* 219(2). https://doi.org/10.1007/s10681-023-03155-8.

Luthar, Zlata, Primož Fabjan, and Katja Mlinarič. 2021. "Biotechnological Methods for Buckwheat Breeding." *Plants* 10(8): 1547. https://doi.org/10.3390/plants10081547.

Malik, M., R. Sindhu, S. B. Dhull, C. Bou-Mitri, Y. Singh, S. Panwar, and B. S. Khatkar. 2023. "Nutritional Composition, Functionality, and Processing Technologies for Amaranth." *Journal of Food Processing and Preservation* 2023: 1753029. https://doi.org/10.1155/2023/1753029.

Martinez-Lopez, Alicia, Maria C. Millan-Linares, Noelia M. Rodriguez-Martin, Francisco Millan, and Sergio Montserrat-de la Paz. 2020. "Nutraceutical Value of Kiwicha (*Amaranthus caudatus* L.)." *Journal of Functional Foods* 65: 103735. https://doi.org/10.1016/j.jff.2019.103735.

Matías, Javier, María José Rodríguez, Verónica Cruz, Patricia Calvo, and María Reguera. 2021. "Heat Stress Lowers Yields, Alters Nutrient Uptake and Changes Seed Quality in Quinoa Grown under Mediterranean Field Conditions." *Journal of Agronomy and Crop Science* 207(3): 481–91. https://doi.org/10.1111/jac.12495.

Mir, Nisar Ahmad, Charanjit Singh Riar, and Sukhcharn Singh. 2018. "Nutritional Constituents of Pseudo Cereals and Their Potential Use in Food Systems: A Review." *Trends in Food Science and Technology* 75(May): 170–80. https://doi.org/10.1016/j.tifs.2018.03.016.

Muñoz, Loreto A., Angel Cobos, Olga Diaz, and José Miguel Aguilera. 2013. "Chia Seed (*Salvia hispanica*): An Ancient Grain and a New Functional Food." *Food Reviews International* 29(4): 394–408. https://doi.org/10.1080/87559129.2013.818014.

Murphy, Kevin S., and Janet Matanguihan. 2015. *Quinoa: Improvement and Sustainable Production*. John Wiley & Sons.

Mustafa, T. A. N., and S. Temel. 2018. "Performance of Some Quinoa (*Chenopodium quinoa* Willd.) Genotypes Grown in Different Climate Conditions." *Turkish Journal of Field Crops* 23(2): 180–86.

Pandey, S., and B. S. Rao. 2016. "Potential Use of Pseudocereals: Buckwheat, Quinoa and Amaranth." In *Food Process Engineering*, 261–90. Apple Academic Press.

Perez, C., E. M. Jones, P. Kristjanson, L. Cramer, P. K. Thornton, W. Förch, and C. Barahona. 2015. "How Resilient Are Farming Households and Communities to a Changing Climate in Africa? A Gender-Based Perspective." *Global Environmental Change: Human and Policy Dimensions* 34(September): 95–107. https://doi.org/10.1016/j.gloenvcha.2015.06.003.

Peterson, A. J., and K. M. Murphy. 2015. "Quinoa Cultivation for Temperate North America: Considerations and Areas for Investigation." *Quinoa: Improvement and Sustainable Production*: 173–92.

Pirzadah, Tanveer, and Reiaz Bilal. 2021. *Buckwheat: Forgotten Crop for the Future: Issues and Challenges.*

Punia, S., and S. B. Dhull. 2019. "Chia Seed (*Salvia hispanica* L.) Mucilage (a Heteropolysaccharide): Functional, Thermal, Rheological Behaviour and Its Utilization." *International Journal of Biological Macromolecules* 140: 1084–90.

Ramzan, Asiya, and Reiaz Ul Rehman. 2021. "*Amaranthus caudatus* L. as a Potential Bioresource for Nutrition and Health." In *Medicinal and Aromatic Plants*, 241–56. Springer International Publishing. https://doi.org/10.1007/978-3-030-589 75-2_9.

Rasha, S., A. A. El-Sheshtawy, and H. E. Ali. 2020. "Phenology, Architecture, Yield and Fatty Acid Content of Chia in Response to Sowing Date and Plant Spacing." *Fayoum Journal of Agricultural Research and Development* 34(1): 314–31.

Rodriguez, J. P., E. Ono, A. M. S. Abdullah, R. Choukr-Allah, and H. Abdelaziz. 2020. "Cultivation of Quinoa (*Chenopodium quinoa*) in Desert Ecoregion." In *Emerging Research in Alternative Crops*, 145–61. Springer.

Rodríguez, Juan Pablo, Hifzur Rahman, Sumitha Thushar, and Rakesh K. Singh. 2020. "Healthy and Resilient Cereals and Pseudo-Cereals for Marginal Agriculture: Molecular Advances for Improving Nutrient Bioavailability." *Frontiers in Genetics* 11(February): 49. https://doi.org/10.3389/fgene.2020.00049.

Roman, V. J. 2021. *Salt Tolerance Strategies of the Ancient Andean Crop Quinoa* (Doctoral Dissertation).

Sadok, Walid, and S. V. Krishna Jagadish. 2020. "The Hidden Costs of Nighttime Warming on Yields." *Trends in Plant Science* 25(7): 644–51. https://doi.org/10.1016/j.tplants.2020.02.003.

Saini, Ramesh, Parchuri Kumar, Reddampalli Venkataramareddy Prasad, Kamatham Sreedhar, Xiaomin Akhilender Naidu, and Young-Soo Shang. 2021. "Omega-3 Polyunsaturated Fatty Acids (PUFAs): Emerging Plant and Microbial Sources, Oxidative Stability, Bioavailability, and Health Benefits – A Review." *Antioxidants* 10(10): 1627.

Schnetzler, Kent A. 2018. "Food Uses and Amaranth Product Research: A Comprehensive Review." In *Amaranth Biology, Chemistry, and Technology*, 155–84. CRC Press. https://doi.org/10.1201/9781351069601-9.

Serrano-Arellano, J., M. Gijón-Rivera, J. L. Chávez-Servín, K. de la Torre-Carbot, J. Xamán, G. Álvarez, and J. M. Belman-Flores. 2015. "Numerical Study of Thermal Environment of a Greenhouse Dedicated to Amaranth Seed Cultivation." *Solar Energy (Phoenix, Ariz.)* 120(October): 536–48. https://doi.org/10.1016/j.solener.2015.08.004.

Sharma, V., S. Chandra, P. Dwivedi, and M. Parturkar. 2015. "Quinoa (*Chenopodium quinoa* Willd.): A Nutritional Healthy Grain." *International Journal of Advanced Research* 3(9): 725–36.

Singh, A., A. Kumari, and H. K. Chaudhary. 2021. "Amaranth, Buckwheat, and Chenopodium: The 'ABC' Nutraceuticals of Northwestern Himalayas." *Agricultural Biotechnology: Latest Research and Trends*: 587–634.

Singh, Mohar, Nikhil Malhotra, and Kriti Sharma. 2020. "Buckwheat (*Fagopyrum* sp.) Genetic Resources: What Can They Contribute Towards Nutritional Security of Changing World?." *Genetic Resources and Crop Evolution* 67(7): 1639–58. https://doi. org/10.1007/s10722-020-00961-0.

Siracusa, Laura, Fabio Gresta, Elisa Sperlinga, and Giuseppe Ruberto. 2017. "Effect of Sowing Time and Soil Water Content on Grain Yield and Phenolic Profile of Four Buckwheat (*Fagopyrum esculentum* Moench.) Varieties in a Mediterranean Environment." *Journal of Food Composition and Analysis: An Official Publication of the United Nations University, International Network of Food Data Systems* 62(September): 1–7. https://doi.org/10.1016/j.jfca.2017.04.005.

Song, Yingjie, Zhuo Cheng, Yumei Dong, Dongmei Liu, Keyu Bai, Devra Jarvis, Jinchao Feng, and Chunlin Long. 2022. "Diversity of Tartary Buckwheat (*Fagopyrum tataricum*) Landraces from Liangshan, Southwest China: Evidence from Morphology and SSR Markers." *Agronomy (Basel, Switzerland)* 12(5): 1022. https://doi.org/10.3390/ agronomy12051022.

Stetter, Markus G., Thomas Müller, and Karl J. Schmid. 2017. "Genomic and Phenotypic Evidence for an Incomplete Domestication of South American Grain Amaranth (*Amaranthus caudatus*)." *Molecular Ecology* 26(3): 871–86. https://doi.org/10.1111/ mec.13974.

Stetter, Markus G., Leo Zeitler, Adrian Steinhaus, Karoline Kroener, Michelle Biljecki, and Karl J. Schmid. 2016. "Crossing Methods and Cultivation Conditions for Rapid Production of Segregating Populations in Three Grain Amaranth Species." *Frontiers in Plant Science* 7(June): 816. https://doi.org/10.3389/fpls.2016.00816.

Stringer, Danielle. 2013. *Investigating the Mechanisms and Effectiveness of Common Buckwheat (Fagopyrum esculentum Moenech) for Acute Modulation of Glycemia.*

Thiyagarajan, Karthikeyan, Fabio Vitali, Valentina Tolaini, Patrizia Galeffi, Cristina Cantale, Prashant Vikram, Sukhwinder Singh, et al. 2016. "Genomic Characterization of Phenylalanine Ammonia Lyase Gene in Buckwheat." *PLOS ONE* 11(3): e0151187. https://doi.org/10.1371/journal.pone.0151187.

Weijuan, Devra I., Selena Jarvis, and Chunlin Ahmed. 2017. "Tartary Buckwheat Genetic Diversity in the Himalayas Associated with Farmer Landrace Diversity and Low Dietary Dependence." *Sustainability* 9(10): 1806.

Woods, M., A. Cobley, S. Verrall, R. Neilson, N. Van Der Velden, and G. Hager. 2020. "Edible Plant Database."

Xingú López, A., A. González Huerta, E. D. L. Cruz Torrez, D. M. Sangerman-Jarquín, G. Orozco De Rosas, and M. Arriaga. 2017. "Chia (*Salvia hispanica* L.) Current Situation and Future Trends." *Revista Mexicana de Ciencias Agrícolas* 8(7): 1619–31.

Yao, Yi-Feng, Xiao-Yan Song, Gan Xie, Ye-Na Tang, Alexandra H. Wortley, Feng Qin, Stephen Blackmore, Cheng-Sen Li, and Yu-Fei Wang. 2023. "New Insights into the Origin of Buckwheat Cultivation in Southwestern China from Pollen Data." *The New Phytologist* 237(6): 2467–77. https://doi.org/10.1111/nph.18659.

Yazar, A., and Ç. İ. Kaya. 2014. "A New Crop for Salt Affected and Dry Agricultural Areas of Turkey: Quinoa (*Chenopodium quinoa* Willd.)." In *Türk Tarım ve Doğa Bilimleri Dergisi, 1(Özel Sayı-2)*, 1440–46. DergiPark (Istanbul University). https://dergipark. org.tr/tr/pub/turkjans/issue/13311/160926

Zhou, Mengjie, Mingxing Huo, Jiankang Wang, Tiantian Shi, Faliang Li, Meiliang Zhou, Junzhen Wang, and Zhiyong Liao. 2022. "Identification of Tartary Buckwheat Varieties Suitable for Forage via Nutrient Value Analysis at Different Growth Stages." *International Journal of Plant Biology* 13(2): 31–43. https://doi.org/10.3390/ijpb13020005.

Zohry, A. E. H. 2020. "Prospects of Quinoa Cultivation in Marginal Lands of Egypt." *Moroccan Journal of Agricultural Sciences* 3.

Zvelebil, M., and M. Pluciennik. 2011. "Historical Origins of Agriculture." *The Role Food, Agriculture, Forestry and Fisheries in Human Nutrition* 41–78. EOLSS Publications.

Chapter 2

Structural and Engineering Aspects of Different Pseudocereals

Arpan Dubey, Prakash Kumar, Rohit Biswas

Pseudocereals are plants that are regarded as grains because of their nutritional profiles, that resemble those of classic cereals. However, the pseudocereals differ from the cereals botanically, as they belong to different plant families from the grass family (Poaceae), to which the cereals belong. The prominent pseudocereals worldwide are quinoa, buckwheat, and amaranth, although other crops, like chia and canihua, are also sometimes included in this category. According to FAO (2021), world production of quinoa and buckwheat stands at 147,037.78 and 1,875,067.97 tonnes, respectively. The major quinoa-producing nations are Peru (>90.00%), Bolivia (3.58%), and Ecuador (2.58%). Buckwheat is largely produced in Russia (49.01%), China (26.79%), Ukraine (5.64%), and the USA (4.39%). The production status of amaranth is not officially monitored by FAO, but some of its key producers are several Asian countries (like India and China), Kenya, and Russia (FAOSTAT, 2023).

Quinoa (*Chenopodium quinoa* Willd) originally belonged to the Andean countries and was a significant crop for the Quechua and Aymara people. Quinoa is a member of the Amaranthaceae family. It is a starch-containing dicotyledonous seed, rich in essential amino acids, with lysine in particular, making it a comprehensive source of high-quality protein (Abugoch James, 2009). Moreover, these proteins don't have a gluten fraction, making them suitable for consumption by people with celiac disease or gluten sensitivity. Quinoa is also a great source of minerals, such as magnesium and iron, along with antioxidants, fibers, and vitamins. Quinoa has a characteristic nutty flavor and can be used as a substitute for

DOI: 10.1201/9781003325277-2

cream soups, salads, and baking (Altuntas, 2018; Graziano et al., 2022). The second major pseudocereal is buckwheat, belonging to the Polygonaceae family, which is represented by nine major species, out of which common buckwheat (*Fagopyrum esculentum* Moench) and Tartary buckwheat (*Fagopyrum tataricum* Gaertn.) are the main ones produced globally (Parde et al., 2003). Buckwheat is a rich source of high-quality dietary fiber, lipids, proteins, and minerals, and contains health-promoting components like phenolic compounds and sterols. Buckwheat is also gluten free, making it suitable for developing gluten-free products for individuals with intolerance associated with gluten. Buckwheat has gained attention as a functional food due to its health-promoting properties that lower blood cholesterol and glucose levels, prevent cancer development, and reduce inflammation. Buckwheat seeds, in the form of raw groats or dehulled seeds, are commonly used in breakfast products, and buckwheat flour is useful for developing various bakery products (Reguera & Haros, 2016). Another major pseudocereal is amaranth (*Amaranthus* spp.). It has a long history of consumption and was one of the most important foods for early civilizations like the Inca, Maya, and Aztecs (Jimena Velarde-Salcedo et al., 2019). Amaranth is considered similar to a combination of beans and rice in terms of its amino acid profile and average protein content. It is rich in soluble fiber and has a protein concentration ranging from 12.5% to 17.6%, with high lysine and methionine levels (Malik et al., 2023). Amaranth's lipid content is highly species-dependent, and it contains beneficial fatty acids like oleic, linoleic, palmitic, and linolenic acids (Thakur et al., 2021).

Amaranth is also nutritionally rich with minerals and vitamins, including calcium, magnesium, niacin, riboflavin, and ascorbic acid. Clinical studies have assessed amaranth's potential health benefits, including cholesterol reduction, antioxidant, anticancer, anti-allergic, and antihypertensive effects (Reguera & Haros, 2016; Malik et al., 2023). Amaranth can be processed using various methods that include expanded grains; it can also be cooked, extruded, toasted, or incorporated into flakes, pastas, and biscuits (Jimena Velarde-Salcedo et al., 2019).

Overall, pseudocereals have gained recognition as nutritious alternatives to traditional grains, and their popularity continues to grow as more people discover their health benefits and culinary versatility. Despite the nutritional superiority associated with pseudocereals, their commercialization is currently limited due to a lack of research on their nutritional composition and insufficient knowledge about processing and utilization methods. Therefore, it is crucial to expand the technology portfolio to enhance functional properties and nutritive value through improved processing techniques. Pseudocereals are highly nutritious crops that offer a rich source of high-quality bioactives, essential amino acids, minerals, and vitamins. They are also gluten free. Considering the significance of these properties, this chapter aims to explore various structural and compositional properties of pseudocereals. The findings will contribute to the development of novel and functional foods in this field.

2.1 Structural Features of Pseudocereals

The structure of pseudocereals refers to their anatomical composition and the arrangement of different tissues within the seeds. Figure 2.1 illustrates various structural components of pseudocereals.

Pseudocereal seeds typically consist of the following main structures:

- **Seed coat (testa):** This is the outermost layer of the seed, providing protection and serving as a physical barrier. It is composed of cells that vary in thickness, color, and texture among different pseudocereal species.
- **Aleurone layer:** This layer lies beneath the seed coat and is rich in proteins, lipids, and enzymes. It plays a crucial role in nutrient storage and seed germination.
- **Endosperm:** The endosperm is the largest part of the seed and serves as the main storage tissue. It contains carbohydrates, proteins, lipids, and other nutrients. In some pseudocereals, such as quinoa, the endosperm is not well developed, and the nutrients are evenly distributed throughout the seed.
- **Embryo:** The embryo is the miniature plant contained within the seed. It consists of the radicle (embryonic root), hypocotyl (connecting structure between the root and shoot), and cotyledons (seed leaves). The embryo is essential for the germination and growth of a new plant.

The structures of pseudocereal seeds contribute to their nutritional composition and functional properties

(Reguera & Haros, 2016). The specific arrangement and composition of tissues vary among different pseudocereal species, influencing their texture, taste, cooking properties, and potential applications in food processing.

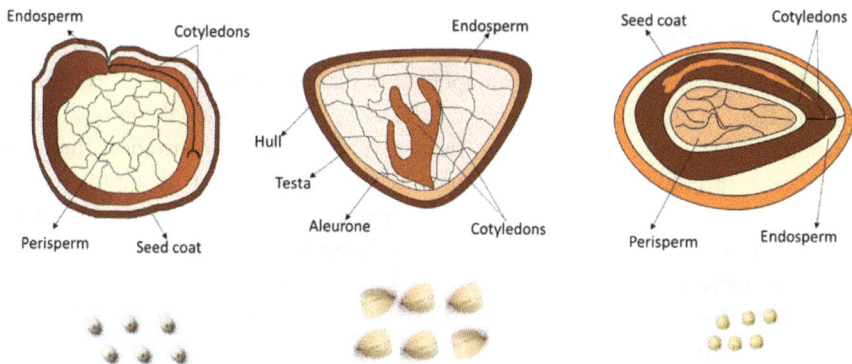

Figure 2.1 **Cross-sectional (above) and actual (below) view of the grain structure of (a) quinoa, (b) buckwheat, and (c) amaranth (adapted from Reguera & Haros (2016)).**

2.2 Chemical Composition of Pseudocereals

2.2.1 Proteins

The protein content of pseudocereals contributes significantly to their nutritional value. Pseudocereal grains are an excellent protein source, characterized by a well-balanced amino acid profile. Specifically, they are rich in lysine and sulfur-containing amino acids. Unlike cereals, which contain prolamin storage proteins that can be harmful to individuals with celiac disease, quinoa, buckwheat, and amaranth primarily consist of albumins and globulins, with minimal or no prolamin storage proteins. In general, compared with quinoa or buckwheat, the protein content of amaranth is reported to be higher (Malik et al., 2023). In pseudocereals, the concentration of essential amino acids is higher, with a more balanced amino acid composition than in most cereals (Wang & Zhu, 2016; Dhull et al., 2020a; Dhull & Punia, 2020a, 2020b). However, the amino acid composition and specific protein content can vary, depending on the variety and cultivation conditions. Most of the protein (about 65%) is found in the seed coat and germ fraction of pseudocereals, whereas 35% of protein is associated with the endosperm fraction. Compared with traditional cereals, quinoa has a higher protein content. Quinoa proteins have a high biological value of 83% and provide all the essential amino acids, making them comparable with milk proteins in terms of their nutritive value (Reguera & Haros, 2016). Quinoa proteins are particularly rich in lysine, which is typically limited in most of the other cereal grains. Moreover, the wide range of amino acids in quinoa, including methionine, contributes to its excellent essential amino acid balance. Buckwheat, another highly nutritious pseudocereal, is known for its high-quality amino acid profile, primarily present in its embryo and aleurone layer regions (Reguera & Haros, 2016). Its proteins offer a higher biological value than those found in cereals, mainly due to their elevated lysine content.

2.2.2 Carbohydrates

Starch is a primary biopolymeric component in plants (Punia et al., 2019a, 2020a, 2020b; Kaur et al., 2020), that play a crucial role in the nutritional composition of pseudocereals such as quinoa, buckwheat, and amaranth. For instance, the most significant carbohydrate form in quinoa is starch, consisting of approximately 58.1–64.2% on a seed dry weight basis. Quinoa starch granules are round in shape and are located mostly in the perisperm of the kernel (Thakur et al., 2021). As far as the composition of the starch is concerned, quinoa starch has only 11% amylose content; that is lower than for other cereals, like wheat (22%), barley (26%), and rice (17%) (Inmaculada González Martín et al., 2014; Reguera & Haros, 2016; Thakur et al., 2021).

The carbohydrate content of amaranth seeds consists mainly of starch (65–75%), followed by dietary fiber (4–5%), and polysaccharide (non-starchy) components. The size of amaranth starch granules varies between 0.8 and 2.5 μm;

that is much lower than for other cereals like maize (5–25 μm), wheat (3–34 μm), and rice (3–8 μm) (Thakur et al., 2021). The smaller amaranth starch granules present increased resistance to amylases, while exhibiting higher swelling capacity, lower gelatinization temperatures, and appreciable water-binding capacity. These characteristics make amaranth starch desirable for its freeze-thaw stability and gelatinization properties in food processing industries. In buckwheat, starch serves as a major storage component in the endosperm, providing energy for plant growth. Buckwheat has starch content ranging from 59 to 70% on the basis of its dry weight, the fluctuations in starch content being attributable to different varieties and the agroclimatic conditions of the growing region. Buckwheat starch has a degree of polymerization in the range of 12–45 units, while amylose content ranges from 15 to 52% (Alvarez-Jubete et al., 2010.). The amylose content in buckwheat starch is significantly higher at 57%, compared with quinoa and amaranth. Regarding monosaccharides and disaccharides, common pseudocereals like quinoa, amaranth, and buckwheat contain fructose, glucose, xylose, arabinose, maltose, and sucrose. Although there are slightly higher amounts of these mono- and disaccharides present in pseudocereals than in other cereals, still their concentration is relatively low. Additionally, pseudocereal kernels contain complex polysaccharides, including cellulose, β-glucans, hemicelluloses, and resistant starch. These polysaccharides contribute to the dietary fiber content, which is comparable with the dietary fiber content in other cereals (Vega-Gálvez et al., 2010).

To summarize the starch characteristics of pseudocereals, amaranth contains starch in smaller concentrations than in cereals, with starch granules mainly located in the perisperm. Quinoa also relies on starch as its primary carbohydrate, with smaller polygonal starch granules and lower amylose concentration than in common cereals. The starch granules of buckwheat are polygonal or round in shape, are significantly smaller than those found in other grains, and exhibit an extraordinarily high amylose concentration, particularly in *F. esculentum* (Rosentrater & Bucklin, 2022).

2.2.3 Lipids

Pseudocereal seeds, including quinoa, buckwheat, and amaranth, have a variable lipid distribution, mainly in the form of lipid bodies. Quinoa contains higher concentrations of lipids in the embryo and perisperm, whereas, in buckwheat and amaranth, the lipids are predominantly concentrated in the endosperm and embryo. Quinoa and amaranth contain approximately two to three times higher lipid concentration than in buckwheat or classic cereals. The degree of saturation for pseudocereal lipids ranges between 75% and 86%, exhibiting higher degree of unsaturation (Wijngaard & Arendt, 2006). In pseudocereals, linoleic acid is the most abundant fatty acid, accounting for 47.5–47.8% dry weight in amaranth, 48.2–56.0% in quinoa, and 36.6–39.0% in buckwheat. Oleic acid is the next most prominent fatty acid, comprising 23.7–32.9% in amaranth, 24.5–26.7%

in quinoa, and 35.2–37.0% in buckwheat. Palmitic acid is also present, ranging from 12.3–20.9% in amaranth and 9.7–11.0% in quinoa, to 15.6–19.7% in buckwheat (Reguera & Haros, 2016; Wijngaard & Arendt, 2006). Amaranth stands out with its high content of squalene, a highly unsaturated open-chain triterpene that serves as the precursor for the entire family of steroids. Squalene levels in amaranth range from 1.9% to 11.2%. Additionally, quinoa contains 3.4–5.8% squalene in its lipid fraction. Although amaranth and quinoa contain higher levels of unsaturated lipids, the protective effect of tocopherols makes them relatively stable against oxidation (Abugoch James, 2009). On the other hand, the lipid composition of buckwheat flour makes it highly susceptible to deterioration as a result of lipid oxidation (Thakur et al., 2021), being particularly rich in unsaturated fatty acids, with linolenic acid being notable. These unsaturated fatty acids have been found to possess significant cardiovascular benefits and to improve insulin sensitivity (Punia et al., 2019b, 2020c; Kidwai et al., 2020). Remarkably, 10% of quinoa's total fatty acid content is accounted for by palmitic acid, which makes it a great source of this fatty acid. Quinoa is a superior source of essential fatty acids including linolenic acid (8.7–11.7%), oleic acid (19.7–29.5%), and linoleic acid (49.0–56.4%), making its lipid profile similar to that of soy oil. On the other hand, the total fat content of amaranth varies between 1.9% and 9.7%, based on genotype. Amaranth is notable for its higher concentrations of fatty acids such as palmitic (19%), linoleic (47%), oleic (26%), and linolenic acid (1.4%) (Thakur et al., 2021). Buckwheat is also rich in unsaturated fatty acids, which are primarily oleic and linoleic acids. Other fatty acids, such as arachidic, eicosenoic, eicosadienoic, and behenic acids are also present in buckwheat, although they make up a smaller proportion of less than 1%.

2.2.4 Vitamins and Minerals

Pseudocereals such as quinoa and amaranth are known for their rich nutritional composition, containing essential vitamins and minerals. They serve as excellent dietary sources of vitamin E, vitamin C, folic acid, and riboflavin. Quinoa is an exceptionally good source of folic acid, having 78.1 μg of folic acid per 100 g, almost twice that of wheat (40 μg/100 g). It is also a good source of vitamin A, vitamin E, thiamine, vitamin C, biotin and riboflavin. Furthermore, quinoa surpasses other grains in terms of mineral content. It provides sufficient amounts of calcium, potassium, and magnesium for a balanced diet. Amaranth is also a great source of minerals as it contains Mg, Fe, Zn, Na, and Ca in the ranges 2300–3360 mg/kg, 72–174 mg/kg, 36.2–40 mg/kg, 160–480 mg/kg and 1300–2850 mg/kg, respectively Becker et al., 1981). Additionally, amaranth contains vitamins like niacin (1.17–1.45 mg/100 g), riboflavin (0.19–0.23 mg/100 g), vitamin C (4.5 mg/100 g) and thiamine (0.07–0.1 mg/100 g) (Becker et al., 1981). Amaranth also contains various tocotrienols and tocopherols, including β- and γ-tocotrienols, and α-, β-, and δ-tocopherols. Quinoa, on the other hand, is abundant in α-tocopherol, which exhibits antioxidant activity, according to some studies a higher concentration of α-tocopherol than β-tocopherol was reported in quinoa (Abugoch James, 2009; Thakur et al., 2021). Quinoa also surpasses wheat, rice, oats, and maize in terms of carotene, riboflavin, tocopherol, and folic acid concentrations, and can provide the daily vitamin and mineral requirements for children between one and three years of age (Vega-Gálvez et al., 2010). Buckwheat has a mineral profile comparable to amaranth, with the mineral concentrations of both pseudocereals being higher than that of quinoa. Notably, macro elements in buckwheat, such as magnesium, potassium, sodium, and calcium, and microelements, like copper, zinc, manganese, and iron, are present in significantly higher levels than in quinoa. Buckwheat is also a great source of thiamine, that is generally present with a strong bond between thiamine and protein. The levels of tocopherols and B vitamins is higher in Tartary buckwheat than in common buckwheat (Reguera & Haros, 2016; Thakur et al., 2021). The γ-tocopherol is the main tocopherol in buckwheat kernels, although some studies report α-tocopherol as the predominant component (Reguera & Haros, 2016). The levels of vitamins and minerals are appreciable in all the pseudo-cereals, which can even have much richer nutrient profiles than those of common cereals.

2.2.5 Bioactive Compounds

Pseudocereals offer a rich source of bioactive compounds with various health benefits. These compounds, such as polyphenols, have been associated with the prevention of degenerative diseases, including cancer (Dhull et al., 2019, 2020b; Kaur et al., 2018) . Buckwheat contains glycosides of flavonols like quercetin, apigenin, and luteolin, whereas quinoa seeds contain glycosides of kaempferol and quercetin, and amaranth seeds contain caffeic acid, *p*-hydroxybenzoic acid, and ferulic acid (Thakur et al., 2021). These polyphenols contribute to the antioxidant and anticarcinogenic properties of these pseudocereals. Additionally, consumption of amaranth and buckwheat seeds has been linked to lower cholesterol and triglyceride levels due to the presence of tocotrienols and squalene compounds. The fagopyritols present in buckwheat grains, representing up to 40% of the total soluble carbohydrate content, have exhibited health- benefiting effects for diabetic patients. Quinoa bioactives include phenolics and other antioxidants, which contribute to improvements in intestinal health (Škrovánková et al., 2020). Furthermore, these pseudocereals also contain other bioactive compounds with notable properties. Phytosterols and saponins, traditionally considered to be antinutrients, are found in the seeds of amaranth, quinoa, and buckwheat, and possess anticarcinogenic and cholesterol-lowering properties. Betalains, which are nitrogen-containing pigments, are abundant in amaranth and quinoa. They can be classified into betacyanins, that impart red-violet colors, or betaxanthins, that result in yellow-orange colors. Moreover, betanins, a type of betacyanin, can be utilized as a natural food colorant and as competent scavengers against low-density lipoprotein (LDL) oxidation and DNA damage (Thakur et al., 2021). Homologs of vitamin E, i.e., tocopherols and tocotrienols, are found in quinoa and amaranth in significant amounts. Gamma-tocopherol is the most abundant form, followed by alpha-tocopherol, beta-tocopherol, and delta-tocopherol. Quinoa grains, available in various color varieties, also contain tocopherols, unsaturated fatty acids, and organic acids, making them promising ingredients for functional foods. The content of free phenols in pseudocereals ranges from 12.4 to 678.1 mg of gallic acid equivalent per 100 g (Škrovánková et al., 2020; Thakur et al., 2021). Amaranth products have the lowest concentrations of phenolic compounds, which are higher in quinoa- and buckwheat-derived products. A higher positive correlation was found among the antioxidant activity and phenolic content of these pseudocereal-derived products. Carotenoids, such as lutein, zeaxanthin, and beta-carotene, are present in varying amounts in quinoa and amaranth kernels. The antioxidant effect of pseudocereals is primarily because of the flavonoid concentrations naturally present in plant-based foods. Tartary buckwheat has four times the flavonoid concentration of common buckwheat. Rutin, a flavonol glycoside, is primarily found in the green parts and flowers of buckwheat plants, with higher concentrations in leaves compared with seeds (Reguera & Haros, 2016). Phytosterols, particularly beta-sitosterol, are present in small amounts in buckwheat, but they are particularly important as they cannot be synthesized by humans (Thakur et al., 2021). Proanthocyanidins are another class of oligomeric flavonoids, that are prevalent in pseudocereal seeds. Their consumption through cereals and pseudocereals has been associated with reductions in chronic ailments, as a result of their anti-inflammatory effects (Tyszka-Czochara et al., 2016).

2.2.6 Antinutrient Factors

Phenolic compounds, known for their health benefits in preventing and treating oxidative stress-related ailments, can also impair protein metabolism, having antinutrient effect (Dhull et al., 2020c, 2020d). These compounds can bind to protein substrates and digestive enzymes, interfering with their function. However, dietary polyphenols have shown potential in mitigating the harmful effects of gluten proteins in individuals with celiac disease. Moreover, buckwheat seeds also contain trypsin inhibitors, which are regarded as being antinutrient in nature. These inhibitors not only inhibit trypsin but can also bind to chymotrypsin, another digestive enzyme. Phytate is another antinutritional component, present in the protein bodies of embryo and aleurone cells of buckwheat seeds (Thakur et al., 2021). Buckwheat kernels can contain a significant amount of tannins, with up to 35–38 g/kg phytic acid in the seed bran. Moreover, phytic acid concentration for amaranth ranges from 2.9 to 7.9 g/kg, whereas, in quinoa, it ranges from 10.5 to 13.5 g/kg. Phytic acid reduces mineral bioavailability by chelating divalent cations of minerals like Fe, Mg, Zn, and Ca (Thakur et al., 2021).

Saponins, natural detergents found in various plants, are another group of antinutritional compounds. Their bitterness and negative effects on animals and humans are structure dependent. Compared to amaranth, that contains 0.9 to 4.91 mg of saponins per kg, quinoa contains higher concentrations (6.27–692.49 mg/ kg). The total saponin content of quinoa seed may range from 0.1 to 5%, in which the epicarp layer contains most of the saponins. The bitter taste associated with quinoa can be attributed to the presence of these saponins. Washing the grains with water (1:8 (w/v) ratio of quinoa to water) can help remove these saponins (Thakur et al., 2021). Compared with other pseudocereals, amaranth contains the lowest saponin concentration (0.09%), causing relatively low toxicity. The bioavailability of zinc and iron is reduced by saponins, as it tends to form complexes with these minerals. The saponin concentration in seven colored-seeded quinoa varieties ranges from 7.51 to 12.12 mg of oleanolic acid equivalents (OAE) per g dry weight. Darker-seeded quinoa varieties tend to have higher phenolic content and greater antioxidant activity compared with lighter-colored varieties. Nine phenolic compounds, including ferulic acid and gallic acid, were identified in different quinoa cultivars, in varying proportions. Free phenolic extracts of quinoa showed greater inhibitory activity against α-glucosidase activity compared with extracts containing bound phenolic components (Tyszka-Czochara et al., 2016).

2.3 Physical and Engineering Properties of Pseudocereals

Pseudocereals have different physical and engineering properties, based on their species and variety. These properties play a crucial role in the processing, handling, and storage of these grains. The size and shape of the pseudocereal seeds affects the flow behavior and packing density, that may in turn affect the processing equipment design. Bulk density, that is the mass of grain per unit volume, includes both the grain particles and the void spaces between them. It is an essential parameter for designing grain handling and storage systems. Different grains have different bulk densities, which affect storage capacities and transportation requirements. Moreover, moisture content of these grains is also a critical parameter that is directly interlinked with its quality, storage requirements, and susceptibility to microbial spoilage. Moreover, the angle of repose is the maximum angle at which a pile of grain can remain in a stable heap. It is directly linked with the surface characteristics and flowability of grains that are important for designing conveying systems. Cohesion refers to the tendency of grains to stick together, whereas friction represents the resistance between grains during flow or movement. Both cohesion and friction influence the flow behavior of grains in handling equipment, such as conveyors and chutes. They also affect the design of grain storage structures, ensuring stability and minimizing grain segregation. Thermal conductivity, specific heat capacity, and thermal diffusivity are important thermal properties of grains. These properties determine the heat transfer characteristics during drying, cooling, and processing operations. Grain moisture content significantly affects their thermal properties. Mechanical properties of grains, such as compressibility, shear strength, and elasticity, are essential for designing equipment and structures involved in grain processing, handling, and storage. These properties affect the forces required for compaction, flow, cutting, and deformation of grains. Understanding the physical and engineering properties of grains is crucial for optimizing processes in the agricultural, food processing, and storage industries. These properties inform the design of equipment, storage facilities, and transportation systems, ensuring the efficient handling and preserving grain quality.

2.3.1 Size and Shape of Pseudocereal Grains

The size of single pseudocereal grains encompasses its length, width, and thickness dimensions. These properties are essential for understanding the type and specification of cleaning, grading, and planting equipment for a particular pseudocereal. The dimensions (length, width, and thickness) of various pseudocereals are listed in Table 2.1. Among the pseudocereals, buckwheat has the largest grain size, whereas amaranth has the smallest. Apart from the general size, it is also recommended to understand the grain size in terms of various diameters like geometrical mean diameter, and arithmetic mean diameter (Rosentrater & Bucklin, 2022). The

Table 2.1   Size and Shape Parameters of the Pseudocereals

Parameter	Quinoa	Buckwheat	Amaranth
Length (mm)	1.68–2.10[a]	3.11–4.55[b]	1.35–1.50[c]
Width (mm)	1.67–2.07[a]	2.73–4.22[b]	1.22–1.37[c]
Thickness (mm)	0.85–1.01[a]	2.43–3.82[b]	0.81–0.93[c]
Geometrical mean Diameter (mm)	1.43±0.02–1.54±0.01[a]	3.41±0.2–4.34±0.28[b]	1.10–1.24[c]
Arithmetic mean diameter (mm)	1.53±0.02–1.62±0.01[a]	3.43±0.22–4.49±0.29[b]	n.r
Sphericity	0.764±0.01–0.786±0.01[a]	0.707±0.04–0.899±0.04[b]	0.81–0.83[c]
Surface area (mm²)	6.46±0.18–7.42±0.14[a]	36.78±491–59.33±770[b]	n.r

[a] (Nadiya Jan et al., 2019)
[b] (Unal et al., 2017)
[c] (Abalone et al., 2004)
n.r: not reported
values reported as: value ± standard deviation

arithmetic mean diameter is the sum of all grain diameters divided by the number of kernels, whereas the geometrical mean diameter is the average of the size of the particles, being analyzed by taking the antilog of the mean logarithm of the grain diameter. The arithmetic and geometrical diameters are again found to be highest for buckwheat and smallest for amaranth. The shape of the grain can be assessed on the basis of the sphericity of the kernel. The shape of the grain will aid in designing the plantation and sorting machinery. The sphericity of any grain, describing how closely it resembles a perfect sphere, is obtained by comparing the surface area of the grain with that of the equivalent sphere. Among all the pseudocereals, amaranth was found to have highest sphericity, although some varieties of buckwheat displayed comparable sphericities (Table 2.1).

It is obvious that the dimensions of whole-grain kernels would have been larger than those of individual processed grains. The processing steps of threshing, dehulling, and cleaning significantly impact the size of the grain. The size of the grain is also dependent on the variety and the moisture content of the grain. The grain size can also differ on the basis of the geographical location of where the grain was cultivated. The size and shape characteristics can also aid in designing the storage and packaging facilities for the pseudocereals.

2.3.2 Porosity, Bulk and True Density of Pseudocereals

Grain bulk is made up of individual grains and the air voids which are interstitially present between them. These porous air voids are generally expressed in terms of porosity or void fraction. Porosity can be expressed as the ratio of the volume occupied by air voids to the total bulk volume. The porosity of the grain bulk is highly dependent on the grain shape, arrangements of the grains, frictional and elastic properties of grains, and overall bulk pressure (Rosentrater & Bucklin, 2022). Porosity of the pseudocereals is listed in Table 2.2.

Table 2.2 Porosity, Thousand-grain Weight, Bulk and True Density of the Pseudocereals

Parameter	Quinoa	Buckwheat	Amaranth
Porosity (%)	27.47–43.20[c]	33.90–44.20[a]	35.26–42.44[b]
Bulk density (kg/m³)	645–721[c]	593–612[e]	812–891[b]
True density (kg/m³)	984–1166[c]	1077–1269[a]	1278–1470[b]
1000-grain weight (g)	2.5–3.1[c]	19.98–21.74[a]	0.79–1.2[d]

[a] (Unal et al., 2017)
[b] (Ilori & Akinyele, 2016)
[c] (Nadiya Jan et al., 2019)
[d] (Reguera & Haros, 2016)
[e] (Parde et al., 2003)

It was observed that buckwheat had the highest porosity, followed by amaranth and quinoa. The porosity of the grain bulk is generally measured with an air compression pycnometer. Average porosity of the clean pseudocereals varied between 27 and 45%, with the smaller-sized grains of quinoa and amaranth exhibiting the lower porosity values. Analysis of the porosity is important as it is helpful in designing the dimensions of the storage bins. Apart from deciding the bin dimensions, porosity also plays a decisive role in optimizing the aeration requirement in the bins. It can benefit in designing the fan arrangements in large silos for grain storage (Rosentrater & Bucklin, 2022). Additionally, porosity could also have significant effect on the application rate of fumigants, as more voids can accommodate a greater fumigant volume.

Bulk density of the grain bulk is the ratio of mass of the gain to the total volume occupied by the grain bulk. It gives an estimate of the total space that would be occupied by the grain bulk. Higher bulk density means lower space requirement as a result of a tighter grain arrangement. Bulk density estimation is primarily important for calculating load requirement in storage and handling equipment, the dimensioning of grain bins, and marketing. Generally, the Winchester Bushel Test is performed for estimating the bulk density of uncompacted grain bulk (Rosentrater & Bucklin, 2022). The bulk density values of different pseudocereals is listed in Table 2.2. By comparing the bulk density of different pseudocereals, it was evident that amaranth, due to its smallest grain size, has the highest bulk density. Additionally, the lowest bulk density was associated with buckwheat, was possibly due to its larger grain size and higher porosity of all the pseudocereals. Bulk density of a pseudocereal also varied because of the variety, moisture content of the grain, and infestation. For instance, the bulk density of quinoa decreased by 8% when the moisture content increased from 5 to 25% (on a weight basis) (Nadiya Jan et al., 2019). Moreover, infestation could also result in dry matter loss during storage, resulting in a decrease in bulk density.

True density, grain density, or absolute density is the ratio of the mass of the grain and the volume of the grain (excluding the volume occupied by air voids). Unlike bulk density, true density is based on the actual volume of the grain. The absolute density of the grain is obtained by using an air compression pycnometer that generally uses an inert gas, such as nitrogen gas. Another method for true density estimation is the toluene or alcohol replacement method. True density of the grain also provides the information about the volume requirement and can also be used for estimating the porosity of the grain bulk (Rosentrater & Bucklin, 2022). Among the pseudocereals, amaranth had the highest true density, followed by quinoa and buckwheat. The weight of the grain is generally expressed as 1000-seed weight and density of the grains could be used to estimate the volume of the average kernel. It must be noted the drying and rehydration of the grains can significantly alter all these bulk properties, and they must be estimated after such operations.

2.3.3 Angle of Repose and Coefficient of Friction of the Pseudocereals

Another important engineering property of any grain is its angle of repose. The angle of repose of of grain can be one of three types. The first is the free-filling angle of repose, that is, when the bulk of grain is poured onto a leveled surface, the pile of bulk forms an angle with the horizontal surface. This angle could increase up to the angle of repose; if it exceeds this, it tends to flow due to its own weight and again reorders itself at the angle of repose. Another type of angle of repose is achieved in the case of unconstrained filling. This happens in the case of filling the grain bulk in silos, containers, wagons, or trolleys, etc. The wall structure restricts the motion of the grain bulk and reduces the kinetic energy of the bulk, limiting the angle of the pile. However, there is no fixed trend or relation between free and constrained angle of repose. The third type of angle of repose is achieved in the case of emptying the grain bulk from the container. In this case, the angle of repose is generally greater than the other two angles of repose, and happens due to the gravitational potential and stored kinetic energy during filling. Knowledge of the angle of repose is essential for understanding and designing auger spaces, bin dimensions, conveyor dimensions, and hopper designs (Rosentrater & Bucklin, 2022). The free angle of repose of the pseudocereals varies between 15° and 33°, with amaranth exhibiting a higher angle of repose than the other two pseudocereals. The range of free angles of reposes of different pseudocereals is listed in Table 2.3. There are limited data available for emptying and filling angles of repose for pseudocereals, although one study reported that, for buckwheat, on average, the filling angle of repose varied between 21.9 and 25.6 and the emptying angle of repose varied between 21.3 and 26.7 (Parde et al., 2003). The angle of repose of any grain bulk is highly dependent on its moisture content; for instance, the angle of repose of amaranth increased by 56.5% with increasing moisture content from 12.48 to 20.09% (on a weight basis) (Ilori & Akinyele, 2016). Other factors that influence the angle of repose are surface characteristics of the grain (friction), the mass distribution of the bulk, and friction between the floor and the grain.

The coefficient of static friction is the ratio of normal force on grain bulk and the lateral shear force required to produce movement of the grain bulk over the wall material. Unlike the angle of repose, that is majorly dependent on the grain properties, the coefficient of static friction depends upon the surface characteristics of both grain and wall material. The coefficient of static friction is measured by using tilt tables, a pull box shear tester, and a triaxial shear tester. The major factors that affect the coefficient of static friction are moisture content of the grain, the bulk pressure, and kernel velocity (Rosentrater & Bucklin, 2022). The values for coefficient of internal friction for pseudocereals for different wall materials are listed in Table 2.3. It is apparent that the coefficient of static friction is higher for rougher surfaces, such as plywood, rubber and concrete. The coefficient of static friction is an important attribute for choosing and designing the wall material of storage and handling equipment.

Table 2.3   Angle of Repose and Coefficient of Static Friction with Different Materials for Various Pseudocereals

Parameter	Quinoa	Buckwheat	Amaranth
Angle of repose (°)	15.05–26.57[d]	21.30–26.70[e]	21.21–33.19[c]
Coefficient of static friction			
• Rubber	0.367–0.435[a]	0.462–0.474[b]	n.r
• Aluminum	n.r	0.431–0.4422[b]	n.r
• Stainless steel	n.r	0.396 - 0.438[e]	0.1406–0.2309[c]
• Galvanized steel	0.25–0.28[d]	0.160–0.29[e]	0.3562–0.5454[c]
• Plywood	0.31–0.36[d]	0.19–0.32[e]	0.4043–0.5869[c]
• Concrete	n.r	0.26–0.43[e]	n.r
• Glass	0.2250–0.25[d]	n.r	0.1477–0.2438[c]

[a] (Altuntas, 2018)
[b] (Unal et al., 2017)
[c] (Ilori & Akinyele, 2016)
[d] (Nadiya Jan et al., 2019)
[e] (Parde et al., 2003)
n.r: not reported

2.3.4 *Thermal, Mechanical and Aerodynamic Properties of Pseudocereals*

Thermal properties are imperative engineering aspects of grain bulk as they can directly impact a variety of mechanical and natural processes. Thermal properties, such as thermal conductivity, specific heat capacity, and thermal diffusivity, significantly affect the drying, cooling, and storage stability of the grain bulk. Hence, it is crucial to understand the thermal behavior of grain bulk in order to design any processing equipment (Rosentrater & Bucklin, 2022). As far as the pseudocereals are concerned there is a lack of research in the field of analyzing the thermal properties.

Mechanical properties of the grain explain the extent of the different forces that a grain could sustain. Mechanical properties are generally expressed as rupture force, hardness, initial cracking force, etc. The mechanical behavior of the grain could be analyzed by using a tensile testing machine or a texture analyzer (Nadiya Jan et al., 2019; Rosentrater & Bucklin, 2022). Rupture force is one of the most important mechanical properties, being the force required for completely breaking the grain. The rupture forces for quinoa and buckwheat were found to be 39.20–51.97 N and 18.72–27.59 N, respectively (Nadiya Jan et al., 2019; Unal et al., 2017). The values of rupture force depend heavily on the moisture content of the grain, and rupture force decreased by 25% when increasing the moisture content of quinoa from 5% to 25% (weight basis) (Nadiya Jan et al., 2019). Similarly, an effect was observed for the initial racking force for quinoa that also decreased by 24% in response to the same increase in the moisture content. These mechanical properties could be useful when designing dehulling, plantation, and milling machinery.

Another important engineering aspect of the pseudocereals is that they significantly affect the flow behavior of the grain during different processes, particularly in drying, storage, and transportation. Nowadays, modern processing facilities use pneumatic conveying systems and fluidized bed operations that present the importance of aerodynamic properties in the processing of pseudocereals. Aerodynamic properties of the grain include terminal velocity, drag coefficient, and flowability, etc. Terminal velocity of the grain is the highest velocity it can achieve when it is subjected to free fall through a fluid medium. The terminal velocity depends upon the size, density, and shape of the grain. Among the pseudocereals, terminal velocity values of 0.6–1.02 and 2.98–3.11 were reported for quinoa and buckwheat, respectively (Reguera & Haros, 2016; Unal et al., 2017). Terminal velocity is a crucial aspect when designing the cleaning and separation equipment like cyclone and specific gravity separators. Comprehensively, by understanding the aerodynamic properties of the pseudocereals, engineers, researchers, and farmers could improve the processing efficiencies, ensure optimum handling, and minimize energy consumption.

2.4 Conclusion

Pseudocereals are a group of plants that are considered grains due to their nutritional profile, similar to those of traditional cereals. However, they differ botanically and belong to different plant families from the cereals. The three major pseudocereals distributed globally are quinoa, amaranth, and buckwheat. Pseudocereals offer a nutritious alternative to traditional grains. They are rich sources of high-quality protein, essential amino acids, phytochemicals, and minerals. Being gluten free adds to their appeal for individuals with dietary restrictions. However, the commercialization of pseudocereals is limited due to a lack of research on their nutritional composition and processing methods. Expanding the technology portfolio and improving processing techniques will be crucial for enhancing their functional properties and nutritive value. The structural features of pseudocereals play a significant role in their nutritional composition and functional properties. Pseudocereal seeds consist of structures such as the seed coat, aleurone layer, endosperm, and embryo. The specific arrangement and composition of these structures vary among different pseudocereal species, affecting their texture, taste, cooking properties, and potential applications in food processing. When examining the chemical composition of pseudocereal kernels, proteins are a key component, characterized by a well-balanced amino acid composition. Pseudocereals are particularly rich in lysine and sulfur-containing amino acids. Carbohydrates, primarily starch, are also present in pseudocereals, with variations in granule size, amylose content, and monosaccharide and disaccharide composition. Pseudocereal lipids are predominantly found in the embryo and endosperm, exhibiting a high degree of unsaturation and containing beneficial fatty acids. Amaranth and quinoa are known for their stability against oxidation, while buckwheat presents a higher risk of deterioration due to its lipid composition. The physical and engineering properties of pseudocereals have significant implications for their processing, handling, and storage. The size and shape of pseudocereal grains influence their flow behavior and packing density, which can impact the design of processing equipment. Different grains have different bulk densities, affecting storage capacities and transportation requirements. Porosity and bulk density play a role in designing storage bins and determining aeration requirements. True density provides information about grain volume requirements and can help estimate porosity. The angle of repose and coefficient of friction are important for designing equipment, determining wall materials, and understanding grain flow behavior. While there is a lack of research on thermal properties of pseudocereals, their mechanical properties, such as rupture force and hardness, are important for designing machinery. The aerodynamic properties of pseudocereals, including terminal velocity, affect their flow behavior in processing and separation equipment. Understanding these physical and engineering properties is crucial for optimizing processes in the agricultural, food processing, and storage industries. They inform the design of equipment, storage facilities, and transportation systems, ensuring efficient handling and preservation of grain quality. In conclusion, pseudocereals are nutritionally dense crops with unique characteristics and health benefits. They offer a rich source of proteins, amino acids, phytochemicals, and minerals, while also being gluten free. Expanding research on their nutritional composition and processing methods would aid in development of more novel pseudocereal-based products. By considering and understanding these physical and engineering properties, stakeholders in the pseudocereal industry can optimize processing efficiencies, ensure proper handling, and minimize energy consumption. Further research is needed to expand knowledge in areas such as thermal properties to enhance the overall understanding of pseudocereal characteristics.

References

Abalone, R., Cassinera, A., Gastón, A., & Lara, M. A. (2004). Some physical properties of amaranth seeds. *Biosystems Engineering, 89*(1), 109–117. https://doi.org/10.1016/j.biosystemseng.2004.06.012

Abugoch James, L. E. (2009). Quinoa (*Chenopodium quinoa* Willd.): Composition, chemistry, nutritional, and functional properties. *Advances in Food and Nutrition Research, 58*, 1–31. https://doi.org/10.1016/S1043-4526(09)58001-1

Altuntas, E. (2018). *Some Selected Engineering Properties of Seven Genotypes in Quinoa Seeds.* www.aaasjournal.org

Alvarez-Jubete, L., Arendt, E. K., & Gallagher, E. (2010). Nutritive value of pseudocereals and their increasing use as functional gluten-free ingredients. *Trends in Food Science and Technology, 21*(2), 106–113. https://doi.org/10.1016/j.tifs.2009.10.014

Becker, R., Wheeler, E. L., Lorenz, K., Stafford, A. E., Grosjean, K., Betschart, A. A., & Saunders, R. M. (1981). A compositional study of amaranth grain. *Journal of Food Science, 46*(4), 1175–1180. https://doi.org/10.1111/j.1365-2621.1981.tb03018.x

Dhull, S. B., & Punia, S. (2020a). Essential fatty acids: Introduction. In *Essential Fatty Acids: Sources, Processing Effects, and Health Benefits*, edited by Dhull, S. B., Punia, S., & Sandhu, K. S., 1–18. CRC Press.

Dhull, S. B., & Punia, S. (2020b). Sources: Plants, animals and microbial. In *Essential Fatty Acids: Sources, Processing Effects, and Health Benefits*, edited by Dhull, S. B., Punia, S., & Sandhu, K. S., 19–56. CRC Press.

Dhull, S. B., Kaur, M., & Sandhu, K. S. (2020b). Antioxidant characterization and in vitro DNA damage protection potential of some Indian fenugreek (*Trigonella foenum-graecum*) cultivars: Effect of solvents. *Journal of Food Science and Technology, 57*, 3457–3466.

Dhull, S. B., Punia, S., & Sandhu, K. S. (Eds.). (2020a). *Essential Fatty Acids: Sources, Processing Effects, and Health Benefits*. CRC Press.

Dhull, S. B., Punia, S., Kidwai, M. K., Kaur, M., Chawla, P., Purewal, S. S., Sangwan, M., & Palthania, S. (2020c). Solid-state fermentation of lentil (*Lens culinaris* L.) with *Aspergillus awamori*: Effect on phenolic compounds, mineral content, and their bio-availability. *Legume Science, 2*(3), e37.

Dhull, S. B., Punia, S., Kumar, R., Kumar, M., Nain, K. B., Jangra, K., & Chudamani, C. (2020d). Solid state fermentation of fenugreek (*Trigonella foenum-graecum*): Implications on bioactive compounds, mineral content and in vitro bioavailability. *Journal of Food Science and Technology, 58*, 1927–1936 .

Dhull, S. B., Punia, S., Sandhu, K. S., Chawla, P., Kaur, R., & Singh, A. (2019). Effect of debittered fenugreek (*Trigonella foenum-graecum* L.) flour addition on physical, nutritional, antioxidant, and sensory properties of wheat flour rusk. *Legume Science, 2*(1), e21.

FAOSTAT. (2023). https://www.fao.org/faostat/en/#compare

Graziano, S., Agrimonti, C., Marmiroli, N., & Gullì, M. (2022). Utilisation and limitations of pseudocereals (quinoa, amaranth, and buckwheat) in food production: A review. In *Trends in Food Science and Technology*, Vol. 125, pp. 154–165. Elsevier Ltd. https://doi.org/10.1016/j.tifs.2022.04.007

Ilori, T. A., & Akinyele, O. A. (2016). Effect of moisture content on selected engineering properties of *Amaranthus cruentus* seed. *Certified Journal, 9001*(7). https://www.researchgate.net/publication/328461498

Inmaculada González Martín, M., Moncada, G. W., Fischer, S., & Escuredo, O. (2014). Chemical characteristics and mineral composition of quinoa by near-infrared spectroscopy. *Journal of the Science of Food and Agriculture, 94*(5), 876–881. https://doi.org/10.1002/jsfa.6325

Jimena Velarde-Salcedo, A., Bojórquez-Velázquez, E., Paulina, A., & De La Rosa, B. (2019). Amaranth.*Whole Grains and their Bioactives: composition and Health*, 209–250

Kaur, L., Dhull, S. B., Kumar, P., & Singh, A. (2020). Banana starch: Properties, description, and modified variations-A review. *International Journal of Biological Macromolecules, 165*(Part B), 2096–2102.

Kaur, P., Dhull, S. B., Sandhu, K. S., Salar, R. K., & Purewal, S. S. (2018). Tulsi (*Ocimum tenuiflorum*) seeds: In vitro DNA damage protection, bioactive compounds and antioxidant potential. *Journal of Food Measurement and Characterization, 12*(3), 1530–1538.

Kidwai, M. K., Singh, A., Malik, T., Dhull, S. B., & Punia, S. (2020). Essential fatty acid bioavailability: A dietary perspective. In *Essential Fatty Acids: Sources, Processing Effects, and Health Benefits*, edited by Dhull, S. B., Punia, S., & Sandhu, K. S., 129–156. CRC Press.

Malik, M., Sindhu, R., Dhull, S. B., Bou-Mitri, C., Singh, Y., Panwar, S., & Khatkar, B. S. (2023). Nutritional composition, functionality, and processing technologies for amaranth. *Journal of Food Processing and Preservation, 2023*, 1753029. https://doi.org/10.1155/2023/1753029

Nadiya Jan, K., Panesar, P. S., & Singh, S. (2019). Effect of moisture content on the physical and mechanical properties of quinoa seeds. *International Agrophysics, 33*(1), 41–48. https://doi.org/10.31545/intagr/104374

Parde, S. R., Johal, A., Jayas, D. S., & White, N. D. G. (2003). Physical properties of buckwheat cultivars. *Canadian Biosystems Engineering, 45*, 3–19.

Reguera, M., & Haros, C. M. (2016). Structure and composition of kernels. In *Pseudocereals: Chemistry and Technology*, 28–48. Wiley Blackwell. https://doi.org/10.1002/9781118938256.ch2

Rosentrater, K. A., & Bucklin, R. (2022). Structural, physical, and engineering properties of cereal grains and grain products. *Storage of Cereal Grains and Their Products*, 135–178. https://doi.org/10.1016/B978-0-12-812758-2.00019-2

Punia, S., Dhull, S. B., Sandhu, K. S., & Kaur, M. (2019a). Faba bean (*Vicia faba*) starch: Structure, properties, and in vitro digestibility—A review. *Legume Science, 1*(1), e18.

Punia, S., Dhull, S. B., Sandhu, K. S., Kaur, M., & Purewal, S. S. (2020a). Kidney bean (*Phaseolus vulgaris*) starch: A review. *Legume Science, 2*(3), e52.

Punia, S., Dhull, S. B., Siroha, A. K., Sandhu, K. S., & Chaudhary, V. (2020c). Mechanism of action of essential fatty acids. In *Essential Fatty Acids: Sources, Processing Effects, and Health Benefits*, edited by Dhull, S. B., Punia, S., & Sandhu, K. S., 89–100. CRC Press.

Punia, S., Sandhu, K. S., Dhull, S. B., Siroha, A. K., Purewal, S. S., Kaur, M., & Kidwai, M. K. (2020b). Oat starch: Physico-chemical, morphological, rheological characteristics and its application-A review. *International Journal of Biological Macromolecules, 154*, 493–498.

Punia, S., Sandhu, K. S., Siroha, A. K., & Dhull, S. B. (2019b). Omega-3 metabolism, absorption, bioavailability and health benefits - a review. *PharmaNutrition, 10*, 100162.

Škrovánková, S., Válková, D., & Mlček, J. (2020). Polyphenols and antioxidant activity in pseudocereals and their products. *Potravinarstvo Slovak Journal of Food Sciences, 14*, 365–370. https://doi.org/10.5219/1341

Thakur, P., Kumar, K., & Dhaliwal, H. S. (2021). Nutritional facts, bio-active components and processing aspects of pseudocereals: A comprehensive review. In *Food Bioscience*, Vol. 42. Elsevier Ltd. https://doi.org/10.1016/j.fbio.2021.101170

Tyszka-Czochara, M., Pasko, P., Zagrodzki, P., Gajdzik, E., Wietecha-Posluszny, R., & Gorinstein, S. (2016). Selenium supplementation of amaranth sprouts influences betacyanin content and improves anti-inflammatory properties via NFκB in murine RAW 264.7 macrophages. *Biological Trace Element Research, 169*(2), 320–330. https://doi.org/10.1007/S12011-015-0429-X

Unal, H., Izli, G., Izli, N., & Asik, B. B. (2017). Comparación de algunascaracterísticasfísicas y químicas de losgranos de trigo sarraceno (*Fagopyrum esculentum* Moench). *CYTA - Journal of Food, 15*(2), 257–265. https://doi.org/10.1080/19476337.2016.1245678

Vega-Gálvez, A., Miranda, M., Vergara, J., Uribe, E., Puente, L., & Martínez, E. A. (2010). Nutrition facts and functional potential of quinoa (*Chenopodium quinoa* Willd.), an ancient Andean grain: A review. *90*(15), 2541–2547.https://doi.org/10.1002/jsfa.4158

Wang, S., & Zhu, F. (2016). Formulation and quality attributes of quinoa food products. *Food and Bioprocess Technology*, *9*(1), 49–68. https://doi.org/10.1007/s11947-015 -1584-y

Wijngaard, H. H., & Arendt, E. K. (2006). Buckwheat. *Cereal Chemistry*, *83*(4), 391–401. https://doi.org/10.1094/CC-83-0391

Chapter 3

Quinoa: Nutrition Profile, Processing and Food Products

Alpana Prajapati, Ajay Kumar Singh

3.1 Introduction

Quinoa (*Chenopodium quinoa* Willd.), pronounced KEEN-wah, is part of the class Dicotyledoneae and belongs to the Chenopodiaceae family. About 250 species come under the genus *Chenopodium*. Quinoa is the seed of the *C. quinoa* plant; botanically, it is not considered a grain and is thus referred to as a pseudo-grain or simply a false grain, with the crops categorized as pseudocereals. Other crops which belong to the pseudocereals are the chenopods, amaranths and buckwheat [1]. It is an allotetraploid species with a chromosome number that is 2n = 4x = 36 because of the random hybridisation between two diploid species of *Chenopodium* millions of years ago, followed by chromosome doubling [2, 3]. It has a flat or oval-shaped seed, pale yellow in colour, but other varieties have seeds of a red or black colour. Quinoa has gained worldwide recognition due to its versatility to flourish in any environmental conditions and its great nutritional and functional qualities. The UN Food and Agriculture Organisation, in its 37th session of the general conference, declared 2013 the International Year of Quinoa.

3.2 Botanical Description

Quinoa is a gynomonoecious (having monoclinous and pistillate flowers on the same plant and an absence of staminate flowers) annual plant. It contains betacyanins, the

DOI: 10.1201/9781003325277-3

stem is erect and it has a tap root (roots that grow vertically downward). Leaves are variously coloured, with an alternate pattern. In India, quinoa shows excellent crop growth, with many cultivars producing large leaves, branches and a height reaching up to 1.5 m. Roots penetrate 1.5 m below the surface, which helps in drought because they can store water and food for later use. Polymorphism is exhibited in the leaves; lower leaves are rhomboidal and upper leaves are lanceolate. The panicle inflorescence has a length of between 15 and 70 cm at the top and the principal axis leads to the generation of two types of secondary axis, the amarantiform and the glomerulate. Female flowers are unisexual (the presence of hermaphrodites at the distal end contains five anthers, five perianth lobes and with two or three stigmatic branches on a superior ovary). In some or all female flowers, males are represented as sterile. The fruits are typically either conical, cylindrical or slightly squashed spheres (ellipsoidal). The mature fruit is a kernel comprised of several layers, viz. (from outside to inside) perigonium, pericarp (saponins are concentrated here) and episperm. Seed colour and size may vary from black to red and yellow, with black being dominant over others.

3.3 Historical Background

Quinoa is a long-established dicotyledonous crop of South America domesticated from 5000 BC to 3000 BC, and indigenous to the Andes region [4, 5]. During the Spanish colonial period, quinoa was specified as a religious and aesthetic crop, and they considered it a non-Christian food. This influenced Spanish people to substitute it with other cereals [6–8]. However, because of its commendable nutritional profile, quinoa production has risen and contributed to food scarcity in underprivileged areas for a couple of decades. In 2018, it came to light that the globalization of quinoa had reached 123 countries. Even though Peru and Bolivia export almost three-quarters of global quinoa supplies, production of quinoa has spread beyond the Andes (Figure 3.1). Consumption of quinoa in Bolivia annually was around 2.37 kg per person, compared with about 1.15 kg per person in Peru [9].

This expansion led to the cultivation of quinoa under difficult climatic conditions which exert adaptation to different environments, resulting in the generation of new varieties.

3.4 Quinoa Production

Quinoa evolved strong tolerance to different stresses, for instance, drought, salinity, humidity (from 40% to 88%), and temperature (-4°C to 38°C). For example, it is revealed that the phenolic content and antioxidant capacity of quinoa seeds are elevated in response to stress [10]. Also, when two Chilean genotypes were grown in two different climatic zones, results showed that in

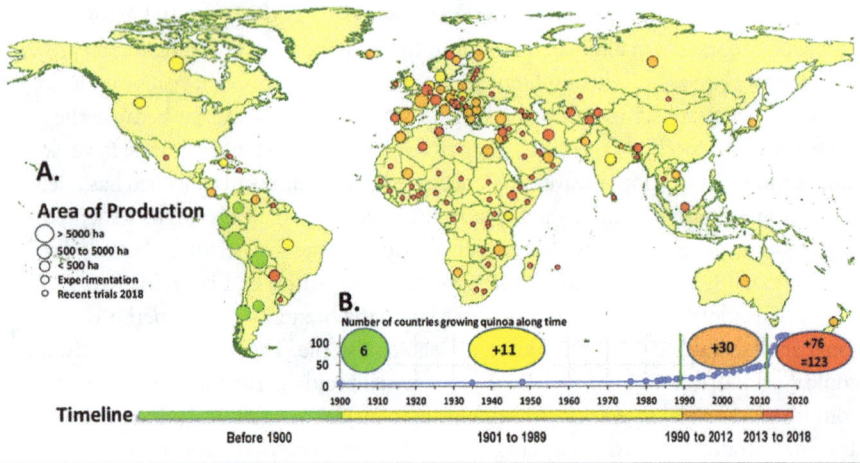

Figure 3.1   Production of quinoa expanded to other countries from 1990 to 2018

cold rainy zones there was an increase in the weight and size of the quinoa seeds, whereas under hot arid conditions there was an increase in proximate components (except for protein) and phenolic compounds [11]. Sowing can be done in various patterns such as broadcasting, sowing in rows, mixed, or by transplanting; spacing (25–50cm between rows) is preferred for quinoa crops as it allows easy hoeing. Seeds should be sown 1–2 cm deep in a finely structured seedbed with moist, well-drained soil. Nitrogenous fertilizers do not affect grain yield and protein content.

3.5 Types of Quinoa

Due to changes in climatic conditions, different varieties of quinoa plants have developed; presently, there are over 1800 varieties available in 120 *Chenopodium* species available. Previously, based on seed colour, they were only classified as white or ivory, red or black. But currently, based on colour, they are classified as black, purple, pink, orange, green, yellow, grey or red (Figure 3.2). The presence of phenolic content and saponins in the pericarp (outer covering) causes such variation in seed colour [12].

3.5.1 White Quinoa

Easily found in local markets and other health or organic food stores. It has a subtle flavour.

Figure 3.2 Types of quinoa based on seed colour

3.5.2 *Red Quinoa*

Due to the moderate amount of phenolic compounds, seed colour ranges from intense orange to dark red, and grains have a great earthy and fruity flavour. However, the colour turns from red to brown after cooking.

3.5.3 *Black Quinoa*

This was developed by cross-breeding quinoa seeds with wild spinach, also called lambs quarters (the Indian name is palak) introduced by farmers in the Colorado

Rockies. Due to the presence of anthocyanins (dark purple coloured), the seeds' colour ranges from dark purple to black. It has an earthy, hearty flavour and is more crunchy than other types of quinoa. Black quinoa cooks faster than other types and is more fibrous. Its crunchiness remains even after cooking but the colour turns inky. It is less available in stores and markets.

Quinoa contains all essential amino acids necessary for the body, being rich in minerals such as calcium, potassium and magnesium, as well as vitamins. It is a good source of dietary fibre and, more importantly, it is a gluten-free low-calorie grain which supports the management of symptoms of celiac disease (a disorder resulting in indigestion, where the symptoms are diarrhoea, bloating, fatigue and anaemia) and other gluten-related diseases. It is a good substitute for various staple foods commonly consumed in India, for instance, rice, wheat and pulses. For diabetic patients who cannot consume rice because of its high glycemic index, quinoa could be the better option. Quinoa is achieving recognition and, because of its nutritional properties and low production cost, health-conscious individuals are switching their choices from rice or wheat to quinoa. Its nutritional quality is appreciated by the National Aeronautics and Space Administration (NASA) and the National Research Council (NRC). NASA has taken quinoa as part of the Controlled Ecological Life Support System (CELSS) [13]. The Food and Agriculture Organization (FAO) intends to promote the cultivation of traditional crops like quinoa and selected it as a crop for greater food security because it supports sustainable production. The health benefits of quinoa indicate that it is a good functional food, with the health benefits possibly being due to the presence of bioactive phytochemicals.

Quinoa contains nutritional as well as anti-nutritional factors, the latter not being generally good for health, but which could be reduced or inactivated to safe health-compatible levels when proper and suitable processing techniques are used, whether in household preparation of food or at industrial-level processing.

Traditionally, quinoa was consumed as a whole grain similar to rice but recently it has also been consumed as pasta or bread made using quinoa flour. There are various recipes to cook quinoa, and more than 100 preparations. In terms of commercialisation, various products are available made with quinoa, like pasta, baby food, cookies, muffins, cereals, snacks, bread, flakes, drinks and diet supplements, although the quinoa content in such food products is up to only 20%. Some people also consume the shoots and leaves of the plant uncooked (in salads) or even cooked as a meal because it is rich in fibre and supports good digestion.

Although quinoa has excellent qualities, different processing methods may affect its nutrient composition; for instance, if the grain is washed with running water to remove anti-nutritionals, the phenolic content may significantly increase. Moreover, if the grains are cooked under pressure after washing, the amounts of phenolic compounds may increase. If the grains are washed followed by cooking at atmospheric pressure the phenolic content does not change as compared to washed-only grains, although, when grains were subjected to washing and then hydrated in water, phenolic content decreases significantly relative to when the grains were only washed. Washed and toasted grains show a considerable decrease in total phenolic content. This may be due to the thermal processing (such as steamed leaves of sweet potato and beans after cooking, whether normal or pressure cooked), resulting in the loss of water-soluble phenolic compounds.

3.6 Nutritional Profile

Presently, people are more aware of their health. To keep pace with the growing population, the availability of complete and nutritious food is crucial. The Food and Agriculture Organization (FAO) released a volume of fact sheets in 2016 which included a fact sheet entitled "Nutritional Benefits of Pulses", in which it is stated that the nutritive value of pulses is amplified when melded with other foods, as this ensures that the body absorbs nutrients more efficiently. For instance, the body cannot produce some ("essential") amino acids, so they need to be provided by pulses in combination with cereals to fulfil protein requirements. Another component that is present in huge amounts in quinoa is starch, as quinoa has a large perisperm enclosed by an embryo layer. The presence of an appropriate amount of dietary fibre improves digestion and stomach-related issues. Quinoa contains proximate compounds along with phytochemical compounds that provide effective health benefits together with antioxidant, anti-obesity, antimicrobial, anti-inflammatory and antihypertensive activities [14]. There are also some anti-nutrients present in quinoa, such as saponins (0.1–5%), phytic acid (1.05–1.35%) and some protease inhibitors; anti-nutritionals decrease the nutritional value, damage sensory properties and lower the quality of quinoa-based food products [15].

3.7 Chemical Composition of Quinoa

Nutritional composition varies among the ecotypes, varieties and locations. Quinoa dry weight is composed mainly of carbohydrates (60–75%), while the dietary fibre component represents 10–13%. Protein content in quinoa (12–16%) is higher than in other cereals, and quinoa is high in unsaturated fatty acids, although the fat content ranges from 5% to 9%. Bioactive compounds, such as polyphenols, are also present in high concentrations (2.7–3.8 g/kg) [16]. Other bioactive compounds in quinoa, which are powerful antioxidants, are flavonoids and carotenoids which can contribute to human health.

3.7.1 Protein

Protein percentage in quinoa seeds ranges from 11% to 19%. Moreover, it contains all nine essential amino acids which are required for good health in humans [17]. Quinoa protein also exhibits gluten-free quality and high digestibility. Quinoa is superior to any cereal (e.g., rice, wheat, barley, and maize) in terms of protein content, with the concentrations of sulphur-containing amino acids (methionine and cysteine) being higher, whereas the aromatic amino acids (tyrosine and phenylalanine) are the limiting amino acids, followed by threonine and then lysine. It is also found that washing quinoa seeds does not significantly affect the protein content. An experiment on animals observed that the Net Protein Utilization (NPU, the amount of protein (in percentage) taken in by the body after eating that food) rate of raw quinoa is 75.7 NPU, whereas the Biological Value (BV) is 82.6, representing the percentage of unused protein from a food retained by the body for protein synthesis, and the True Digestibility (TD) is 91.7, representing the percentage of protein absorbed from the gastrointestinal tract [18]. The protein quality in quinoa is somewhat like casein, so it would not be surprising to replace dairy products with this pseudocereal.

3.7.2 Carbohydrates

The carbohydrate content of the quinoa seed dry matter ranges from 67% to 74%, consisting of free sugars (glucose, fructose, sucrose and maltose), starch at about 55–65%, located as simple units (smaller in size than 3 μm) at the perisperm in a spherical shape. Dietary fibres make up about 10%. Dietary fibre is the carbohydrate fraction that helps in digestion due to its resistance towards digestive enzymes during absorption in the small intestine; later, partial or full digestion is done in the large intestine, thus providing sound digestive health [19]. The starch present in quinoa seed is submicron-sized and composed of amylopectin which is a polygonal and spherical form [20]. The viscosity of quinoa starch is higher than cereal flour, so it could be used as a condenser, with a high concentration of minerals.

3.7.3 Lipids

Quinoa seed contains about 6.9% lipids, which is superior to the value for other grains [15]. Quinoa seed oil has an unsaturated fatty acid concentration 20 times higher than that of rice oil. Palmitic oil is the major source of saturated fatty acids in quinoa, whereas the main unsaturated fatty acids make up 87–88% of the total fatty acids, consisting of linoleic acid (18:2n-6; 49–57%), oleic acid (18:1n-9; 20–30%) and α-linolenic acid (18:3n-3; 8.5–12%) [21, 22]. Quinoa is also rich in oil and lipid-soluble antioxidants like α- and γ-tocopherol, which provide the natural antioxidant potential that increases the availability of oil at the cell membrane level and protects fatty acids from free radical oxidation [23].

3.7.4 Dietary Fibre

Fibre is an indigestible component of plants that consists of two principal elements: soluble fibre and insoluble fibre. Soluble fibres are easily dissolved in water, fermented in colon to generate gases, are physiologically active and have prebiotic properties. On the other hand, insoluble fibres do not dissolve in water, provide bulking mass and are metabolically inert or they can be prebiotic and fermented metabolically in the large intestine. Bulking fibres help to ease defecation. High-fibre-rich whole grains lower the risk of cardiovascular disease and type 2 diabetes. Dietary fibre in quinoa is about 2.6–10% of the total dry weight of the grain, consisting of 78% and 22% of the insoluble fibres and soluble fibres, respectively [24].

3.7.5 Gluten

Gluten is a structural protein found in certain grains, mainly in wheat, barley and rye, where it provides elasticity and gives food a chewy texture. Although it is generally not a problem, some people are not able to tolerate gluten, as it causes them to suffer bloating, diarrhoea and stomach pain. In some people, gluten triggers an immune response, causing celiac disease, an autoimmune disease. Quinoa is one of the healthiest grains, containing high concentrations of antioxidants; the absence of gluten makes quinoa seed products suitable for celiac disease patients.

3.8 Microconstituents

3.8.1 Minerals

Dietary minerals are necessary as they play an important role in body function by regulating electrolyte balance, glucose homoeostasis, transmission of nerve impulses and function as enzyme cofactors. Ash (residues remaining after burning a solid substance) determines the inorganic and mineral availability. Quinoa contains a wide range of minerals, and the ash content of its seeds is higher than that of any other cereal crop, such as maize, rice or wheat.

3.8.2 Vitamins

These are essential nutrients for proper body function. Quinoa is rich in the vitamin B complex like folic acid (B_6) and riboflavin (B_2). Thiamine (B_1) is present in similar concentrations as others, but the amount of niacin (B_3) is less.

3.8.3 Saponins

Saponin content in quinoa grain was determined spectrophotometrically to be 3.33% [25]. Saponins are plant-derived organic chemicals, normally triterpenes, which are present in the seed's outer layer, which can be toxic; when blended in water, saponins produce a foam. If the saponin concentration is <0.11%, it is sweet in taste; if >0.11%, it tastes bitter. Quinoa seed also contains saponin triterpenes, mostly as aglycones, for instance, monodesmosidic of hederagenin, bidesmosidic of oleanolic acid, serjanic acid and phytolaccagenic acid. The tridesmosidic saponin of hederagenin is also present, as are various other aglycones [26]. Saponin content may vary among different growing stages; at the time of branching, saponin content is low, whereas, at the blooming stage, saponin content is high. They are removed by processing, either by the dry method (toasting and subsequent rubbing to remove the outer layer) or by the wet method (washed and rubbed in cold water). For large-scale or commercial purposes, abrasive dehulling is carried out to remove saponins but the drawback is that some saponins may still be intact with the perisperm. Saponins present in quinoa have the potential for use for medical purposes as they induce intestinal permeability and help in the absorption of some drugs [27]. It has been reported that quinoa saponins may have the potential to function as an adjuvant for mucosal administration of vaccines.

3.9 Phytochemicals in Quinoa

Quinoa is a good source of several bioactives (secondary plant metabolites), such as carotenoids, phytosterols, squalene, fagopyritols, phyecdysteroids and phenolic compounds, including dietary fibres.

Phenolic compounds are grouped into two main categories: phenolic acids and polyphenols. Phenolic acids, such as gallic acid, ferulic acid, coumaric acid and ellagic acid, contain one aromatic ring, and have strong antioxidant properties, as well as including anti-allergenic, anti-inflammatory, antiviral, anticarcinogenic, and hypocholesterolaemic effects. On the other hand, polyphenols, such as tannins, lignans, flavonoids, etc., contain more than three aromatic rings and are mainly present in the pericarp; these bioactives protect plants from insects and pests. In food, polyphenols impart colour and flavour; if present in excess, they cause bitterness.

Flavonoids and total phenolic content (TPC) in different varieties of quinoa have been reported in the range of 36.2–144.3 mg/100g and 16.8–59.7 of grain, respectively [28]. On average, TPC in quinoa is 71.7 mg gallic acid equivalents (GAE)/100 g, higher than wheat (53.1 mg GAE/100 g) and amaranth (21.2 mg GAE/100 g) [29]. Phenolic compounds vary with the type of seeds; in red quinoa seeds, TPC is 50% greater, with flavonoid content 90% higher, and antioxidant activity 150% higher than in yellow quinoa seed. TPC in black, red, and white quinoa seeds is approximately 68.20, 63.46 and 46.69 mg/100 g, respectively [30].

3.10 Health Benefits of Quinoa

Quinoa is also known as "the mother food" or the "superfood" and is considered a gift from God, even for medicinal purposes. Major health issues, such as allergenicity, celiac disease and diabetes, can be prevented by quinoa seed consumption. Below are some studies which have been carried out to highlight the benefits of quinoa on health.

3.10.1 Allergenicity

In a study, it was observed that consuming quinoa did not affect wheat protein antibodies and it was advised that quinoa could be consumed by patients who were allergic to wheat proteins [31]. Quinoa has some anti-allergic properties, which were revealed by a case study. A 52-year-old French man who consumed quinoa with fish and bread developed a structured reaction consisting of dysphonia, dysphagia, angioedema and urticaria; it was confirmed, when samples of the ingested food were tested, that samples having quinoa were reactive towards immunoglobulin E (IgE) in serum [32].

3.10.2 Celiac disease

The Glycaemic Index (GI) is the value of carbohydrates present in food on a scale of 0 to 100 according to their impact on blood sugar levels two hours after eating. Foods which have GI values less than 55 produce a gradual rise in insulin and blood sugar levels; that is, low-GI diets show improved glucose levels and lipid levels, while also maintaining weight because they help to control appetite [33, 34]. It has been reported that individuals with high dietary fibre intakes have lower fasting insulin [35] and longevity increases as complex carbohydrates are ingested [36]. Celiac disease can be ameliorated by a low-GI diet.

3.10.3 Diabetes

Diabetes is a metabolic disorder in which the body suffers high sugar levels for a long time, and is caused when there is inadequate insulin production or insulin resistance. It is not restricted to any specific age group, but middle-aged people are more vulnerable, particularly in low- to middle-income countries, so it is important to identify glycaemic control aimed at cost-effective interference. Alpha-glucosidase inhibitors in quinoa act on the brush-bordered epithelial cells of the small intestine and inactivate enzymes responsible for digestion of complex carbohydrates, consequently hindering secretion of glucose into the blood and improved glucose tolerance at subsequent meals [37].

3.11 Biotic and Abiotic Factors Affecting Nutritional Profile

Factors which affect the nutritional quality of any crop can be stresses, biotic or abiotic, where biotic stresses include living organisms, like microbes, insects, weeds and native plants, whereas abiotic stresses are negative effects of non-living factors on living organisms at appropriate environments, like ultraviolet light, temperature extremes, flood, drought and soil salinity. The major issues that affect the nutritional profile of quinoa are discussed below.

3.11.1 Biotic Factors

Quinoa is affected by various diseases like mildew, damping off, blight, and mosaic caused by a variety of pathogens.

3.11.1.1 Downy Mildew

This is a plant disease caused by *Peronospora farinosa*, an oomycete that thrives under cool and moist conditions. It causes severe damage to the crop and reduces the crop yield by 33–58% even in the most resistant cultivars [38]. Quinoa showed a high resistance level towards downy mildew in northern India.

3.11.1.2 Insects and Pests

Pests cause damage ranging from 8% to 40% [39]. Birds also affect the crop as they damage the inflorescence, but this is not serious because of the presence of saponins, which resist damage by initiating chemical defence and confer protection against pests and birds [40].

3.11.1.3 Damping-off

This disease of seedlings or germinating seeds is primarily caused by fungi. Seeds become colonized, growth is stunted and plant tissues rot near the soil surface soon after germination. The main causative agents of damping off in quinoa are *Ascochyta caulina*, *Pythium* spp., *Fusarium* spp., *Fusarium avenaceum*, and *Alternaria* spp. Such fungi are found under greenhouse conditions on the infected parts of quinoa seedlings and are isolated using a pathogenicity test [41].

3.11.1.4 Other Fungal Diseases

Stalk rot is caused by *Phoma exigua* var. *foveata*, where leaves appear brown, with glossy abrasions on the stem. *Phoma cava* causes reddish-brown to black lesions visible on the base of the stem; in serious cases, it spreads all over the stem under cool conditions between 15 and 25°C. Grey mould is a fungal disease caused by *Botrytis cinerea* which attacks soft plant tissues, causing the growth of fuzzy grey-brown mould. Leaf spot disease is caused by *Ascochyta hydalospora* affects seeds and is a trash-borne disease that infects both winter- and spring-sown crops, appearing as pod spots with little sunken, grey centres and a brown margin, developing into elongated stem lesions.

3.11.1.5 Bacterial Blight

The pathogenic agents are bacteria, mainly *Pseudomonas* spp., quinoa is affected by *Pseudomonas syringae,* a Gram-negative bacillus, causing severe yellowing, browning, spotting, withering, or dying of some parts of a plant or even the entire plant.

3.11.1.6 Chlorotic Mosaic

This is caused by a viral pathogen caused by *Chenopodium* mosaic virus.

3.11.1.7 False Nodule

The causative agents are the plant-parasitic nematodes *Nacobbus* spp. and *Theca vermiculatus* spp.

3.11.2 Abiotic Stresses

3.11.2.1 Temperature

Temperature is an important factor in crop productivity, with, for instance, low temperature being the limiting factor for species distribution of quinoa.

3.11.2.2 Drought

Water is a necessity for any plant, and quinoa has a high capability to grow under stressful conditions like drought. But excess irrigation affects yield; in the seedling stage, damping-off and stunting may occur, leading to a decrease in plant height but not in the yield [42]. The low water requirement of quinoa crops makes them suitable to grow under water-deficient conditions and in places where water availability is limited so that farmers do not have to depend on seasonal rains.

3.12 Processing

After harvesting, quinoa undergoes post-harvest operations to ensure quality. This procedure includes drying or stacking, threshing, winnowing and storage of grains [43, 44].

3.12.1 Drying and Stacking

Drying and stacking of the harvested plants involves arrangement of plants after drying into stacks or piling harvested crops for storage. It can be done by different methods, such as Taucas (panicles ordered towards the same side), Arcos (stacking in crosses), or Chucus (mounds in a cone shape).

3.12.2 Threshing

This involves separating the grains from the panicle by beating the grains to separate them from the straw. It can be done manually or by using machines. This should be done immediately after harvesting to get high-quality harvested grains.

3.12.3 Winnowing

This is the method for separating grains from the chaff or straw. Winnowing is achieved by tossing and lifting the threshed grain that blows out the lighter chaff to one side, for the heavier seeds to fall vertically. There are three methods by which this process is done. First is the traditional method in which wind is used, achieving low yields of about 400 kg per day. Second is a manual method with some modifications, where machines regulate air flow, passing 600 kg of yield per hour, and the third is fully machine-based, which has a flow capacity of 500 to 800 kg yield per hour which is the highest of the three methods.

3.12.4 Storage

This is the last step in post-harvest processing, storing grains for further use under appropriate conditions. Grains should be stored in a cold, dry and dark place to protect them from moisture and pests. Grains are mainly stored in mylar or polyethylene bags.

To meet the quality standards of quinoa seeds, further processing needs to be done. A preliminary step involves cleaning to remove impurities, washing to remove saponin, drying and then sorting to separate grains based on physical appearance, i.e., size and colour [43, 44]. As a result of the high nutritional profile and low cultivation cost, quinoa production is increasing steadily. Various industrial-scale advancements in processing were introduced to postharvest processing to relieve traditional methods used in small-scale production. The quinoa grain production chain is divided into three phases: agricultural production, processing and value-added food product development.

Although saponins are a stumbling block to the consumption of quinoa because of its bitter taste, it also has valuable qualities like antioxidant activity and anti-inflammatory effect, and it may also reduce blood cholesterol levels. To overcome this problem, researchers have created the "Quinoa Alliance" programme that focuses on creating a second step in quinoa processing [45]. Quinoa cultivation and its processing are not common, so the specific machinery is not well developed to overcome technological problems that restrict the processing as well as the market for such technologies. Adapted technologies are being used by some companies which are less efficient and cause problems during the processing chain. For example, rice scarification (to break seed dormancy by scratching or mechanical rubbing grains) is now used in quinoa processing where the quinoa grain runs against a metallic net to ensure saponin is removed, with increasing friction accelerating the process. The obstacles discovered here in post-harvest processing gave rise to innovation in technology to overcome the drawbacks. Consequently, some inefficiencies were noted: significant loss of raw materials; diminished grain quality; increase in production costs due to high consumptions of water, electrical energy and gas; high operational labour costs and generation of wastewater residue with high saponin content. It was later determined that these inefficiencies led to water pollution, as well as the recovery of unusable saponin which loses its commercial value. Table 3.1

Table 3.1 The Solution to Some Inefficiencies Occurs by Using the Traditional Method of Quinoa Processing [46]

Source of inefficiencies	Their solution
Adaption of technologies which are originally for the processing of other grains, used appropriately for quinoa (e.g., scarification).	Developing and implementing an efficient system for dry cleaning, and using intrinsic coarse feature of quinoa for scarification.
Using technologies that did not focus on the recovery of high commercial value sub-products. (e.g., saponins)	Installing highly efficient saponin recovery system to retrieve an important sub-product having high economic value in the market.
Employing drying systems with inadequate air flow allows re-humidification of some part of the product.	Using turbines in drying system for adequate airflow.
Excessive and unnecessary number of unit operations in the process.	Installing robust drying and washing steps and eliminating unwanted processing steps.
Using technologies that operate in small batches, rather than a continuous process.	Operating in continuity requires fewer operators.
Using washing systems, with variable and large residence time range, leads to variability in products (hardly all grains are washed for a precise amount of time).	Creating a washing system that simulates the laminar trajectory of the grain, using a turbulent flow, thus creating a correlative process. Also, reduces residence time.

displays the cause of inefficiencies during quinoa processing and the solutions that have been adopted to make processing more effective according to the results of the "Quinoa Alliance" programme.

Processing of quinoa after harvest is important to maintain the shelf-life of grains, to preserve their quality and protect them from hazards. However, the quality of food may be affected by several processing techniques. The primary objective of processing is to remove saponins (a toxic metabolite) from the grains, and further process changes according to the food product to be used for instance, soup, flour, etc. [46].

It is suggested that, for the extraction of bioactive compounds from quinoa bran, techniques could be used such as albumin with chemically assisted electronic fractionation, while phenolic acids, triterpene saponins and flavonoids could be extracted using an ultrasound-assisted extraction method [47].

Thermal processing of quinoa flour can degrade the saponins, enhancing the organoleptic and pharmacological properties, whereas roasting and extrusion directly affect the chemical composition of flour and reminiscence of saponin substances takes place that can be visualized in Nuclear Magnetic Resonance (NMR) spectra. Steam preconditioning moderately affects the chemical profile, and this comparison of processing methods and their effect on quinoa flour is determined by high-performance liquid chromatography (HPLC) [48].

3.13 Utilization of By-products of Quinoa

3.13.1 Saponins

As the consumption of quinoa increases because of its nutritional value, researchers are interested in developing new technologies. The most significant innovation in quinoa processing is unit operations for the removal of saponins; as mentioned before, saponins give a bitter taste to the product so they must be removed to achieve desirable organoleptic properties. However, this by-product of quinoa can be used in other industries.

The physiological and biological properties of saponin made it useful for commercial applications in agriculture such as the preparation of bio-insecticides and food, as well as in the cosmetics and pharmaceutical sector. For instance, the ability to foam at low concentrations makes saponins a good natural foaming agent. Even though it is an antinutritional component, it also has some positive health effects, such as antifungal, anticarcinogenic, and cholesterol-lowering properties, which attract the pharmaceutical industry.

In studies, it has been suggested that phytochemicals, such as saponins, can be transformed if the alkali transformation method is used [49]. In an experiment, it was observed that the alkali-transformed simple quinoa saponins show remarkable inhibitory activity against bacteria.

3.13.2 Husks

The husk is the dried outer covering of seeds or some fruits. In a study, husks have been used to develop molluscicides (pesticides against molluscs). The golden apple snail (GAS) is the most dangerous species to crops, threatening rice agriculture; due to advances in processing, novel molluscicides have been developed against GAS from quinoa husk. It is better than synthetic molluscicides because, even at the highest concentration, it shows no toxicity towards the fish involved in rice cultivation. It was observed that the quinoa husk-based molluscicide can kill GAS completely, without affecting fish. By application of 11 ppm quinoa saponin, within 48 h, GAS mortality reaches 94%, without affecting the growth of the rice shoots [50].

3.13.3 Biomass

In certain rural areas, using agricultural waste as a source of energy with which to cook food is still practised, because modern energy sources are limited for farmers. Quinoa pellets could be an alternative energy source in rural areas instead of conventional fuel sources. Pellets have been produced by combining starch with ash, the latter being produced by burning the biomass of the crop. Quinoa biomass residues from the crop are available in large quantities and they can be used as feed for animals, as fodder can not be grown at elevated altitudes. Its seeds and husks are also used as animal feed; this is just the dry matter yield that has a high protein concentration and good digestibility. The quinoa bran can substitute 30% of normal feed consumption, using feed with antiparasitic potential in its by-products. When feeding quinoa by-products to non-ruminant livestock, make sure that it does not exceed 30% in the diet; on the other hand, the presence of leaves, stalks or saponin from the harvest may not be involved in feeding ruminant livestock. Using quinoa by-products for feedstock needs more research because cost and competition issues may arise. Saponin is an excellent foaming and emulsifying agent and extends its applications from the food industry to other industries like cosmetics and pharmaceuticals [51].

3.13.4 Effect of Processing on Phytochemical Content

Domestic processing of quinoa seeds involves washing and soaking seeds in water which may influence the nutritional quality of seeds. For instance, it has been reported that domestic methods, like soaking then germination, increases vitamin C by 15% in domestically processed Indian quinoa seeds, compared with 46% in industrially processed quinoa [52].

Germinated seeds exhibit a 134% surge in total phenolic content and a 90% rise in total antioxidant activity compared with raw quinoa seeds [53]. Phenolic compound content and antioxidant capacity can be increased by only washing seeds under running water, also minimising the saponin concentration by 2.75%

(because saponins are soluble in water) and their bitter taste from the grains. But cooking or toasting does not affect the saponin content. While hydration temperature plays a significant role in releasing saponin, washing then hydrating seeds in water increases the reduction in saponin content by 3.63% compared with only-washed grains. Hydration helps to penetrate their interior and release more saponin by diffusion [54]. Using the dry methods (like toasting and abrasive rubbing of the grains) to remove saponins results in a reduction in vitamins and minerals.

During the puffing process, depletion in amino acid content occurs, and ultimately a loss of proteins. Exposure to high temperature for a long time during the process can adversely affect nutritional quality [43, 44, 55].

3.14 Food Products

Quinoa attracts considerable recognition. It is consumed in different forms; some use it as a whole grain and some consume its processed version, for instance, in bread, biscuits and processed food. The quality of the product may vary after processing, depending on the type of process the raw material goes through.

Quinoa, although not a cereal crop, can be milled into flour to make toasted and baked products for instance, for which it can be substituted for all-purpose flour and can be used to make bread, pasta, cookies, noodles, pancakes and various other products [56]. Dehulled quinoa flour contains 15.6% protein, 4.6% lipid, 8.9% total dietary fibre, in which 1.2% is soluble fibre, and rich in minerals and vitamins, mainly potassium and vitamin B_3 (niacin) [57]. It is valuable to include quinoa flour in extruded snacks in place of rice, maize or oats. In a study, it was estimated that, on the incorporation of 60% quinoa, rather than oats, rice or maize, the protein content was 15.99%, lipid 4.24%, carbohydrate 77.83% and ash 1.94 %, and the product shelf life was 80 days when stored at ambient temperature. The fatty acids present were omega-6 (58%), omega-3 (4.4%) and oleic fatty acid (23%) [58]. 'Kancolla' is a variety of quinoa which is sweet in taste, and consumed just as rice and wheat.

Boiled quinoa grains can be a good replacement for rice or can be used for the thickening of soups or porridge. Sprouted quinoa, which is green quinoa seeds sprouted, is used in salads. Puffed quinoa or popcorn quinoa seed pops can also be made. Quinoa grains have been used to make soups, drinks, desserts and dry snacks.

Following are the traditional preparations made in South America using quinoa.

1. Lawa: prepared like porridge, called Mazamorra (semi-thick), with water, lime, raw flour and animal fat
2. P'esque: grains cooked in water, without salt, served with grated cheese or milk, depending on availability
3. Quinoa soup: produced by shallow-cooked quinoa, tubers, fresh or dried meat, and vegetables

4. Mucuna: quinoa flour-based steam-cooked balls with seasoning like humitas or tamales in the centre
5. Kispiña: steamed buns in different sizes and shapes
6. Phiri: roasted and slightly dampened coarse quinoa flour
7. Tacti o tactacho: fried buns, like doughnuts, made with llama fat and flour
8. Phisara: quinoa grain cooked and slightly roasted
9. Juchacha: ground quinoa and katahui-based Andean soup, along with roasted barley flour
10. Kaswira de ajara: oil-fried flattened bread, made with black quinoa or ajara and katahui (lime)
11. El Ullphu, Ullphi: roasted flour mixed in water with sugar according to taste and served cold
12. Kaswira de quinua: oil-fried flattened bread, made with white quinoa and katahui (lime)
13. Quichiquispiña: steamed and pan-fried bread, made with quinoa flour and katahui
14. Turuchaquispiña o Polonca: large steamed bread loaves cooked in a clay pot, made with lightly ground quinoa (chama) in a K'ona, and some katahui
15. K'apikispiña: ground quinoa in a K'ona, cooked in a clay pot to make the steamed bun, a common feast among all Saints but are larger in size than Mululsitoquispiña
16. Mululsitoquispiña: steamed bread, made with flour and katahui, cooked in a clay pot
17. Q'usa: a macerated (mashed) cold drink, also called quinoa chicha
18. Chiwa: juvenile leaves of quinoa known as chiwa in Aymara and lliccha in Quechua, used in salads and soups. The young leaves are rich in minerals (iron, calcium and phosphorus) and vitamins.

3.15 New Food Products Developed with Quinoa

Quinoa can be taken as a school breakfast for children; in combination with legumes, it will help to improve dietary quality. It is also cost-effective because other processed or semi-processed food in the market is usually expensive and ulti-mately unaffordable for most of the population.

The grains show flexibility for cooking purposes, and other parts are also used in cooking, including leaves, stems and roots. Grains are mostly used for cook-ing soups and stews, and toasted seeds and tender leaves are used in soups, crepes and tortillas. Quinoa flour is mostly used in baked goods (bread loaves, cookies, pancakes and muffins), and nutrition bars, confections, granola (breakfast cereal), and various fermented and non-fermented beverages. Moreover, grains and other by-products like husks are used to feed livestock.

Processed or semi-processed products are generally made with cereals, quinoa flour being used in producing almost all types of food products. It is a replacement for cereals for individuals with coeliac disease and those who are gluten intolerant, while also developing products with different organoleptic characteristics than other common noodles.

3.15.1 Quinoa Flakes

Produced by pressing dried quinoa grain (15% moisture) between two converging rollers, and is consumed mainly for breakfast. These are significant sources of minerals and nutrients. The texture and appearance may vary because plasticity characteristics vary with different quinoa.

3.15.2 Quinoa Extruded Products

The extruded food products are manufactured by using extruders, which is a method in which the food is prepared at room temperature and used for combining and sculpting foods. Some examples of extruded food products are pasta, noodles, vermicelli, etc. Low-temperature extrusion is often used to develop pasta and vermicelli requiring temperatures below 100°C.

3.15.3 Puffed Quinoa

The grains are exposed to high temperature and pressure, then taken away immediately from the heat to a sudden pressure drop that pops quinoa: moisture in the grains is evaporated, so that the weight of the product is dropped and expands into a lighter product. Puffed quinoa can then be consumed as it is, or used to make an energy bar or as a breakfast cereal.

Several studies have determined the effect of quinoa-based food on humans as well as animals. Quinoa was used as an ingredient in several food products and given to consumers of different ages and the results were observed (Table 3.2).

3.16 Conclusion

Quinoa, an ancient crop, was originally cultivated in pre-HispanicAndean communities, and has now been identified as an excellent source of nutrients and biologically active compounds like phenolic acids, saponins, flavonoids, etc. It also contains all essential amino acids, minerals, vitamins, omega-3 fatty acids and it is gluten free, being a good substitute for barley, wheat and rye for coeliac patients. In addition, quinoa reduces the risk of chronic diseases. The nutritional quality is influenced by the variety of quinoa, and the different growing locations. Moreover, the antinutrients in quinoa grains, like saponins, could be easily reduced or inactivated to safe levels when relevant industrial processing or household preparation processes are practised. The replacement of refined wheat flour with quinoa flour bring up several problems, including alteration

Table 3.2 The Effect of Quinoa-Supplemented Diet on Different Health Problems

Quinoa Component	Study	Model System	Observation	Reference
Hydrolyzed quinoa	Hydrolyzed quinoa (2000 mg/kg body weight) supplemented in animal diet and control with non-supplemented animals.	Adult Wistar rat (60-day-old males)	No change in TC between the control and treated group. However, a significant reduction ($p<0.05$) in TG levels were observed in the quinoa-supplemented group compared with the control group.	[59]
Quinoa cereal bar	Cereal bar weighing 25 g and containing 9.75 g of quinoa, two bars fed daily to the subject for 30 days	Nine male and 13 female students (aged 18–45 years)	TG, TC and LDL-C level reduced significantly from 101.83, 175.56 and 152.63 mg/dL, respectively, to 89.31, 158.28 and 121.31 mg/dL, respectively.	[60]
Quinoa flakes and grains	In line with their gluten tolerance, subjects were fed 50 g quinoa daily for six weeks. It was recommended to consume flakes for breakfast and replace rice with quinoa and use it in salads as a side dish.	Two male and 17 female celiac patients (aged >18 years)	No notable difference in LDL-C, TG and TC levels. However, a significant drop in HDL-C from 1.82 to 1.68 mmol/L ($p<0.05$) was observed.	[61]
Roasted and raw quinoa	Experimental animals were divided into four groups, in which three groups were fed with 63 g/kg body weight of raw, roasted or restricted raw quinoa for 40 days. The fourth group was treated as the no-quinoa control.	40 male Wistar rats	Significant increases occurred in TG, TC and HDL-C in treated groups compared to control. But there was no significant difference among the values for the three plus-quinoa groups.	[62]

(Continued)

Table 3.2 (Continued)   The Effect of Quinoa-Supplemented Diet on Different Health Problems

Quinoa Component	Study	Model System	Observation	Reference
Quinoa seeds	Four groups of experimental animals were set up: 1. Control (C) 2. Control plus 31% of fructose (CF) 3. Group fed with quinoa seeds in diet (Q) 4. Group fed with quinoa plus 31% of fructose in diet (QF). consisting of 6 animals in each group.	24 male Wistar rats	A marked decrease ($p<0.05$) in TC and TG in the group fed with quinoa (Q) compared with the control (C). There was no effect of the fructose-infused diet on TC (CF vs C or QF vs Q). However, there was a significant decrease in LDL-C and HDL-C level ($p<0.008$) and ($p<0.05$), respectively, in Q compared with C	[63]
Quinoa flakes	One group of subjects was fed 25 g of quinoa flakes (QF) and the other group was fed 25 g of cornflakes (CF) every day for four consecutive weeks	35 postmenstrual and obese women	There was a significant decrease in LDL-C level from 129±35 to 121±26 mg/dL and in TC level from 191±35 to 181±28 mg/dL in the QF group compared with the CF group. No significant difference was observed in serum glucose levels between CF and QF	[64]

(Continued)

Table 3.2 (Continued) The Effect of Quinoa-Supplemented Diet on Different Health Problems

Quinoa Component	Study	Model System	Observation	Reference
Quinoa flakes	Five groups of experimental animals: 1. Fed with white wheat bread (GP) 2. Fed with wheat bread and 0.1g cholesterol (GPC) 3. Fed with wheat bread with 15% quinoa and 0.1g cholesterol (GP15C) 4. Fed with wheat bread with 20% quinoa and 0.1 g cholesterol (GP20C) 5. Fed with 0.1 g cholesterol (GC)	36 adult Wistar rats	Observed significant decreases ($p<0.05$) in TG, TC and LDL-C levels in GP20C compared with GP from 67.66 to 36.06, 85.76 to 68.38 and 36.83 mg/dL to 18.59 mg/dL respectively. However, a significant increase in HDL-C level from 35.40 mg/dL to 42.48 mg/dL was observed in the GP20C group compared with GP and GPC.	[65]
Quinoa flour	Three groups of animals were fed protein from different sources: casein (control), quinoa flour or amaranth flour for 15 days.	24 albino male Wistar rats (weighing 100–120g)	Significant reduction ($p<0.01$) in TC and HDL-C levels were observed in both quinoa- and amaranth-based diets compared with the group with the casein-based diet. No weight gain was observed in any group. However, a significant controlled level of postprandial cholecystokinin hormone was observed in the quinoa-based diet compared with the control.	[66]

(Continued)

Table 3.2 (Continued) The Effect of Quinoa-Supplemented Diet on Different Health Problems

Quinoa Component	Study	Model System	Observation	Reference
Quinoa extract	The experimental treatments were divided into four groups based on consuming a high-fat (HF) or a low-fat (LF) diet for three weeks. The HF group supplemented or not supplemented with either pure 20-hydroxyecdysone (HF20E) or with quinoa extract (HFQ)	48 male C57BL/J mice (6-weeks-old)	TG and TC levels in plasma showed no significant difference between any group. But remarkable differences (70%, $p<0.001$) in weight gain were observed between the HF and LF groups. However, the quinoa and 20E treatments in the HF group did not show a reduction in weight gain.	[67]
Mixed quinoa	Five groups of animals were fed diets containing 1. High-fat basal diet (control) 2. Milled quinoa in ratios (60:15:15:10) 3. Defatted soybean (50:20:20:10) 4. Carrot powder (40:35:25: 10) 5. Resistant starch (30:30:30:10) for 30 days	30 albino male rats (weighing 45–55g)	LDL-C, TC and TG levels decreased and HDL-C level significantly increased ($p<0.05$) about 81.01%, 56.03%, 54.29% and 55.81% respectively in the milled quinoa-fed group compared with the control.	[68]

(Continued)

Table 3.2 (Continued) The Effect of Quinoa-Supplemented Diet on Different Health Problems

Quinoa Component	Study	Model System	Observation	Reference
Quinoa saponin extract (QS)	Three groups of 3T3-L1 preadipocyte cells were treated and grown in different media: 1. Dulbecco's modified Eagle's medium (DMEM), 2. 12.5 µg/mL QS in DMEM (QS-12.5) 3. 25 µg/mL QS in DMEM (QS-25).	Preadipocyte cell 3t3-L1 (derived cell line from mouse) (*in vitro*)	Both QS and control inhibited adipocyte differentiation. compared with the control significantly (*p*<0.05). TG reduction was observed in QS-25 and QS-12.5 by 25.01% and 18.38%, respectively. Diet-induced obesity due to protein expression was suppressed by both QS-25 and QS-12.5 in comparison with the control.	[69]
Quinoa saponin extract	Saponin extract from milled and defatted quinoa seeds was incubated with macrophage cells RAW 264.7.	RAW 264.7 macrophage cells stimulated from Lipopolysaccharide (*in vitro* murine cell lines)	Quinoa saponin extract significantly decreased concentration-dependent inhibition of nitric oxide (NO) generation. Inhibition in production of IL-G (67–90%) and TNFα production (33–80%) due to treated macrophage cells.	[70]

(Continued)

Table 3.2 (Continued)   The Effect of Quinoa-Supplemented Diet on Different Health Problems

Quinoa Component	Study	Model System	Observation	Reference
Quinoa leachate (QL)	Experimental animals fed with QL were divided into four groups depending on their diet 1. Control diet 2. Normal diet plus orally fed QL of 250 mg/kg body weight. 3. Normal diet plus orally fed QL of 500 mg/kg body weight. 4. Normal diet plus metformin at 300 mg/kg of body weight. (positive control) for 15 weeks.	C57Bl/6 J male mice diet-induced obese (5-weeks-old)	Under fasting conditions, blood glucose concentration significantly decreased ($p<0.05$) in subjects fed with either Metformin or QL at 500 mg/kg body weight.	[71]
Quinoa-based bread	White wheat bread (WWB) as control, Quinoa breakfast (QB) or buckwheat breakfast (BB) were fed to subjects, followed by regular lunch after four hours. It was a random cross-over study.	12 healthy subjects (six women and six men) aged between 22 and 40 years and 12 individuals with type 2 diabetes (seven women and five men) aged 40–68 years)	After breakfast, iAUC (incremental area under the curve) value for glucose significantly decreased ($p<0.001$) in healthy subjects who were fed BB compared with QB. In diabetic patients, however, QB and BB subjects had lower iAUC values compared with WWB. But, after lunch, the iAUC value was induced in QB and BB fed diabetic and healthy individuals. QB subjects showed a higher glycaemic index (89.4) than BB subjects (26.8) because at 30 min it was at its peak.	[72]

(Continued)

Table 3.2 (Continued) The Effect of Quinoa-Supplemented Diet on Different Health Problems

Quinoa Component	Study	Model System	Observation	Reference
Quinoa milk beverage	Subjects fed with 50 g of anhydrous glucose in 250 mL water were considered to be the control group, the second group was with 312.5 mL of quinoa milk containing 50 g carbohydrates, the third group was fed with 25 mL commercial rice milk, and the fourth group consumed 25 mL commercial cow's milk.	12 healthy individuals aged 20–50 years (six females and six males)	The glycaemic indices after being fed cow's milk, rice milk or quinoa milk were 39, 79 and 52, respectively. It was concluded from the study that quinoa milk can replace current milk products because it does not cause any known adverse effect in humans plus has elevated protein content and lowers the glycaemic index.	[73]

Note: TC: Total cholesterol; TG: Triglycerol; LDL-C: Low-density lipoprotein cholesterol; HDL-C: High-density lipoprotein cholesterol.

in organoleptic properties that need to be corrected. Quinoa is a strategic crop complementing the diet of rural/marginal area populations where protein and energy malnutrition affects large populations in developing countries. Quinoa grains show versatility for culinary uses; in addition to grains, other parts used are leaves in soups and salads, and flour in making soups, pancakes, tortillas, biscuits, cookies, etc. The excellence of the nutritional profile and multifunctional characteristics in food products makes it ideal to be used in the food industry. Other industrial potentials of quinoa include utilization of grains as fillers in the plastic industry, as dusting powders and complementary protein for human and animal foods. An initiative should be taken to amplify agriculture production with high grain quality, higher yields, low saponin content and large seed size. To embrace quinoa in India, necessary information about the crop must be proclaimed to the farmers and the consumers and efficient post-harvest technologies and a decent marketing system are required. If we solemnly endorse the demand for healthy and ethical food products to pursue sustainability, then we must learn from the setbacks we see so far in this chapter and act accordingly. Quinoa is likely to outbuild its underemployed status to promote health and would be considered a supergrain of the twenty-first century.

3.17 Future Recommendations

Quinoa seeds have attracted dietary significance due to their extraordinary nutritional values, their absence of gluten and presence of complete protein, bioactive compounds, micronutrients, minerals and vitamins. It is a supergrain not only because of its nutritional quality but because its farming pattern is also excellent; it is stress tolerant, make cultivation easier and more efficient for farmers. The malnutrition increases as population rises; it is challenging for government to provide nutritious food in the rural areas. Its exceptional potential attracts government officials to believe that it can be used to overcome food scarcity and can take part in food security. Since FAO has declared 2013 an 'International year of quinoa', quinoa is spread all over the world, New food products are developed using pseudocereals, researchers are taking interest to study about such super crops, it will develop interest in the upcoming researchers to develop more technology to solve any occurred inefficiency in processing. individuals should learn more about underutilized crops and their benefits to make use of it.

References

1. Martínez-Villaluenga, C., Peñas, E. and Hernández-Ledesma, B. (2020). Pseudocereal grains: Nutritional value, health benefits and current applications for the development of Gluten-Free Foods. *Food Chem. Toxicol. Int. J. Publ. Br. Ind. Biol. Res. Assoc.* 137: 111178.

2. Kolano, B, Siwinska, D., Gomez Pando, L., Szymanowska-Pulka, J. and Maluszynska, J. (2011). Genome size variation in *Chenopodium quinoa* (Chenopodiaceae). *Plant Syst. Evol.* 298(1): 251–255.
3. Palomino, G.. and Hernández, L. T. and de la Cruz Torres, E. (2008). Nuclear Genome size and chromosome analysis in *Chenopodium quinoa* and C - Berlandierisubsp: Nuttalliae. *Euphytica* 164: 221.
4. Núñez, L. (1974). *La Agricultura Prehistoricaen Los Andes Meridionales*; Editorial Orbe, Santiago de Chile, Chile.
5. González, J.A., Eisa, S.S.S., Hussin, S.A.E.S. and Prado, F.E. (2015). Quinoa: An incan crop to face global changes in agriculture. In *Quinoa: Improvement and Sustainable Production*, Murphy, K., Matanguihan, J., Eds.; John Wiley & Sons, Inc., Hoboken, NJ, USA, pp. 1–18. ISBN 978-1-118-62805-8. 10.1002/9781118628041.ch1
6. Bazile, D., Martínez, E., Fuentes, F., Chia, E., Namdar-Irani, M., Olguín, P., Saa, C., Thomet, M. and Vidal, A. (2015). Quinoa in Chile. In *State of the Art Report of Quinoa in the World in 2013*, Bazile, D., Bertero, D., Nieto, C., Eds.; FAO & CIRAD, Rome, Italy, pp. 401–421. ISBN 978-92-5-108558-5. http://www.fao.org/3/a-i4042e.pdf (accessed on 12 October 2019).
7. Cauda, C., Micheletti, C., Minerdo, B., Scaffidi, C. and Signoroni, E. (2013). *Quinoa in the Kitchen*; Slow Food Editore, Bra, Italy. ISBN 9788884993397.
8. Bojanic, A. (2011). *Quinoa: An Ancient Crop to Contribute to World Food Security*; FAO Fiat Panis, Rome, Italy. http://www.fao.org/3/aq287e/aq287e.pdf (accessed on 28 October 2019).
9. Alandia, G., Rodriguez, J.P., Jacobsen, S.-E., Bazile, D. and Condori, B. (2020). Global expansion of quinoa and challenges for the Andean region. *Glob. Food Sec.* 26: 100429. ISSN 2211-9124.
10. Panuccio, M.R., Jacobsen, S.E., Akhtar, S.S. and Muscolo, A. (2014). Effect of saline water on seed germination and early seedling growth of the halophyte quinoa. *AoB Plants* 6: plu047. 10.1093/aobpla/plu047.
11. Miranda, M., Vega-Galvez, A., Martinez, E.A., Lopez, J., Marin, R., Aranda, M. and Fuentes, F. (2013). Influence of contrasting environments on seed composition of two quinoa genotypes: Nutritional and functional properties. *Chil. J. Agric. Res.* 73: 108–116. 10.4067/S0718-58392013000200004.
12. Vega-Gálvez, A., Miranda, M., Vergara, J., Uribe, E., Puente, L. and Martínez, E.A. (2010). Nutrition facts and functional potential of quinoa (*Chenopodium quinoa* Willd.), an ancient Andean grain: A review. *J. Sci. Food Agric.* 90(15): 2541–2547.
13. Schlick, G. and Bubenheim, D.L. (1993) Quinoa: An emerging "new" crop with potential for CELSS. NASA Technical Paper 3422, pp. 1–9.
14. Guixing, R., Cong, T., Xin, F., Shengyuan, G., Gang, Z., Lizhen, Z., Zou, L. and Peiyou, Q. (2023). Nutrient composition, functional activity and industrial applications of quinoa (*Chenopodium quinoa* Willd.) "*Food Chemistry*" 410: 135290.
15. Wu, V.-G. (2015). Nutritional properties of quinoa. *Quinoa Improv. Sustain. Prod.* 10.1002/9781118628041.ch11.
16. Vega-Gálvez, A., Miranda, M., Vergara, J., Uribe, E., Puente, L. and Martínez, E.A. (2010). Nutrition facts and functional potential of quinoa (*Chenopodium quinoa* Willd.), an ancient Andean grain: A review. *J. Sci. Food Agric.* 90(15): 2541–2547.
17. Rao, N. and Shahid, M. (2012). Quinoa-a promising new crop for the Arabian Peninsula. *Am. Eurasian J. Agric. Environ. Sci.* 12: 1350–1355.

18. Ruales, J. and Nair, B.M. (1992). Nutritional quality of the protein in quinoa (*Chenopodium quinoa* Willd.) seeds. *Plant Foods Hum. Nutr.* 42(1): 1–11. 10.1007/BF02196067.
19. Brownawell, A.M., Caers, W., Gibson, G.R., Kendall, C.W.C., Lewis, K.D., Ringel, Y. and Slavin, J.L. (2012). Prebiotics and the health benefits of fiber: Current regulatory status, future research, and goals. *J. Nutr.* 142(5): 962–974.
20. Brenda, C.J., Torres-Vargas, O.L. and Rodríguez-García, M.E. (2019). Physicochemical characterization of quinoa (*Chenopodium quinoa*) flour and isolated starch. *Food Chemistry* 298: 124982.
21. Ando, H., Chen, Y.C., Tang, H., Shimizu, M., Watanabe, K. and Mitsunaga, Y. (2002). Food components in fractions of quinoa seed. *Food Sci. Technol. Res.* 8(1): 80–84. 10.3136/fstr.8.80.
22. Miranda, M., Vega-Gálvez, A., Uribe, E., López, J., Martinez, E.A., Rodríguez, M.J., Quispe, I. and Di Scala, K. (2011). Physico-chemical analysis, antioxidant capacity and vitamins of six ecotypes of Chilean quinoa (*Chenopodium quinoa* Willd.). *Proc. Food Sci.* 1: 1439–1446. 10.1016/j.profoo.2011.09.213.
23. Filho, A.M.M.,Pirozi, M.R., Borges, J.T.D.S., Pinheiro Sant'Ana, H.M.,Chaves, J.B.P. and Coimbra, J.S.D.R. (2017). Quinoa: Nutritional, functional, and antinutritional aspects. *Crit. Rev. Food Sci. Nutr.* 57(8): 1618–1630. 10.1080/10408398.2014.1001811.
24. Bastidas, E.G., Roura, R., Rizzolo, D.A.D., Massanés, T. andGomis, R. (2016). Quinoa (*Chenopodium quinoa* Willd.), from nutritional value to potential health benefits: An integrative review. *J. Nutr. Food Sci.* 6: 497. 10.4172/2155-9600.1000497.
25. Nickel, J.,Spanier, L.P., Botelho, F.T., Gularte, M.A. and Helbig, E. (2016). Effect of different types of processing on the total phenolic compound content, antioxidant capacity, and saponin content of *Chenopodium quinoa* Willd. grains. *Food Chem.* 209. 10.1016/j.foodchem.2016.04.031.
26. Kuljanabhagavad, T. and Wink, M. (2009). Biological activities and chemistry of saponins from *Chenopodium quinoa* Willd. *Phytochem. Rev.* 8(2): 473–490. 10.1007/s11101-009-9121-0.
27. Estrada, A., Li, B. and Laarveld, B. (1998). Adjuvant action of *Chenopodium quinoa* saponins on the induction of antibody responses to intragastric and intranasal administered antigens in mice. *Comp. Immunol. Microbiol. Infect. Dis.* 21(3): 225–236.
28. Repo-Carrasco-Valencia, R., Hellström, J.K., Pihlava, J.M. and Mattila, P.H. (2010). Flavonoids and other phenolic compounds in Andean indigenous grains: Quinoa (*Chenopodium quinoa*), kañiwa (*Chenopodium pallidicaule*) and kiwicha (*Amaranthus caudatus*). *Food Chem.* 120(1): 128–133.
29. Alvarez-Jubete, L., Wijngaard, H., Arendt, E.K. and Gallagher, E. (2010). Polyphenol composition and in vitro antioxidant activity of amaranth, quinoa buckwheat and wheat as affected by sprouting and baking. *Food Chem.* 119(2): 770–778.
30. Brend, Y., Galili, L., Badani, H. et al. (2012). Total phenolic content and antioxidant activity of red and yellow quinoa (*Chenopodium quinoa* Willd.) seeds as affected by baking and cooking conditions. *Food Nutr. Sci.* 3(8): 1150.
31. Asao, M. and Watanabe, K. (2010). Functional and bioactive properties of quinoa and amaranth. *Food Sci. Technol. Res.* 16(2): 163–168.
32. Astier, C., Moneret-Vautrin, D.A.,Puillandre, E. and Bihain, B.E. (2009). First case report of anaphylaxis to quinoa, a novel food in France. *Allergy* 64(5): 819–820.
33. Thomas, D. and Elliott, E.J. (2009). Low glycaemic index, or low glycaemic load, diets for diabetes mellitus. *Cochrane Database Syst. Rev.* (1): CD006296.

34. Thomas, D.E., Elliott, E.J. and Baur, L. (2007). Low glycaemic index or low glycaemic load diets for overweight and obesity. *Cochrane Database Syst. Rev.* (3): CD005105.
35. Lee, D., Hwang, W., Artan, M., Jeong, D.E. and Lee, S.J. (2015). (̣ects of nutritional components on aging. *Aging Cell* 14(1): 8–16.
36. Pereira, M.A., Jacobs, D.R. Jr, Pins, J.J., Raatz, S.K. and Gross, M.D. (2002). (̣ect of whole grains on insulin sensitivity in overweight hyperinsulinemic adults. *Am. J. Clin. Nutr.* 75(5): 848–885.
37. Ranilla, L.G., Apostolidis, E., Genovese, M.I., Lajolo, F.M. and Shetty, K. (2009). Evaluation of indigenous grains from the Peruvian Andean region for antidiabetes and antihypertension potential using in vitro methods. *J. Med. Food* 12(4): 704–713.
38. Danielsen, S., Jacobsen, S.E., Echegaray, J. and Ames, T. (2001). Impact of downy mildew on the yield of quinoa. In *CIP Program Report 1999–2000*; International Potato Center, Lima, Peru, pp. 397–401.
39. Ortiz, R.V. and Zanabria, E. (1979). Plagas. Quinua y Kaniwa. CultivosAndinos. In: Tapia, M.E. (Ed.), *Serie Libros y MaterialesEducativos*, vol. 49. Instituto Interamericano de CienciasAgricolas, Bogota, Columbia, pp. 121–136.
40. Risi, J. and Galwey, N.W. (1984). The Chenopodium grains of the Andes: Inca crops for modern agriculture. *Adv. Appl. Biol.* 10: 145–216.
41. Monika, D. and Karel, V. (2018). Seedling damping-off of *Chenopodium quinoa* Willd. *Plant Prot. Sci.* 40: 5–10. 10.17221/3119-PPS.
42. Oelke, E.A., Putnam, D.H., Teynor, T.M. and Oplinger, E.S. (1992). Alternative field crops manual. University of Wisconsin Cooperative Extension Service, University of Minnesota Extension Service, Centre for Alternative Plant and Animal Products.
43. Bojanic, A. (2011). Quinoa: An ancient crop to contribute to world food security; FAO; Fiat Panis, Rome, Italy. http://www.fao.org/3/aq287e/aq287e.pdf (accessed on 28 October 2019).
44. Quiroga, C., Escalera, R., Aroni, G., Bonifacio, A., González, J.A., Villca, M., Saravia, R., Ruiz, A. Traditional processes and technological innovations in quinoa harvesting, processing and industrialization. In: *State of the Art Report of Quinoa in the World in 2013*, Bazile, D., Bertero, D., Nieto, C., Eds.; FAO & CIRAD: Rome, Italy, pp. 258–296. ISBN 978-92-5-108558-5. http://www.fao.org/3/a-i4042e.pdf (accessed on 12 October 2019).
45. Birbuet, J.C. andMachicado, C.G. (2009). Technological progress and productivity in the quinoa sector. http://hdl.handle.net/10419/45661 (accessed on 1 November 2019).
46. López, L.M., Capparelli, A. and Nielsen, A.E. (2011). Traditional post-harvest processing to make quinoa grains (*Chenopodium quinoa* var. *quinoa*) apt for consumption in Northern Lipez (Potosí, Bolivia): Ethnoarchaeological and archaeobotanical analyses. *Archaeol. Anthropol. Sci.* 3(1): 49–70.
47. Yang, F., Guo, T., Zhou, Y., Han, S., Sun, S. and Luo, F. (2022). Biological functions of active ingredients in quinoa bran: Advance and prospective. *Crit. Rev. Food Sci. Nutr.* 10.1080/10408398.2022.2139219.
48. Brady, K., Ho, C.T., Rosen, R.T., Sang, S. and Karwe, M.V. (2007). Effects of processing on the nutraceutical profile of quinoa. *Food Chem.* 100(3): 1209–1216.
49. San Martín, R., Ndjoko, K. and Hostettmann, K. (2008). Novel molluscicide against *Pomacea canaliculata* based on quinoa (*Chenopodium quinoa*) saponins. *Crop Prot.* 27(3–5): 310–319.

50. Sun, X., Yang, X., Xue, P., Zhang, Z. and Ren, G. (2019). Improved antibacterial effects of alkali-transformed saponin from quinoa husks against halitosis-related bacteria. *BMC Complement. Altern. Med.* 19(1): 1–10.

51. Bazile, D., Martínez, E., Fuentes, F., Chia, E., Namdar-Irani, M., Olguín, P., Saa, C., Thomet, M. and Vidal, A. (2015). Quinoa in Chile. In *State of the Art Report of Quinoa in the World in 2013*, Bazile, D., Bertero, D., Nieto, C., Eds.; FAO & CIRAD, Rome, Italy, pp. 401–421. http: //www.fao.org/3/a-i4042e.pdf (accessed on 12 October 2019).

52. Kaur, I., Tanwar, B., Reddy, M. and Chauhan, A. (2016). Vitamin C, total polyphenols and antioxidant activity in raw, domestically processed and industrially processed Indian *Chenopodium quinoa* seeds. *J. Appl. Pharm. Sci.* 6(4): 139–145.

53. Gomez-Caravaca, A.M., Iafelice, G., Verardo, V., Marconi, E. and Caboni, M.F. (2014). Influence of pearling process on phenolic and saponin content in quinoa (*Chenopodium quinoa* Willd.). *Food Chem.* 157: 174–178.

54. Vega-Gálvez, A., Martín, R.S., Sanders, M., Miranda, M. and Lara, E. (2010). Characteristics and mathematical modeling of convective drying of quinoa (*Chenopodium quinoa* Willd.): Influence of temperature on the kinetic parameters. *J. Food Process. Preserv.* 34(6): 945–963.

55. Tanwar, B., Goyal, A., Irshaan, S., Kumar, V., Sihag, M.K., Patel, A. and Kaur, I. (2019). Quinoa. In *Whole Grains and Their Bioactives*, Johnson, J., Wallace, T., Eds.; John Wiley & Sons, Ltd., Chichester, pp. 269–305.

56. Vilcacundo, R. and Hernández-Ledesma, B. (2017). Nutritional and biological value of quinoa (*Chenopodium quinoa* Willd.). Curr. Opin. Food Sci. 14: 1–6.

57. Ranhotra, G.S., Gelroth, J.A., Glaser, B.K., Lorenz, K.J. and Johnson, D.L. (1993). Composition and protein nutritional quality of Quinoa. *American Association of Cereal Chemists, Inc.* 70(3).

58. Alajil, O., Hymavathi, T.V., Maheswari, K.U. and Rudra, S.G. (2020). Nutritional, physico-chemical and sensory attributes of quinoa based extrudates. *Vegetos* 33(3): 390–400.

59. Meneguetti, Q.A., Brenzan, M.A., Batista, M.R., Bazotte, R.B., Silva, D.R. and Garcia Cortez, D.A. (2011). Biological effects of hydrolyzed quinoa extract from seeds of *Chenopodium quinoa* Willd. *J. Med. Food* 14(6): 653–657.

60. Farinazzi-Machado, F.M.V., Barbalho, S.M. and Oshiiwa, M. (2012). Use of cereal bars with quinoa (*Chenopodium quinoa* Willd.) to reduce risk factors related to cardiovascular diseases. *Food Sci. Technol. (Campinas)* 32(2): 239–244.

61. Zevallos, V.F., Herencia, L.I., Chang, F., Donnelly, S., Ellis, H.J. and Ciclitira, P.J. (2014). Gastrointestinal effects of eating quinoa (*Chenopodium quinoa* Willd.) in celiac patients. *Am. J. Gastroenterol.* 109(2): 270–278.

62. Machado, F.M.V.F., Barbalho, S.M. and Oliveira, F.D.D.L. (2014). Efeitos da suplementação de quinoa (*Chenopodium quinoa* Willd.) crua e torrada no perfilmetabólico de ratos Wistar. *J. Health Sci. Inst* 32(1): 59–63.

63. Paśko, P., Zagrodzki, P. and Bartoń, H. (2010). Effect of quinoa seeds (*Chenopodium quinoa*) in diet on some biochemical parameters and essential elements in blood of high fructose-fed rats. *Plant Foods Hum. Nutr.* 4: 333–338.

64. De Carvalho, F.G., Ovídio, P.P., Padovan, G.J., Jordão Junior, A.A., Marchini, J.S. and Navarro, A.M. (2014). Metabolic parameters of postmenopausal women after quinoa or corn flakes intake – A prospective and double-blind study. *Int. J. Food Sci. Nutr.* 65(3): 380–385.

65. Gewehr, M.F., Pagno, C.H., Danelli, D., Melo, L.Md., Flôres, S.H. and Jong, E.Vd. (2016). Evaluation of the functionality of bread loaves prepared with quinoa flakes through biological tests. *Braz. J. Pharm. Sci.* 52(2): 337–346.
66. Mithila, M.V. and Khanum, F. (2015). Effectual comparison of quinoa and amaranth supplemented diets in controlling appetite: A biochemical study in rats. *J. Food Sci. Technol.* 52(10): 6735–6741.
67. Foucault, A.S., Mathe, V., Lafont, R. et al. (2012). Quinoa extract enriched in 20 hydroxyecdysone protects mice from diet-induced obesity and modulates adipokines expression. *Obesity (Silver Spring)* 20(2): 270–277.
68. Hejazi, M.A. (2016). Preparation of different formulae from quinoa and different sources dietary fiber to treat obesity in rats. *Nat. Sci.* 14(2): 55–65.
69. Yao, Y., Zhu, Y., Gao, Y., Shi, Z., Hu, Y. and Ren, G. (2015). Suppressive effects of saponin-enriched extracts from quinoa on 3T3-L1 adipocyte differentiation. *Food Funct.* 6(10): 3282–3290.
70. Yao, Y., Yang, X., Shi, Z. and Ren, G. (2014). Anti-inflammatory activity of saponins from quinoa (*Chenopodium quinoa* Willd.) seeds in lipopolysaccharide-stimulated RAW 264.7 macrophages cells. *J. Food Sci.* 79(5): H1018–H1023.
71. Graf, B.L., Poulev, A., Kuhn, P., Grace, M.H., Lila, M.A. and Raskin, I. (2014). Quinoa seeds leach phytoecdysteroids and other compounds with anti-diabetic properties. *Food Chem.* 163: 178–185.
72. Gabrial, S.G., Shakib, M.C.R. and Gabrial, G.N. (2016). Effect of pseudocereal-based breakfast meals on the first and second meal glucose tolerance in healthy and diabetic subjects. *Open Access Maced. J. Med. Sci.* 4(4): 565.
73. Pineli, L.D.L.D.O., Botelho, R.B., Zandonadi, R.P. et al. (2015). Low glycemic index and increased protein content in a novel quinoa milk. *LWT Food Sci. Technol.* 63(2): 1261–1267.

Chapter 4

Amaranth: Nutritional Profile, Processing and Food Products

Vikas Kumar, Danish Shafi Mir, Nitin Sharma

4.1 Introduction

There has been increasing research and development in recent years into basic plant materials that can serve as both food and medication. The food and beverage industries make extensive use of various cereal grains. Pseudocereals are an umbrella term for a diverse group of plants. The seeds of these plants are used as food, just like cereal grains, and are often ground into flour before being ingested. They taste and have nutritional content comparable to that of cereal grains. Although these are not our standard breakfast cereals, they do offer similar composition and nutritional value. Our forefathers all around the world and throughout the ages relied on pseudocereals as their main source of nutrition. Certain pseudocereals are more common in particular parts of the world. In the world's poorest regions, pseudocereals remain a staple food. Quinoa, amaranth, chia, and buckwheat are the pseudocereals with the most research behind them (Morales et al. 2020). They show much promise as a natural supply of several bioactive compounds. Peptides and protein hydrolysates isolated from these plants have been shown to have several health benefits (Morales et al. 2020). In 2021, bioactive peptides derived from amaranth protein with inhibitory properties toward cholesterol esterase and pancreatic lipase activities have been isolated by Fisayo Ajayi et al. (2021).

Due to climate change, the challenge of world hunger, and changes in crop profiles in Europe and other regions worldwide, it is becoming desirable to

DOI: 10.1201/9781003325277-4

hunt for novel plants with a high nutritional potential. Amaranth has various beneficial health properties in addition to its value as a food crop (Karamac et al. 2019; Malik et al. 2023). Other significant benefits of amaranth include good crop performance, resilience to drought, and high photosynthetic rate. It has been widely used as a food crop since the times of the Aztecs, Mayans, and Incans (Kulczyski et al. 2017). It was widely introduced to other nations as a crop, vegetable, weed, or cereal in the sixteenth and seventeenth centuries. In addition to being eaten, amaranth seeds also serve a spiritual purpose. Kamierczak et al. (2011) reported its widespread use in ceremonial and religious contexts. Despite its potential, not enough has been done to domesticate and cultivate this plant yet. Because of the many positive effects amaranth could have on the economy, including those for farmers, food manufacturers, and consumers, this needs to be emphasized strongly. There are around 70 different annual species in the family Amaranthaceae (Szwejkowskaet al. 2012; Kulczyskiet al. 2017; Moszaket al. 2018; Park et al. 2020; Aderibigbe et al. 2022). Some *Amaranthus* species are considered weeds, while others are cultivated extensively as pseudocereals, vegetables, and ornamentals. *Amaranthus cruentus*, *Amaranthus hypochondriacus*, and *Amaranthus caudatus* are the most cultivated *Amaranthus* species for human consumption, while *Amaranthus blitum*, *Amaranthus gangeticus*, *Amaranthis mangostanus*, and *Amaranthus tricolor* are cultivated as vegetables in India (Szwejkowska et al. 2012). In Africa, amaranth leaves are consumed as medicine (Aderibigbe et al. 2022). *Amaranthus* species thrive in the warm, humid climates found in many parts of the world. In Poland, amaranth seeds are employed in the production of flour, cereals, confectionery, expanded grains, bread, pasta, and noodles (Januszewska-Jówiak et al. 2008). This chapter focused on the nutritional properties of amaranth, which may assist researchers in formulating a variety of food products with both health and medicinal benefits.

4.2 Chemical Composition of Amaranth Seeds

Amaranths are used for both human and animal food due to their high nutrient content in both the grain and the leaves (Mustafa et al. 2011). High protein and minerals are found in the crop compared with other regularly used true cereal grains such millet, sorghum, rice, wheat, and maize (Mustafa et al. 2011). In addition, it contains an optimal ratio of essential amino acids to unsaturated fatty acids. The protein amaranth contains is rich in the essential amino acid lysine, which is deficient in most cereal proteins (Amare et al. 2015). Amaranth protein tends to contain a significant amount of sulfur-containing amino acids, even though these amino acids are usually found in low quantities in pulse crops. It is reported that the protein content of amaranth grain is nearly at the FAO-/WHO-recommended levels due to its well-balanced amino acid composition (Januszewska-Józwiak et al.

2008; O'Brien and Price 2008; Murya and Pratibha 2018; Karamac et al. 2019). Amaranth has been identified as a potentially game-changing crop to address the food needs of the world's population (Mekonnen et al. 2018; Malik et al. 2023). Amaranth greens are rich in protein (14 to 43 g kg^{-1} in fresh matter), lysine (40 to 56 g kg^{-1}), and carotenoids (60 to 200 mg kg^{-1}) (Szwejkowska et al. 2012). The concentrations of oxalates and nitrates in fresh matter ranged from 4.10 to 9.2 g kg^{-1} and 3.0 to 16.5 g kg^{-1}, respectively (Prakash and Pal 1991). Several *Amaranthus* species are among the richest food sources for vitamins C, B_6, folate, and carotene outside of the green leafy vegetables and cereals (Musa et al. 2011). The nutritional value of amaranth grain is displayed in Table 4.1. Nutritional and chemical composition varies among different species of *Amaranthus* (Kadoshnikov et al. 2005).

Proteins, lipids, carbohydrates, vitamins, and minerals are the primary biological constituents in amaranth (Park et al. 2020). The protein content of amaranth seeds, which averages about 18%, varies widely between cultivars, climates, soil types, and fertilization practices (Januszewska-Józwiaket al. 2008; Moszaket al. 2018). Albumins make up the greatest proportion of amaranth proteins. The primary carbohydrate found in amaranth grain is starch (Moszaket al. 2018). The polysaccharide content of amaranth seeds ranges between 45 and 65% (Januszewska-Józwiak et al. 2008). The soluble (mostly pectins) and insoluble fibre fractions are significant components of amaranth. Lignin, cellulose, and hemicelluloses, the components of the insoluble fraction, improve digestion. Seeds typically include a fibre content of 2–8% of dry weight (Kaźmierczak et al. 2011), although this varies from crop to crop. The most common unsaturated fatty acids are linoleic acid (62%), oleic acid (20%), linolenic acid (1%), and arachidonic acid (2%). Saturated fatty acids can be found in *Amaranthus* at low concentrations (palmitic (13%), stearic (2.6%), arachidic (0.7%), and myristic acids (0.1%) (Kaźmierczak et al. 2011)). The lipid fraction of amaranth includes compounds with biological significance, such as tocopherols, tocotrienols, and sterols (Jamka et al. 2021).

Seeds and leaves of amaranth have been found to be rich sources of squalene (2 and 8%, respectively), as well as a wide variety of vitamins (particularly the B group) and minerals (Szwejkowska et al. 2012; Obiedzinska et al. 2012; Park et al. 2020). On average, *Amaranthus* seeds contain 3.3% of various minerals by weight (Januszewska-Józwiak et al. 2008). Iron (Fe), phosphorus (P), and calcium (Ca) are the most abundant, followed by potassium (K) and magnesium (Mg). Other minerals found in *Amaranthus* include copper (Cu), zinc (Zn), sodium (Na), chromium (Cr), manganese (Mn), nickel (Ni), lead (Pb), cadmium (Cd), and cobalt (Co). *Amaranthus* seeds and leaves include trace amounts of polyphenols, saponins, haemagglutinins, phytin, nitrates, and oxalates. Betacyanins are present and active in *Amaranthus*, contributing to its astringent effect on flavour. Betacyanins, of which betanidin is the best known, are red or purple betalain pigments. Many species of *Amaranthus* have been shown to contain these chemicals (Fleming 2000). Recently, betalains have been identified as being highly bioactive natural chemicals that can have positive effects on human health.

Table 4.1 Nutritional Content of the *Amaranthus* spp. (USDA 2016)

Nutrient	Value per 100 g
Water	11.29 g
Energy	371 Kcal
Protein	13.56 g
Total lipid (fat)	7.02 g
Ash	2.88 g
Carbohydrates	65.25 g
Fiber, total dietary	6.7 g
Sugars, total	1.69 g
Starch	57.27 g
Minerals	
Calcium, Ca	159 mg
Phosphorus, P	508 mg
Iron, Fe	7.61 mg
Zinc, Zn	2.87 mg
Magnesium, Mg	248 mg
Manganese, Mn	3.333 mg
Phosphorus	557 mg
Sodium	4 mg
Zinc	2.87 mg
Selenium	18.7 mg
Vitamins	
Ascorbic acid	4.20 mg
Thiamin	0.12 mg
Riboflavin	0.200 mg
Niacin	0.92 mg
Folate, total	82 mcg

(Continued)

Table 4.1 (Continued)   Nutritional Content of the *Amaranthus* spp. (USDA 2016)

Nutrient	Value per 100 g
Vitamin A	2 IU
Pantothenic acid, vit. B55	1.46 mg
Vitamin E (α-tocopherol)	1.19 mg
Vitamin E (β-tocopherol)	0.96 mg
Vitamin E (δ-tocopherol)	0.69 mg
Vitamin B_6	0.591 mg
Choline	69.8 mg
Betaine	67.6 mg
Lutein	28 mcg
Essential amino acids	
Arginine	1.060 g
Alanine	0.799 g
Asparticacid	1.261g
Tryptophan	0.181 g
Threonine	0.558 g
Isoleucine	0.582 g
Serine	1.148 g
Leucine	0.879 g
Lysine	0.747 g
Methionine	0.226 g
Phenylalanine	0.542 g
Glycine	1.636 g
Proline	0.698 g
Tyrosine	0.329 g
Valine	0.679 g
Histidine	0.389 g

(Continued)

Table 4.1 (Continued) Nutritional Content of the *Amaranthus* spp. (USDA 2016)

Nutrient	Value per 100 g
Fatty acid	
Palmitic acid (C16:0)	1.154 g
Stearic acid (C18:0)	0.223 g
Oleic acid (C18:1)	1.671 g
Linoleic acid (C18:2) ω-6	2.736 g
Linolenic acid (C18:3) ω-3	0.042 g
Squalene in Amaranthus oil	2.4 to 8.00 %

4.3 Antinutrient Content of *Amaranthus* Grain and Effects of Processing

Amaranthus grain poses a risk due to the nutrient-inhibitory effects of oxalates and phytates. Dobos (1992) observed that saponin concentrations in several *Amaranthus* species were extremely low, on average 0.09%. Nitrates are mostly found in the leaves and not in the grains, similar to the situation in common cereals, Amaranth has only trace quantities of protease inhibitors (trypsin and chymotrypsin inhibitors). These antinutrients have protective functions in plants; phytic acid, for example, stores phosphorus for later use. However, they block nutrients from being absorbed in humans (Dhull et al. 2019, 2020a, 2020b). Since humans cannot degrade phytate, it cannot provide inositol or phosphate in the diet. Phytate interacts with proteins to generate complexes that reduce their bioavailability (Kaur et al. 2018, Dhull et al 2021). In addition, phytate and oxalate have been linked to poorer starch digestion. Cuadrado et al. (2019) revealed that oxalate binds calcium and reduces absorption, whereas saponins form complexes with zinc and iron. Several processing techniques have been utilized to reduce the antinutrient content of *Amaranthus* grains.

The amaranth grain hull contains significant amounts of tannins, which can be greatly reduced by removing the hull. However, this process does not eliminate phytic acid from the grain. It is impossible to remove phytic acid by sanding away the outer layers of the grain or by using water extraction since the acid is dispersed uniformly throughout the grain. The impact of five processing procedures (defatting, blanching, germination, fermentation, and autoclaving) on the protein digestibility and antinutrient profile of *Amaranthus viridis* grain was evaluated by Babatunde and Gbadamosi (2017). Protein digestibility was shown to increase by 82% after germination at 30 °C for 72 h, significantly more than the 76% increase occurring after fermentation. It was found that autoclaving and blanching were superior to other heat treatments for lowering tannin and oxalate content, respectively. Njoki (2015) studied the effects of dry heating processes (roasting at 160 °C for 10 min or popping at 190 °C for 15 s) and wet heating techniques (boiling whole grains at a water:seed ratio of 4:1 and boiling *Amaranthus* flour at a water:flour ratio of 6:1). The protein digestibility of flour and whole grain was increased by 24% and 15%, respectively, by boiling, whereas roasting decreased protein digestibility. The protein digestibility of flour and whole grain saw an increase of 24% and 15%, respectively, in response to boiling, while protein digestibility decreased with roasting. Boiling was more effective than roasting and popping at decreasing tannin, phytate, and oxalate concentrations. Steeping *Amaranthus* grain for 5 h and then germinating it for 24 h were shown to be the most effective processing times based on dry matter loss and decrease of antinutrient levels (Kanensi et al. 2011). Clearly, the antinutrient content of amaranth grains can be lowered through appropriate processing, so that their value for food security need not be compromised.

4.4 Development of Food Products Using Amaranth

The food industry has identified amaranth as a "supergrain" with vast unexplored opportunities for new product innovation. Unlike wheat, it does not contain gluten, making it a useful ingredient in creating gluten-free products. Amaranth grain has the potential to provide consumers with the necessary nutrients while also benefitting from value-added processing technologies. Amaranth grain may be malted to make beer or boiled and eaten as porridge at home (Mugalavai 2013). Popped or roasted amaranth grain may be ground into flour, which can then be used alone or in combination with other ingredients to create a wide variety of baked goods and cereals (Emire and Arega 2012). The combination of unpopped grain with powdered fish or the flour of other cereals such as maize, sorghum, and millet can be used to create a thin porridge (O'Brien and Price 2008; Janet 2015). The development of several food products with amaranth seeds is discussed below.

4.4.1 Development of amaranth/wheat-based bread

Amaranth grain has the versatility to serve as a standalone ingredient or as part of a blended flour in the production of breakfast dishes, baked goods, gluten-free foods, and extruded goods. Together with wheat, it can be used to make leavened bread (Elizabeth 2010). Flat cakes made with amaranth flour are popular in Latin America and the Himalayas (Teutonico and Knorr 1985). Emire and Arega (2012) used the commercial bread-producing machinery of the Kality Food Sharing Corporation in Addis Ababa, Ethiopia, to create amaranth-/wheat-based breads with a variety of blend formulas. For the manufacturing of composite flour, wheat was replaced with amaranth at different percentages (5%, 10%, 15%, 20%, and

30%). As the percentage of amaranth in the dough increased, the baked bread's nutritional value skyrocketed. Certain rheological qualities and sensory features of loaves baked at 220 °C for 18 min were enhanced by the addition of up to 10% amaranth. Bread was baked using a combination of 0–55% grain amaranth flour and wheat flour in another investigation (Jerome 2001). It was found that the bread's quality declined at a 15% (w/w) substitution level of grain amaranth flour. Amaranth flour was found to considerably enhance protein, fat, ash, dietary fibre, and mineral contents by Sanz-Penella et al. (2013); however, this rise was accompanied by larger quantities of phytates, which may reduce the bioavailability of the minerals. Increases in crumb hardness and elasticity were also seen with increasing amaranth concentration, suggesting that the greatest amount of amaranth flour that may be used in a product is 20%, while retaining its quality and nutritional value (Sanz-Penella et al. 2013).

4.4.2 Development of Weaning Foods

Low-income mothers in sub-Saharan Africa, in both rural and urban areas, commonly use complementary meals, such as porridge made from maize, millet, and sorghum grains (Akinsola et al. 2017). The manufactured gruel from these grains is insufficient to meet the nutritional needs of newborns since it is deficient in calories, high-quality protein, and dietary minerals. The porridge's low energy level is because it is too thick for youngsters to consume without difficulty, so that it must be watered down before being given to them. Weaning meals' protein and other vital elements have been boosted with amaranth (Anigo et al. 2009; Akinsola et al. 2017). Amaranth and sorghum are used together in various proportions in order to get a nutrient-rich supplement meal (Okoth 2011). Malted flour made from amaranth and sorghum grain was utilized to create better flour with fewer anti-nutritional elements. The protein level was greatest in the mixture of 90% amaranth and 10% sorghum grain. This is within the range of what an infant of 12–23 months old, with poor breast milk consumption, needs from supplemental meals (9.1 g d⁻¹) in terms of protein (WHO 2007). Porridge made from a mixture of maize, amaranth grain, and chickpea in varying proportions was developed as a supplementary diet for the rural population of Ethiopia (Zebdewos et al. 2015). The porridge manufactured from 70% amaranth and 30% chickpea outperformed the porridge made from whole maize and a combination of all three grains in terms of iron, zinc, and calcium content, as well as protein and phytate levels. Amaranth flour was proven to improve the food's protein quality. Mburu et al. (2012) soaked and pre-gelatinized amaranth grains to extract flour, which they used to make a porridge-like material by mixing 15 g of the flour with 100 mL hot water. The resultant gruel has all sensory components within an acceptable range, was rich in minerals, such as magnesium, manganese, phosphorus, iron, and zinc, and phytochemicals such as tocopherols, protein, riboflavin, and niacin, and it was claimed to meet babies' daily nutrient needs if eaten three times a day. It is important to spread awareness about amaranth-based supplementary meals to both rural and urban families due to their beneficial protein-energy balance.

4.4.3 Development of Amaranth-based Biscuits

Breads, biscuits, and other treats made in bakeries are the most widely consumed types of processed food worldwide (Caleja et al. 2017). Biscuits are high-energy, low-fibre snacks that are often produced from wheat flour and fat. When consumed in large quantities, this can be harmful to health (Caleja et al. 2017). Thus, the use of alternative flours that are high in vital nutrients is necessary to increase the total nutritious contribution of biscuits. Biscuits made from amaranth/wheat composite flours have been created in a number of research projects (Sindhuja et al. 2005; Renu and Anirban 2015; Virginia and Ajit 2014). Utilization of amaranth flour to replace wheat flour in a 35% volume ratio by Sindhuja et al. (2005) led to the improvement of organoleptic qualities including its colour, taste, flavour, and appearance in the product. After experimenting with wheat/grain amaranth composite flours, Renu and Anirban (2015) found that their cookies were just as tasty when made with 100% grain amaranth. Including up to 20% grain amaranth with wheat flour for baking biscuits was revealed to have the optimum sensory characteristics (Virginia and Ajit 2014). Gluten-free biscuits were made by using amaranth, buckwheat, and quinoa by Schoenlechner et al. (2008). A sensory panel favoured the biscuits made with buckwheat and amaranth because of their crunchier texture compared to those made with quinoa and amaranth. These results clearly showed the utility of amaranth flour as a substitute for wheat flour for making gluten-free biscuits.

4.4.4 Development of Functional Beverages

The role of food and drink in the prevention and treatment of disease has received increased attention in recent years. The nutritional advantages of functional foods and beverages are augmented by additional physiological benefits (Corbo et al. 2014). Drinks rank as the top functional food category in terms of popularity, as they offer an economical option, align with consumer preferences for attractive packaging and presentation, are easy to distribute, and can be stored for an extended period of time, provided that they are appropriately refrigerated (Panghal 2017a, 2017b; Goyal et al. 2023). In addition, they are a fantastic method of providing people with essential nutrients and bioactive chemicals. It was predicted that $105.5 billion was made in revenue worldwide from sales of functional drinks (Pandal 2017). The food industry is increasingly interested in exploring how it may derive functional foods from existing food products. A functional drink made from extruded amaranth grains is available on the market now (Milan-Carrillo et al. 2012). There was a considerable improvement in sample acceptance when amaranthwas included. Grain amaranth was fermented for varying amounts of time to create a non-alcoholic beverage called Kunu that was compared to the better-known sorghum-based Kunu (Isaac-Bamgboye et al. 2019). Amaranth-kunu, when fermented for 48 h, was shown to have a higher level of acceptability than sorghum Kunu. Moreover, amaranth/Kunu has higher protein, fat, and ash content than sorghum-based Kunu. Arguelles-Lopez et al. (2018) used extrusion and germination methods to generate an amaranth chai-based beverage. Protein content in germinated amaranth grains and chia seeds was found to be greater than that of extruded seeds. Amaranth and chia flour were mixed in a ratio of 70:30 to create the drink. Several key athletic performance variables were measured in cyclists who were given either an amaranth-based beverage or a commercially available beverage. This was done in a randomized cross-over design (Espino-Gonzalez et al. 2018). Except for time-trial performance, none of the other factors tested showed any statistically significant variation. When compared with a commercially available beverage, performance was improved when amaranth was used instead. An amaranth drink can be prepared by combining 88.8 g L^{-1} amaranth grains, 41.3 g L^{-1} sugar, and 0.35 g L^{-1} sodium chloride, which can be further flavored with grape or orange extract, yielding a drink with 1.5% protein content, 10% carbohydrate concentration, and an electrolyte concentration of 0.35 g L^{-1} (sodium chloride).The amaranth beverage had more calories (52.48 kcal 100 mL^{-1}) than most commercial drinks (24 kcal 100 mL^{-1}). Because of their ability to prevent and treat chronic illnesses, amaranth grains are excellent raw beverage materials. Thus, amaranth in functional beverages might boost their nutritional content, giving them a preferable option to the sugary and caffeinated drinks that are now trendy but do not provide numerous medicinal benefits.

4.4.5 Utilization of Amaranth in Animal Nutrition

The need for animal feed components is rising. As a significant component of human diets, there is a high demand for resources like maize. Hence, it is imperative to explore alternative sources of energy and protein. *Amaranthus* species can potentially fulfill this role as a crop. Cereals, like sorghum and millet, employed in animal diets serve only as energy sources and must be supplemented with other protein sources. Amaranth, which has a similar calorific and protein content to these cereals, may be used in their place or as an alternative supplement. Amaranth has similar calorific content as other cereals but twice the amount of protein. In several regions, both the amaranth grain and leaves have been utilized as a fodder or silage crop for feed for various animals, including cattle, chickens, pigs, and rabbits. To keep up with the ever-increasing demand for animal feed, it is crucial to recognize the need to utilize alternative sources of energy and protein, like amaranth (Peiretti 2018).

Nonetheless, it has been shown that *Amaranthus* contains growth-inhibiting chemicals, the activity of which can be reduced by cooking. In a 14-day experiment, in which rats were fed amaranth grains that had been subjected to various pretreatments (raw, dry-heating, moist heating, or soaking in water), only the group fed moist-heated grains gained weight; the other rats gained weight for the first six days and then lost weight (Pond et al. 1991). Pre-heated amaranth grains can be used as a feed supplement for broiler chickens. Heat treatment can diminish or eliminate antinutritional components present in the grain. Research showed that broiler chicks can be fed extruded grain amaranth without losing weight. Fasuyi and Nonyerem (2007) employed sun-dried *A. cruentus* leaves to create a potentially nutrient-dense feed meal that could be added to broiler finisher diets. In order to reduce the antinutritional components in *Amaranthus* leaf, which might prevent the body from absorbing necessary nutrients, the leaves were soaked in water and then dried in the sun. The study found that a 10% inclusion of *A. cruentus* leaf in bird feed was enough to increase broiler chicken performance without creating any harmful side effects. According to Fasuyi and Akindahuns (2009), steam-pelleted diets, including amaranth grain, increased feed intake, growth performance, and fat deposition in broilers. Tillman and Waldroup (1986) recommended including up to 40% *A. cruentus* grain in broiler diets, but only if the amaranth is adequately extruded or autoclaved (Peiretti 2018). Seguin et al. (2013) showed that ensiling *A. hypochondriacus* produced digestible silage for ruminants. It has been demonstrated that ensiling decreased the oxalate content of amaranth. Olorunnisomo and Ayodele (2009) found that *A. cruentus* may produce high-quality fodder, which the West African dwarf sheep used in the study digested successfully. Alegbejo (2014) reported that *A. retroflexus* and *A. hybridus* have been successfully used as fodder for sheep and calves.

4.4.6 Amaranthus *Oil (Squalene) and its Industrial Application*

The highly valued compound squalene, which is recognized for its biological significance and is present in a wide variety of plant and animal cells. According to some sources, the *Amaranthus* plant is the most squalene-dense plant on the planet (Alvarez et al. 2010; Venskutonis and Kraujalis 2013). The plant derived oil squalene is suitable for use in cosmetics (moisturisers, sunscreens, and makeup, etc.) and other personal care items (Ofitserov 2001). *Amaranthus* squalene has a high antioxidant content and protects the skin from free-radical damage (Huang et al. 2009). The use of *Amaranthus* oil promotes skin regeneration and assists in restoring the skin's protective barrier. It boosts the skin's moisture retention and decreases its evaporation rate, all the while activating cellular defence mechanisms (Huang et al. 2009). After analyzing the oil of 104 different *Amaranthus* genotypes and 30 species, by He and Cai (2002), it was observed that the amount of squalene present was significantly higher in the grains, with an average of 4.2%, as compared to the leaves, which had an average of only 0.26%. Squalene is reported to be present in olive oil (564 mg 100 g^{-1}), soybean oil (9.9 mg 100 g^{-1}), grape seed oil (14.1 mg 100 g^{-1}), peanut oil (27.3 mg 100 g^{-1}), corn oil (24.7 mg 100 g^{-1}), sunflower seed oil (13.8 mg 100 g^{-1}), and *Amaranthus* (5942 mg 100 g^{-1}) (Nergiz and Celikkale 2011; Lozano-Grande et al. 2018). *Amaranthus* has a high squalene concentration but is under-utilized in the commercial sector. Olive oil is the most common source for

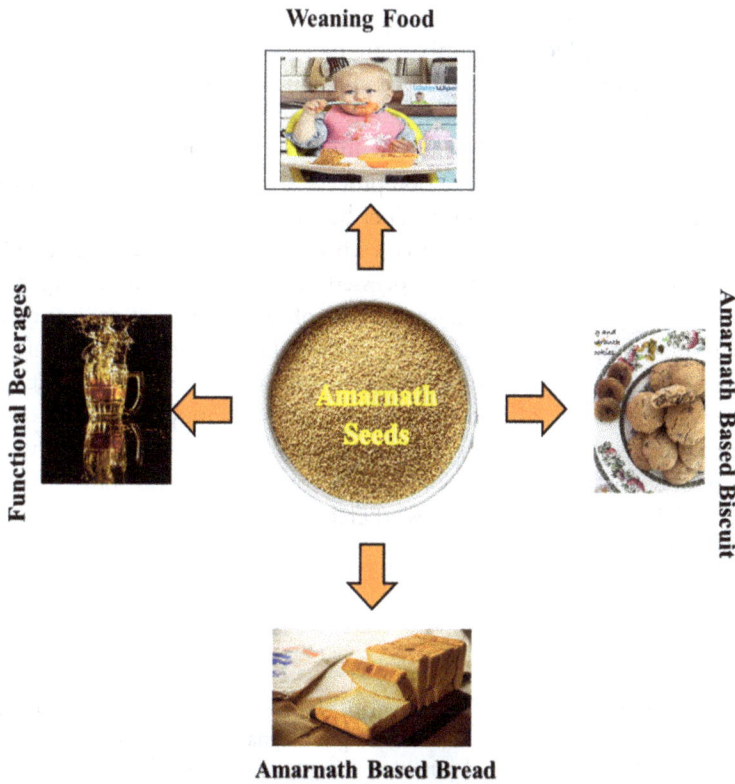

Figure 4.1 Schematic representation of food products developed from *Amaranthus* seeds.

commercial extraction. Squalene is expected to experience a substantial surge in demand within the next ten years, according to industry experts, making it imperative to investigate other sources of plants, such as *Amaranthus*, to create cutting-edge methods for extraction to ensure high-quality and high-yield production at an industrial scale (Figure 4.1).

4.5 Conclusion

Grain *Amaranthus*, the new crop, has been the centre of attention ever since the 1970s because of claims of its special characteristics and adaptability. This rediscovered crop provides agricultural benefits and is known for its resilience under stress conditions, such as high irradiance, high temperatures, and drought. Popular foods like bread, cookies, and pasta can serve as ideal vehicles for protein enrichment due to their high consumption rates. Several experts agree that *Amaranthus* is safe for those with coeliac disease, because it contains no gluten. Nevertheless, the absence of gluten prevents grain amaranth from being included in composite flour for leavened products. The unique aroma and flavour of amaranth, which has been described as spicy, with a slightly pungent and bitter aftertaste, is another factor restricting acceptability of composite breads. Yet, sensory and non-sensory elements also play a role in consumers' willingness to try new foods. Non-sensory elements include things like production processes, customer attitudes, awareness of health and the environment, and product beliefs, in addition to the obvious ones like price and ease of preparation. Amaranth-containing bread can be launched on the market as being produced and processed according to norms of organic agriculture due to increased consumer awareness of health and the environment and due to product specialities. Therefore, the potential for its use in the food, feed, and even some industrial sectors is eventually determined by the characteristics proven by numerous, above-all fundamental, scientific data. With proper production and storage techniques, amaranth grain has the potential to make a significant contribution to food security and nutrition. In addition, it can provide raw and intermediate materials for various industries, thereby increasing production possibilities and employment opportunities.

References

Aderibigbe, O. R., O. O. Ezekiel, S. O. Owolade, J. K. Korese, B. Sturm, O. Hensel. 2022. Exploring the potentials of underutilized grain Amaranthus (*Amaranthus* spp.) along the value chain for food and nutrition security: A review. *Critical Reviews in Food Science and Nutrition* 62(3):656–669.

Ajayi, F. F., P. Mudgil, C. Y. Gan, S. Maqsood. 2021. Identification and characterization of cholesterol esterase and lipase inhibitory peptides from Amaranthus protein hydrolysates. *Food Chemistry: X* 12:100165.

Akinsola, A. O., O. O. Onabanjo, M. A. Idowu, and B. I. O. AdeOmowaye. 2017. Traditional complementary foods: A critical review. *Greener Journal of Agricultural Sciences* 7(9):226–242.

Alegbejo, J. O. 2014. Nutritional value and utilization of Amaranthus (*Amaranthus* spp.) – A review. *Bayero Journal of Pure and Applied Sciences* 6 (1):136–143.

Alvarez, J. L., M. Auty, E. K. Arend, and E. Gallagher. 2010. Baking properties and microstructure of pseudocereal flours in gluten-free bread formulations. *European Food Research and Technology* 230(3):437–445.

Amare, E., C. Mouquet-Rivier, A. Servent, G. Morel, A. Adish, and G. Haki. 2015. Protein quality of Amaranthus grains cultivated in Ethiopia as affected by popping and fermentation.. *Food and Nutrition Sciences* 6(1): 38–48.

Anigo, K. M., D. A. Ameh, S. Ibrahim, and S. K. Danbauchi. 2009. Nutrient composition of commonly used complementary foods in Northwestern Nigeria. *African Journal of Biotechnology* 8(18):4211–4216.

Arguelles-Lopez, O. D., C. Reyes-Moreno, R. Roberto Gutierrez Dorado, M. F. Sanchez-Osuna, J. Lopez-Cervantes, E. O. Cuevas-Rodriguez, J. Milan-Carrillo, and J. X. K. Perales-Sanchez. 2018. Functional beverages elaborated from Amaranthus and chia flours processed by germination and extrusion. *Biotecnia* 20 (3):135–145.

Babatunde, T. O., and S. O. Gbadamosi. 2017. Effect of different treatments on invitro protein digestibility, antinutrients, antioxidant properties and mineral composition of *Amaranthus viridis* seed. *Cogent Food and Agriculture* 3(1):1–14.

Caleja, C., L. Barros, A. L. Antonio, M. B. P. P. Oliveira, and I. C. F. R. Ferreira. 2017. A comparative study between natural and synthetic antioxidants: Evaluation of their performance after incorporation into biscuits. *Food Chemistry* 216:342–346.

Corbo, M. R., A. Bevilacqua, L. Petruzzi, F. P. Casanova, and M. Sinigaglia. 2014. Functional beverages: The emerging side of functional foods. *Comprehensive Reviews in Food Science and Food Safety* 13(6):1192–1206.

Cuadrado, C., K. Takacs, E. E. Szabó, and M. Pedrosa 2019. Non-nutritional factors: Lectins, phytic acid, proteases inhibitors, allergens. In *Legumes: Nutritional Quality, Processing and Potential Health Benefits*, ed. M. A. Martín-Cabrejas, 152–176. London: Royal Society of Chemistry.

Dhull, S. B., M. Kaur, and K. S. Sandhu. 2020a. Antioxidant characterization and *in vitro* DNA damage protection potential of some Indian fenugreek (*Trigonella foenum-graecum*) cultivars: Effect of solvents. *Journal of Food Science and Technology* 57(9): 3457–3466.

Dhull, S. B., S. Punia, M. K. Kidwai, M. Kaur, P. Chawla, S. S. Purewal, M. Sangwan, and S. Palthania. 2020b. Solid-state fermentation of lentil (*Lens culinaris* L.) with *Aspergillus awamori*: Effect on phenolic compounds, mineral content, and their bio-availability. *Legume Science* 2(3): e37.

Dhull, S. B., S. Punia, R. Kumar, M. Kumar, K. B. Nain, K. Jangra, and C. Chudamani. 2021. Solid state fermentation of fenugreek (*Trigonella foenum-graecum*): Implications on bioactive compounds, mineral content and in vitro bioavailability. *Journal of Food Science and Technology* 58(5): 1927–1936

Dhull, S. B., S. Punia, K. S. Sandhu, P. Chawla, R. Kaur, and A. Singh. 2019. Effect of debittered fenugreek (*Trigonella foenum-graecum* L.) flour addition on physical, nutritional, antioxidant, and sensory properties of wheat flour rusk. *Legume Science* 2(1):e21.

Dobos, G. 1992. Koerner Amaranthusals neve kulturpflanze in Oesterrcich. Introducktion und Zuechterische Aspeke. PhD thesis, University of Natural Resources and Applied Life Sciences.

Elizabeth, A. A. 2010. Ancient grains: Opportunities and challenges for Amaranthus, quinoa, millet, sorghum and *tef* in gluten-free food products. IFT Annual Meeting, Chicago, IL.

Emire, S. A., and M. Arega. 2012. Value added product development and quality characterization of Amaranthus (*Amaranthus caudatus* L.) grown in East Africa. *African Journal of Food Science and Technology* 3 (6):129–141.

Espino-González, E., M. J. Muñoz-Daw, J. M. Rivera-Sosa, M. L. De la Torre-Díaz, G. E. Cano-Olivas, J. C. De Lara-Gallegos, and M. C. Enríquez-Leal. 2018. Influence of an Amaranthus-based beverage on cycling performance: A pilot study. *Biotecnia* 20 (2):31–36.

Fasuyi, A. O., and A. D. Nonyerem. 2007. Biochemical, nutritional and haematological implications of *Telfairia occidentalis* leaf meal as protein supplement in broiler starter diets. *African Journal of Biotechnology* 6:1055–1063.

Fasuyi, A. O., and A. O. Akindahuns. 2009. Nutritional evaluation of *Amaranthus cruentus* leaf meal-based broiler diets supplemented with cellulase/glucanase/xylanase enzymes. *American Journal of Food Technology* 4(3):108–118.

Fleming, T. 2000. *PDR for Herbal Medicines*, 2nd ed.; Medical Economics Company, Montvale, pp. 75–76.

Goyal, C., P. Dhyani, D. C. Rai, S. Tyagi, S. B. Dhull, P. K. Sadh, J. S. Duhan, and B. S. Saharan. 2023. Emerging trends and advancements in the processing of dairy whey for sustainable biorefining. *Journal of Food Processing and Preservation* 2023: 1–24.

He, H. P., Y. Cai, M. Sun, and H. Corke. 2002. Extraction and purification of squalene from Amaranthus grain. *Journal of Agriculture and Food Chemistry* 50 (2):368–372.

Huang, Z. R., Y. K. Lin, and J. Y. Fang. 2009. Biological and pharmacological activities of squalene and related compounds: Potential uses in cosmetic dermatology. *Molecules* 14(1):540.

Isaac-Bamgboye, F. J., M. O. Edema, and O. F. Oshundahunsi. 2019. Nutritional quality, physicochemical properties and sensory evaluation of Amaranthus-Kunu produced from fermented grain Amaranthus (*Amaranthus hybridus*). *Annals of Food Science and Technology* 20 (2):322–331.

Jamka, M., A. Morawska, P. Krzyzanowska-Jankowska, J. Bajerska, J. Przysławski, J. Walkowiak, A. Lisowska. 2021. Comparison of the effect of Amaranthus oil vs. rapeseed oil on selected atherosclerosis markers in overweight and obese subjects: A randomized double-blind cross-over trial. *International Journal of Environmental Research and Public Health* 18:8540.

Janet, O. A. 2015. Nutritional value and utilization of Amaranthus (*Amaranthus* spp.) – A review. *Bayero Journal of Pure and Applied Science* 6(1):136–143.

Januszewska-Jóźwiak, K., Synowiecki, J. 2008. Charakterystykaiprzydatnośćskładnik ówszarłatu w biotechnologiiżywności. *Biotechnologia* 3:89–102.

Jerome, A. A. 2001. The effect of Amaranthus grain flour on the quality of bread. *International Journal of Food Properties* 4 (2):341–351.

Kázmierczak, A., I. Bolesławska, J. Przysławski. 2011. Szarłat—Jegowykorzystanie w profi laktyceileczeniuwybranychchoróbcywilizacyjnych. *NowinyLekarskie* 80:192–198.

Kadoshnikov, S. I., I. G. Kadoshnikova, and D. M. Martirosyan. 2005. Investigation of fractional composition of the protein in Amaranthus. *Book: Non-Traditional Natural Resources, Innovation Technologies and Products* 12:81–104. Moscow: Russian Academy of Natural Sciences.

Kanensi, O. J., S. Ochola, N. K. Gikonyo, and A. Makokha. 2011. Optimization of the period of steeping and germination for Amaranthus grain. *Journal of Agriculture and Food Technology* 1(6):101–105.

Karamac, M., F. Gai, E. Longato, G. Meineri, M. A. Janiak, R. Amarowicz, P. G. Peiretti. 2019. Antioxidant activity and phenolic composition of Amaranthus (*Amaranthus caudatus*) during plant growth. *Antioxidants* 8:173.

Kaur, P., S. B. Dhull, K. S. Sandhu, R. K. Salar, and S. S. Purewal. 2018. Tulsi (*Ocimum tenuiflorum*) seeds: *In vitro* DNA damage protection, bioactive compounds and antioxidant potential. *Journal of Food Measurement and Characterization* 12(3): 1530–1538.

Kaźmierczak, A., I. Bolesławska, J. Przysławski. 2011. Szarłat—Jegowykorzystanie w profi laktyceileczeniuwybranychchoróbcywilizacyjnych. *NowinyLekarskie* 80:192–198.

Kulczyński, B., A. Gramza-Michałowska, M. Grdeń. 2017. Amarantus—Wartośćodż ywczaiwłaściwościprozdrowotne. *Bromat. Chem. Toksykol.* 1:1–7.

Lozano-Grande, M. A., S. Gorinstei, E. Espitia-Rangel, G. Dávila-Ortiz, and A. L. Martínez-Ayala. 2018. Plant sources, extraction methods, and uses of squalene. *International Journal of Agronomy* 2018:1–13.

Malik, M., R. Sindhu, S. B. Dhull, C. Bou-Mitri, Y. Singh, S. Panwar, and B. S. Khatkar. 2023. Nutritional composition, functionality, and processing technologies for amaranth. *Journal of Food Processing and Preservation* 2023:1753029. https://doi.org/10.1155/2023/1753029.

Mburu, M. W., N. K. Gikonyo, G. M. Kenji, and A. M. Mwasaru. 2012. Nutritional and functional properties of a complementary food based on Kenyan Amaranthus grain (*Amaranthus cruentus*). *African Journal of Food, Agriculture, Nutrition and Development* 12(2):1–19.

Mekonnen, G., M. Woldesenbet, T. Teshale, and T. Biru. 2018. *Amaranthus caudatus* production and nutrition contents for food security and healthy living in MenitShasha, MenitGoldya and Maji Districts of Bench Maji Zone, Southwestern Ethiopia. *Nutrition and Food Science International Journal* 7(3):1–7.

Milan-Carrillo, J., A. Montoya-Rodrıguez, R. Gutierrez-Dorado, X. Perales-Sanchez, and C. Reyes-Moreno. 2012. Optimization of extrusion' process for producing high antioxidant instant Amaranthus (*Amaranthus hypochondriacus* L.) flour using response surface methodology. *Applied Mathematics* 3(10):1516–1525.

Morales, D., M. Miguel, M. Garcés-Rimón. 2020. Pseudocereals: A novel source of biologically active peptides. *Critical Reviews in Food Science and Nutrition* 61:1537–1544.

Moszak, M., A. Zawada, M. Grzymisławski. 2018. Właściwościorazzastosowanieole jurzepakowegoioleju z amarantusa w leczeniuzaburzeńmetabolicznychzwiązan ych z otyłością (The properties and the use of rapeseed oil and Amaranthus oil in the treatment of metabolic disorders related to obesity). *Forum ZaburzeńMetab* 9:53–64.

Mugalavai, V. K. 2013. Effect of Amaranthus: Maize flour ratio on the quality and acceptability of ugali and porridge (Kenyan cereal staples). *ARPN Journal of Agricultural and Biological Sciences* 5(1):1–7.

Murya, N. K., and A. Pratibha. 2018. Amaranthus grain nutritional benefits: A review. *Journal of Pharmacognosy and Phytochemistry* 7(2):2258–2262.

Musa, A., J. A. Oladiran, M. I. S. Ezenwa, H. O. Akanya, and E. O. Ogbadoyi. 2011. Effect of heading on some micronutrients, anti-nutrients and toxic substances in *Amaranthus cruentus* grown in Minna, Niger State, Nigeria. *American Journal of Food and Nutrition* 1(4):147–154.

Mustafa, A. F., P. Seguin, and B. Gelinas. 2011. Chemical composition, dietary fibre, tannins and minerals of grain Amaranthus genotypes. *International Journal of Food Sciences and Nutrition* 62(7):750–754.

Nergiz, C., and D. Çelikkale. 2011. The effect of consecutive steps of refining on squalene content of vegetable oils. *Journal of Food Science and Technology* 48(3):382–385.

Njoki, J. W. 2015. Impact of different processing Techniques on nutrients and Antinutrients content of grain Amaranthus (*Amaranthus albus*). A. thesis, submitted in partial fulfillment for the Degree of Master of Science in Food Science and Nutrition in the Jomo Kenyatta University of Agriculture and Technology.

O'Brien, G. K., and M. L. Price. 2008. Amaranthus: Grain and vegetable types. Echo Technical Note.

Obiedzinska, A., and B. Waszkiewicz-Robak. 2012. Olejtłoczonenazimnojakozywno' s'cfunkcjonalna Zywnośc. *Nauka Technol. Jakóśc* 1:27–44.

Ofitserov, E. N. 2001. Amaranthus: Perspective raw material for food-processing and pharmaceutical industry. *Chemical Compatibility and Sinulation* 2(5):14–18.

Okoth, J. K., S. Ochola, N. K. Gikonyo, and, A. Makokha. 2011. Optimization of the period of steeping and germination for Amaranthus grain. *Journal of Agriculture and Food Technology* 1(6):101–105.

Olorunnisomo, O. A., and O. J. Ayodele. 2009. Effects of intercropping and fertilizer application on the yield and nutritive value of maize and Amaranthus forages in Nigeria. *Grass and Forage Science* 64(4):413–420.

Pandal, N. 2017. Nutraceuticals: Global markets.Available at: https://www.bccresearch.com/market-research/food-and-beverage/nutraceuticals-markets-report-fod013f.html

Panghal, A., V. Kumar, S. B. Dhull, Y. Gat, and N. Chhikara. 2017a. Utilization of dairy industry waste-whey in formulation of papaya RTS beverage. *Current Research in Nutrition and Food Science Journal* 5(2):168–174.

Panghal, A., K. Virkar, V. Kumar, S. B. Dhull, Y. Gat, and N. Chhikara. 2017b. Development of probiotic beetroot drink. *Current Research in Nutrition and Food Science Journal* 5(3). http://doi.org/10.12944/CRNFSJ.5.3.10.

Park, S. J., A. Sharma and H. J. Lee 2020. A review of recent studies on the antioxidant activities of a third-millennium food: *Amaranthus* spp. *Antioxidants* 9:1236.

Peiretti, P. G. 2018. Amaranthus in animal nutrition: A review. *Livestockresearch for Rural Development* 30 (5):1–20.

Pond, W. G., J. W. Lehmann, R. Elmore, F. Husby, C. C. Calvert, C. W. Newman, B. Lewis, R. L. Harrold, and J. Froseth. 1991. Feeding value of raw or heated grain Amaranthus germplasm. *Animal Feed Science and Technology* 33(3–4):221–236.

Prakash, D., and M. Pal. 1991. Nutritional and antinutritional composition of vegetable and grain Amaranthus leaves. *Journal of the Science of Food and Agriculture* 57(4):573–583.

Renu, P. R., and D. Anirban. 2015. Development and shelf-life study of Amaranthus cookies. 7th Indo-Global Summit and Expo on Food & Beverages. *Journal of Food Processing and Technology*. ISSN: 2157–7110.

Sanz-Penella, J. M., M. Wronkowska, M. Soral-Smietana, and C. M. Haros. 2013. Effect of whole Amaranthus flour on bread properties and nutritive value. *LWT - Food Science and Technology* 50(2):679–685.

Schoenlechner, R., S. Siebenhandl, and E. Berghofer. 2008. Pseudocereals. In *Gluten-Free Cereal Products and Beverages*, eds. E. K. Arendt and F. D. Bello, 149–176. London: Elsevier.

Seguin, P., A. F. Mustafa, D. J. Donnelly, and B. Gélinas. 2013. Chemical composition and ruminal nutrient degradability of fresh and ensiled Amaranthus forage. *Journal of the Science of Food and Agriculture* 93(15):3730–3733.

Sindhuja, A., M. L. Sudha , and A. Rahim. 2005. Effect of incorporation of Amaranthus flour on the quality of cookies. *European Food Research and Technology* 221(5): 597–601.

Szwejkowska, B., and S. Bielski. 2012. Wartość prozdrowotna nasion szarłatu (*Amaranthus cruentus* L.). *Postępy Fitoterapi* 4:240–243.

Teutonico, R. A., and D. Knorr. 1985. Amaranthus: Composition, properties and applications of rediscovered food crop. *Journal of Food Technology* 39:49–60.

Tillman, P. B., and P. W. Waldroup. 1986. Processing grain Amaranthus for use in broiler diets. *Poultry Science* 65(10):1960–1966.

Venskutonis, P., and P. Kraujalis. 2013. Nutritional components of Amaranthus seeds and vegetables: A review on composition, properties, and uses. *Comprehensive Reviews in Food Science and Food Safety* 12 (4):381–412.

Virginia, P., and P. Ajit. 2014. Development of nutritious snacks by incorporation of Amaranthus seeds, watermelon seeds and their flour. *Indian Journal of Community Health* 26(5):86–94.

WHO. 2007. Protein and amino acid requirements in human nutrition. Report of a joint WHO/FAO/UNU expert consultation. World Health Organization Technical Report Series 935. WHO, Geneva. http://www.who.int/nutrition/publications/nutrient/requirements/WHO_TRS_935/en/

Zebdewos, A., P. Singh, G. Birhanu, S. J. Whiting, C. J. Henry, and A. Kebebu. 2015. Formulation of complementary food using Amaranthus, chickpea and maize improves iron, calcium and zinc content. *African Journal of Food, Agriculture, Nutrition and Development* 15:290–295.

Chapter 5

Buckwheat: Nutritional Profile, Processing, and Food Products

Nitin Sharma, Smita Mall, Vikas Kumar

5.1 Introduction

Pseudocereal crops, like buckwheat, amaranth, and quinoa, are gaining in popularity nowadays due to their exceptional nutritional and biological values. These crops are rich in dietary fibers, essential amino acids, minerals like calcium, iron, and zinc, vitamins, and other bioactive compounds, such as saponins, polyphenols, and phytosterols (Alvarez-Jubete et al. 2010; Martínez-Villaluenga et al. 2020; Sidorova et al. 2020; Malik et al. 2023). These grains are considered the "grain of the twenty-first century", and their potential impact on human health is the subject of considerable interest and research. Although scientific evidence supporting the health benefits of pseudocereals is limited, several studies have suggested their beneficial effects on obesity, metabolic syndrome, and complications associated with type 2 diabetes (Punia and Dhull 2019; Martínez-Villaluenga et al. 2020). For instance, research studies have shown that pseudocereals can improve insulin sensitivity, reduce blood glucose levels, and promote antidiabetic effects.

Grain products are an important component of the global diet, providing a vital source of complex carbohydrates. There are two primary categories of grains: cereals and pseudocereals. Cereals comprise wheat, rice, corn, oats, barley, and rye, while pseudocereals comprise quinoa, amaranth, buckwheat, and chia. Although the production volume of pseudocereals is much lower than that of cereals, they have gained more popularity in recent years owing to their high nutritional value. Due

DOI: 10.1201/9781003325277-5

to their gluten-free nature, pseudocereals can be considered an excellent alternative for individuals with celiac disease or gluten intolerance (Ciudad-Mulero 2019). Quinoa, amaranth, and buckwheat grains produce various food products such as bread, pasta, cookies, muffins, bars, snacks, soups, drinks, and cereals (Taylor and Awika 2017). However, buckwheat must be addressed in the twenty-first century, primarily due to the focus on high-yielding varieties of other cereals like rice, wheat, and maize. This has significantly decreased the cultivated area (Chettry et al. 2021; Chrungoo et al. 2021). Global buckwheat production has experienced a consistent rise and will reach approximately 4 million tons by 2021 (Zou et al. 2021).

Buckwheat (*Fagopyrum*) contains several health-promoting properties that could make it a potential functional food (Krkoskova and Mrazova 2005). Phenolic compounds, such as rutin and quercetin, present in buckwheat have been reported to decrease inflammation, thereby protecting against chronic diseases. In addition, several phenolic compounds that have been shown to have antioxidant (Kaur et al. 2018; Dhull et al. 2019), anticancer, and antidiabetic properties have also been reported in buckwheat. Phytosterols present in buckwheat are responsible for reduction in the cholesterol level of consumers and an increase in heart health (González-Sarrías et al. 2013; Dhull et al. 2020a). Buckwheat has become increasingly popular in recent years. Its potential therapeutic effects on chronic diseases, such as anti-oxidative, cardioprotective, anticancer, hepatoprotective, antihypertension, antitumour, anti-inflammatory, antidiabetic, neuroprotective, cholesterol-lowering, and cognition-improving activities, have attracted the attention of numerous food scientists and researchers (Lv et al. 2017; Kwon et al. 2018; Ge and Wang 2020).

Further research into buckwheat's therapeutic potential is required in light of these promising findings (Zou et al. 2021). The present chapter focuses on the nutritional profile and processing of buckwheat. This chapter also discusses the various food products developed from buckwheat.

5.2 Plant Description

Fagopyrum (buckwheat) is a genus of plants belonging to the Polygonaceae family, which is composed of approximately 1,200 species (Sanchez et al. 2011). Russia and China are the main producers of buckwheat in the world (Li and Zhang 2001), but its consumption has become increasingly popular in the United States, Canada, and Europe (Stember 2006). Buckwheat is cultivated in several northern countries, from China and India to Korea, Bhutan, and Nepal. It is also grown in Kazakhstan, Tajikistan, Russia, Ukraine, Lithuania, Estonia, Belarus, Moldova, Poland, Serbia, Croatia, Slovenia, Austria, Italy, the United States of America, and Canada (Yilmaz et al. 2020; Huda et al. 2021; Podolska et al. 2021; Raguindin et al. 2021). Buckwheat is highly adaptable and can be cultivated in diverse habitats and regions worldwide, including high-altitude areas with minimal rainfall and lower temperatures. This grain can be grown on nutrient-deficient soils and is more pest resistant than other cereal crops (Wijngaard et al. 2007; Ge and Wang 2020; Živkovi et al. 2021). Common buckwheat (*Fagopyrum esculentum* Moench) and tartary buckwheat (*Fagopyrum tataricum* Gaertn.) are the most widely grown species of buckwheat (Huda et al. 2021).

Buckwheat has a short growing season of about 10–12 weeks and can thrive under poor soil conditions (Huda et al. 2021; Chettry et al. 2021; Chrungoo et al. 2021). Buckwheat is a versatile pseudocereal, grown over centuries for its grain (for food and feed) and greens (vegetables and fodder). It does not belong to the Poaceae (grass) family, which includes all true cereals, such as barley and wheat, but shares similarities in chemical composition and edibility (Wijngaard et al. 2005; Brasil et al. 2020). It is a dicotyledonous plant, whereas wheat and barley are monocotyledonous plants, indicating different seed structures, a pair of S-shaped cotyledons instead of the usual single cotyledon, and the distribution of their primary storage compounds differing from that of other plants. Its triangular seeds are not botanically linked to wheat, and its name is probably derived from its comparable uses to wheat, such as in making flour and other baked goods. Despite these differences, buckwheat remains a popular and versatile crop, known for its nutritional benefits and gluten-free properties (Wijngaard et al. 2005, 2007).

The most economically significant species of buckwheat is the common buckwheat (*F. esculentum*), which accounts for 90% of global buckwheat production. It is primarily cultivated in temperate regions of the Northern Hemisphere (Phiarais et al. 2005; Huda et al. 2021). Common buckwheat is typically characterized by producing dimorphic flowers that come in either pin type (long pistil and short stamen) or thrum type (short pistil and long stamen), resulting in self-incompatibility (Woo et al. 2010, 2016). Tartary buckwheat (*F. tataricum*), also known as "bitter buckwheat" due to much higher concentrations of bitter substances in the grains, has several advantages over other species, such as high grain yield, self-pollination, and greater tolerance to adverse climatic conditions, being mainly a high-altitude crop (Zhang et al. 2012; Chrungoo et al. 2021). Additionally, tartary buckwheat contributes to various health benefits due to its higher rutin content (Suzuki et al. 2021). Grains of both common buckwheat and tartary buckwheat are gluten free and are often used as a substitute for wheat and other grains in recipes.

5.3 Nutritional and Bioactive Composition

Buckwheat is a rich source of proteins, ranging from 8.5% to 18.8%, depending on various factors such as cultivar, source, and climate (Dziadek et al. 2016); see its nutritional content summarised in Table 5.1. This protein content is higher than that of classic cereal grains (Bobkov 2016). The protein in buckwheat comprises globulins (43.3–64.5%), albumins (12.5–18.2%), prolamins (0.8–2.9%), and glutelins (8.0–22.2%), along with 15% residual proteins (Chrungoo et al. 2016). Buckwheat lacks the 30 kDa prolamin proteins responsible for coeliac disease, as

Table 5.1   Nutritional Content of Buckwheat

Nutrient	Value per 100 g
Water	9.75 g
Energy	356 Kcal
Protein	11.1 g
Total lipid (fat)	3.4 g
Ash	2.1 g
Carbohydrates	71.5 g
Fiber, total dietary	10 g
Sugars, total	2.6 g
Starch	61.6 g
Minerals	
Calcium, Ca	18 mg
Phosphorus, P	347 mg
Iron, Fe	2.2 mg
Zinc, Zn	2.4 mg
Magnesium, Mg	231 mg
Manganese, Mn	1.3 mg
Potassium, K	460 mg
Sodium, Na	1 mg
Selenium, Se	8.3 µg
Vitamins	
Thiamin	0.12 mg
Riboflavin	0.425 mg
Niacin	7.02 mg
Folate, total	30 mg
Vitamin B$_6$	0.21 mg
Biotin	18.6 mg

(Continued)

Table 5.1 (Continued) Nutritional Content of Buckwheat

Nutrient	Value per 100 g
Pantothenic acid	1.23
Essential amino acids	
Arginine	0.982 g
Alanine	0.748 g
Aspartic acid	1.13 g
Glutamic acid	2.05 g
Tryptophan	0.192 g
Threonine	0.506 g
Isoleucine	0.498 g
Serine	0.685 g
Leucine	0.832 g
Lysine	0.672 g
Methionine	0.172 g
Cystine	0.229 g
Phenylalanine	0.52 g
Glycine	1.03g
Proline	0.507 g
Tyrosine	0.241 g
Valine	0.678 g
Histidine	0.309 g

(USDA 2019)

confirmed through gel electrophoresis and enzyme-linked immunosorbent methods (Petr et al. 2003). It also provides significant amounts of essential macronutrients, such as phosphorus, potassium, magnesium, and calcium, although its iron, manganese, and zinc levels are comparatively lower (Zhu 2016). The highest concentrations of micronutrients in buckwheat are reported in the seed coat, hull, and aleurone layers, as opposed to cereal grains (Orožen et al. 2012). After enzymatic digestion, minerals such as zinc, copper, and potassium become easily absorbable (Klepacka et al. 2020). Lipid content in buckwheat varies between 1.5% and 3.7%, but these lipids are crucial for various physiological processes (Ruan et al. 2020;

Punia et al. 2019, 2020; Kidwai et al. 2020). Buckwheat is rich in essential amino acids such as arginine and lysine, constituting around 13.36% of its protein content (Watanabe 1998; Christa and Soral-Śmietana 2008). Buckwheat also provides vitamins (B_1, B_2, B_3, B_5, B_6, C, and E), inositol, and imino sugars (Comino et al. 2013; Gimenez-Bastida et al. 2015; Christa and Soral-Śmietana 2008). It is rich in the essential trace element, selenium (0.0099 to 0.1208 mg/g), which has been shown to enhance the immune system, reduce inflammation, and protect against cancer and AIDS (Zheng et al. 2011).

Buckwheat groats and hulls are rich in flavonoids such as rutin, orientin, vitexin, quercetin, isovitexin, and isoorientin, according to Zhang et al. (2012) illustrated in Figure 5.1. The plant's flowers, stalks, leaves, and grains are rich sources of rutin, with the highest concentrations found in the leaves and seeds (Ahmed et al. 2014; Das et al. 2019). Specifically, the leaves have a rutin concentration of 0.08–0.10 mg/g, while the seeds range from 0.12–0.36 mg/g, as reported by Park et al. (2008) and Brunori et al. (2010). Aside from rutin, buckwheat contains the flavonoids hyperoside (a quercetin galactoside), quercitrin, and catechins (Kalinova and Vrchotova 2009). Compared with other foods, the concentration of free and bound phenolic acids is highest in buckwheat (Li et al. 2016). The phenolic acids in buckwheat mostly consist of derivatives of benzoic acid and cinnamic acid (Mir et al. 2018). Figure 5.1 illustrates several phenolic acids such as *p*-hydroxybenzoic acid, syringic acid, protocatechuic acid, vanillic acid, ferulic acid, *p*-coumaric acid, gallic acid, caffeic acid, chlorogenic acid, and salicylic acid found in buckwheat. Buckwheat grains also contain fagopyrin, a photosensitive polyphenolic compound (Kim and Hwang 2020), with a naphthodianthrone skeleton similar to hypericin compounds. However, due to their adverse reaction properties and photosensitizing effects, fagopyrins found in buckwheat seeds and leaves are deemed unsafe. Buckwheat embryos contain a higher concentration of fagopyritols, which are galactosyl derivatives of D-chiro-inositol in the bran fraction (Cheng et al. 2019). These fagopyritols play a crucial role in preventing the desiccation of buckwheat plants during seed development. D-fagomine is an imino-sugar associated with buckwheat (groats, bran, leaves, and flour) as a dietary food component with biological activity and antimicrobial properties (Amezqueta et al. 2012). These phytocompounds are responsible for several medicinal properties, such as antibacterial, trypsin inhibitors, anticancer, hypotensive, and antidiabetic activities (Gimenez-Bastida et al. 2015; Dhull et al. 2020b, Dhull et al. 2021; Sofi et al. 2023).

5.4 Food Products Developed from Buckwheat

Buckwheat has wide applications in the food industry, being used to prepare various food items such as bread, chapatis, biscuits, cakes, dhindo, wine, buckwheat tea, and more. This makes it suitable for being used as a staple diet in India. Adding 30% buckwheat flour to bread has been found to increase the antioxidant capacity

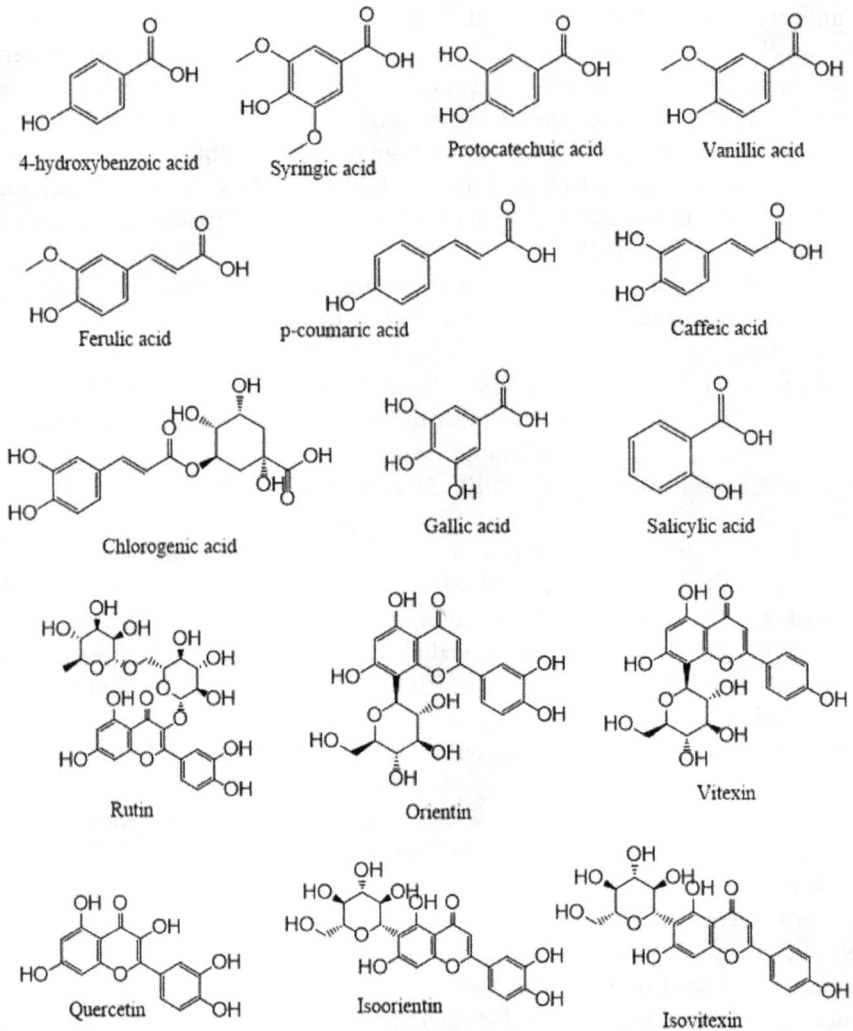

Figure 5.1 Chemical structures of phenolic acids and flavonoids reported from buckwheat.

of serum (Luitel et al. 2017). Buckwheat kernels are rich in soluble protein but have leucine as the first limiting amino acid (Wei et al. 2003). The biosynthesis of flavonoids in buckwheat makes it a food that promotes health (Taguchi 2016). Buckwheat kernels are rich in minerals such as K, Fe, and Zn in the albumins, Ca, Mg, and Mn in the globulins, and Na in prolamins and glutelins (Wei et al. 2003). Buckwheat protein has one of the highest amino acid scores of proteins in plant-based foods (Qin et al. 2010). Buckwheat products are also high in resistant starch, which is linked to several health benefits (Skrabanja et al. 2004). The many commercial uses of the buckwheat plant make it a highly valuable crop for generating both nourishment and income. As consumers place greater emphasis on the taste, health benefits, and nutritional value of their food products, the production of functional cereal (and pseudocereal) items has significantly increased in recent years (Kowalska et al. 2012). A few examples of processed products that use buckwheat in varying concentrations are presented in the following subsections.

5.4.1 Buckwheat Flour

Buckwheat flour is a versatile ingredient that can be incorporated into various baked products to improve their nutritional and sensory properties. Studies have shown that buckwheat flour can successfully replace a portion of wheat flour in biscuits, bread, and other baked goods without compromising their acceptability. Sensory evaluation of wheat-based biscuits made with 30% or 40% buckwheat flour incorporation (Filipcev et al. 2011a; Filipciv et al. 2011b) showed high acceptability. Similarly, adding 30% buckwheat flour to the wheat bread recipe was highly acceptable, resulting in improved antioxidant and sensory properties of the bread (Chlopicka et al. 2012). Addition of 30% buckwheat flour to the bread resulted in better sensory results (Wronkowska et al. 2008; Wronkowska et al. 2012). Similarly, incorporating 10% to 40% buckwheat flour into regional bakery products like Turkish bread and round rolls enhanced their sensory characteristics.

Studies have also shown that substituting wheat flour with buckwheat flour can increase the antioxidant properties of baked goods without significant decreases in sensory quality. In addition, whole-grain buckwheat flour effectively improves wheat bread's antioxidant properties (Selimovic et al. 2014). Improved physical properties have been achieved with the addition of buckwheat hemicellulose to wheat-buckwheat bread (Hromadkova et al. 2007). Despite the dark colour of crumbs, bread with buckwheat hemicellulose was highly accepted. Incorporating buckwheat flour also increased the attractiveness and colour of bread (Lin et al. 2009b). However, some challenges are associated with the incorporation of buckwheat flour into bakery products. Formulating gluten-free products with added buckwheat ingredients is more challenging. The optimal proportion of buckwheat flour in gluten-free cakes is 30%, 10% in gluten-free cookies, and 20% in muffins (Loredana et al. 2015).

Sensory qualities of bakery items produced with buckwheat flour, such as spaghetti pasta, noodles, and pancakes, have been studied. Buckwheat flour improves the bulkiness, hardness, and adhesiveness of durum semolina spaghetti, while also influencing sensory qualities such as scent, flavour, glossy appearance, and colour (Chillo et al. 2008). Incorporating buckwheat flour (20%) into instant noodles leads to higher-quality noodles, particularly in terms of texture and colour (Choy et al. 2013). In gluten-free noodle production, common buckwheat is preferred over tartary buckwheat due to its better sensory properties, including higher scores in all sensorial descriptors such as colour, taste, aroma, and mouthfeel (Ma et al. 2013). Buckwheat flour can also be used in breadmaking to add nutty and earthy flavours while improving the nutritional profile of the bread. Buckwheat flour can make cakes, muffins, and other baked goods. The sensory properties of these baked goods can be influenced by the amount of buckwheat flour used, with higher amounts resulting in a more distinct buckwheat taste and aroma. However, the texture of the baked goods may become denser and heavier with increasing amounts of buckwheat flour (Beitane et al. 2014). Overall, buckwheat flour can be a valuable ingredient in bakery products, providing unique flavours and nutritional benefits. However, the amount of buckwheat flour used should be carefully considered to achieve the final product's desired sensory properties and texture.

5.4.2 Buckwheat Honey

Buckwheat honey is a dark-coloured honey with a distinct flavour and aroma that is different from other types of honey. It is produced by bees that gather nectar from the buckwheat plant's flowers, a common crop in many regions of the world. The flavour of buckwheat honey is often described as rich, malty, and with a slightly bitter aftertaste. Sensory evaluation of buckwheat honey reveals a complex aroma (Zhou et al. 2002), which is a combination of malty, burnt sugar, vanilla, buttery, floral, and fruity-ester notes. This is due to presence of volatile compounds such as 3-methyl butanal, maltol, furaneol, vanillin, various esters, 2-phenyl ethanol, β-damascenone, and phenylacetaldehyde. In addition to its unique flavour and aroma, buckwheat honey is also known for its health benefits due to its high concentration of antioxidants, which are responsible for its anti-inflammatory properties.

5.4.3 Buckwheat Sprouts

Buckwheat sprouts are a popular food product that is gaining attention due to the numerous health benefits. The sprouts are made by sprouting the seeds of the gluten-free buckwheat plant. Buckwheat sprouts are an excellent source of flavonoids and anthocyanins, which are powerful antioxidants because of the higher amounts of rutin, isoorientin, vitexin, isovitexin, cyanidin 3-*O*-glucoside, and cyanidin 3-*O*-rutinoside (Liu et al. 2007; Kim et al. 2007; Nakamura et al. 2013). Incorporating buckwheat sprouts into the diet can be an effective way to get maximum benefit. Nonetheless, more information must be provided on incorporating buckwheat sprouts into food formulations. Only a few studies have explored their use as an ingredient in food products (Sun Lim et al. 2001; Kim et al. 2004). Sturza et al (2020) found that incorporating tartary buckwheat sprouts into steamed bread formulations, instead of buckwheat flour, was possible. Still, the strong bitterness and astringency of the sprouts limited their addition to relatively low levels (8%) due to a negative effect on consumer acceptance with greater amounts. In conclusion, buckwheat sprouts are a nutritious and flavourful food product that can be a valuable addition to a healthy diet. Further research is needed to explore the optimal ways to incorporate buckwheat sprouts into food formulations and maximize their potential health benefits.

5.4.4 Buckwheat Beer

Buckwheat beer is a gluten-free beer made from buckwheat malt and other gluten-free ingredients. It specifically targets consumers with coeliac disease or gluten intolerance, who cannot consume traditional gluten-containing beers. Studies have shown that buckwheat beer can achieve high sensory acceptability and quality comparable with barley beer (Podeszwa 2013). Buckwheat malt, produced by malting and kilning buckwheat groats, is an essential ingredient in buckwheat beer. Buckwheat malt and beer may be significantly affected by the biochemical features of the buckwheat groats (Nic Phiaris et al. 2010). In the production of buckwheat beer, the malting and mashing processes are optimized to ensure the desired sensory characteristics of the beer (Dezelak et al. 2014). The extensive mashing regime and the use of dried yeast also contribute to the unique taste and bitterness of buckwheat beer. Researchers found that buckwheat beer had higher carbonation and flavour purity ratings than beers brewed with quinoa or other gluten-free pseudocereals. Nonetheless, the high polyphenol and amino acid content of buckwheat can increase the perception of bitterness in beer. Overall, buckwheat beer offers a potential alternative for consumers with coeliac disease or gluten intolerance seeking a gluten-free beer option that maintains high sensory acceptability and quality. Further research and optimization of the malting and brewing processes may lead to even better-tasting and more cost-effective gluten-free beer options for consumers.

5.5 Conclusions

Buckwheat has the best nutritional profile of all pseudocereals, as it contains many phytochemicals, vitamins, and minerals. Buckwheat is a low-cost and protein-rich raw material, with a higher protein content than cereals, which could be used to combat protein-related deficiency in developing nations. Buckwheat, containing bioactive components, has health and pharmacological significance. Pharmaceutical companies may employ the bioactive components from buckwheat, that have been identified, to treat various disorders. The most intriguing trend in the food industry is the development of functional foods with added health benefits. Due to their nutritional value and suitability for those with gluten intolerance, buckwheat-related food products, with excellent sensory and techno-functional characteristics, have gained popularity in the past decade. Further research must be carried out in order to enhance the taste and texture of gluten-free buckwheat products.

References

Ahmed, A., Khalid, N., Ahmad, A., Abbasi, N. A., Latif, M. S. Z., & Randhawa, M. A. (2014). Phytochemicals and biofunctional properties of buckwheat: A review. *J Agri Sci* 152: 349–369.

Alvarez-Jubete, L., Arendt, E.K., Gallagher, E. 2010. Nutritive value of pseudocereals and their increasing use as functional gluten-free ingredients. *Trends Food Sci Technol* 21(2): 106–113.

Amezqueta, S., Galan, E., Fuguet, E., Carrascal, M., Abian, J., & Torres, J. (2012). Determination of d-fagomine in buckwheat and mulberry by cation exchange HPLC/ESI–Q-MS. *Anal Bioanal Chem* 402: 1953–1960.

Beitane, I., Krumina- Zernture, G., Murniece, I. 2014. Sensory, colour and structural properties of pancakes prepared with pea and buckwheat flours. *Foodbalt* 2014: 234–238.

Brasil, V.C.B., Guimarães, B.P., Evaristo, R.B.W., Carmo, T.S., Ghesti, G.F. 2020. Buckwheat (*Fagopyrum esculentum* Moench) characterization as adjunct in beer brewing. *Food Sci Technol* 41: 265–272.

Brunori, A., Baviello, G., Zannettino, C., Corsini, G., Sandor, G., & Vegvari, G. (2010). The use of Tartary buckwheat whole flour for bakery products: Recent experience in Italy. *Annals of the University Dunarea de Jos Galati Fascicle VI – Food Technology* 34: 33–38.

Cheng, F., Ge, X., Gao, C., Li, Y., & Wang, M. (2019). The distribution of D-chiro-inositol in buckwheat and its antioxidative effect in HepG2. *J Cereal Sci* 89: 102808.

Chettry, U., Chrungoo, N.K. 2021. Beyond the cereal box: Breeding buckwheat as a strategic crop for human nutrition. *Plant Foods Hum Nutr* 76(4): 399–409.

Chillo, S., Laverse, J., Falcone, P.M., Protopapa, A., Del Nobile, M.A. 2008. Influence of the addition of buckwheat flour and durum wheat bran on spaghetti quality. *J Cereal Sci* 47(2): 144–152.

Chlopicka, J., Pasko, P., Gorinstein, S., Jedryas, A., Zagrodzki, P. 2012. Total phenolic and total flavonoid content, antioxidant activity and sensory evaluation of pseudocereal breads. *LWT Food Sci Technol* 46(2): 548–555.

Choy, A.L., Morrison, P.D., Hughes, J.G., Marriott, Ph. J. Small, D.M. 2013. Quality and antioxidant properties of instant noodles enhanced with common buckwheat flour. *J Cereal Sci* 57(3): 281–287.

Christa, K., Soral-Śmietana, M. 2008. Buckwheat grains and buckwheat products – Nutritional and prophylactic value of their components – A review. *Czech J Food Sci* 26(3): 153–162.

Chrungoo, N. K., Dohtdong, L., & Chettry, U. (2016). Diversity in seed storage proteins and their genes in buckwheat. In *Molecular breeding and nutritional aspects of buckwheat* (pp. 387–399). Academic Press.

Chrungoo, N.K., Chettry, U. 2021. Buckwheat: A critical approach towards assessment of its potential as a super crop. *Ind J Genet Plant Breed* 81(1): 1–23.

Ciudad-Mulero, M., Fernández-Ruiz, V., Matallana-González, M.C., Morales, P. 2019. Dietary fiber sources and human benefits: The case study of cereal and pseudocereals. *Adv Food Nutr Res* 90: 83–134.

Comino, I., de Lourdes Moreno, M., Real, A., Rodríguez-Herrera, A., Barro, F., & Sousa, C. (2013). The gluten-free diet: testing alternative cereals tolerated by celiac patients. *Nutrients* 5(10): 4250–4268.

Das, S. K., Avasthe, R. K., Ghosh, G. K., & Dutta, S. K. (2019). Pseudocereal buckwheat with potential anticancer activity. *Bulletin of Pure and Applied Sciences Section B-Botany* 38(2): 94–95.

Dezelak, M., Zarnkow, M., Becker, T., Kosir, I.J. 2014. Processing of bottom-fermented gluten-free beer-like beverages based on buckwheat and quinoa malt with chemical and sensory characterization. *J Inst Brew* 120: 360–370.

Dhull, S.B., Kaur, M., Sandhu, K.S. 2020a. Antioxidant characterization and in vitro DNA damage protection potential of some Indian fenugreek (*Trigonella foenum-graecum*) cultivars: Effect of solvents. *J Food Sci Technol* 57(9): 3457–3466.

Dhull, S.B., Punia, S., Kidwai, M.K., Kaur, M., Chawla, P., Purewal, S.S., Sangwan, M., Palthania, S. 2020b. Solid-state fermentation of lentil (*Lens culinaris* L.) with *Aspergillus awamori*: Effect on phenolic compounds, mineral content, and their bioavailability. *Legume Sci* 2(3): e37.

Dhull, S.B., Punia, S., Kumar, R., Kumar, M., Nain, K.B., Jangra, K., Chudamani, C. 2021. Solid state fermentation of fenugreek (*Trigonella foenum-graecum*): implications on bioactive compounds, mineral content and in vitro bioavailability. *J Food Sci Technol* 58(5): 1927-1936.

Dhull, S.B., Punia, S., Sandhu, K.S., Chawla, P., Kaur, R., Singh, A. 2019. Effect of debittered fenugreek (*Trigonella foenum-graecum* L.) flour addition on physical, nutritional, antioxidant, and sensory properties of wheat flour rusk. *Legume Sci* 2(1): e21.

Dziadek, K., Kopeć, A., Pastucha, E., Piątkowska, E., Leszczyńska, T., Pisulewska, E., Witkowicz, R., & Francik, R. (2016). Basic chemical composition and bioactive compounds content in selected cultivars of buckwheat whole seeds, dehulled seeds and hulls. *J Cereal Sci* 69: 1–8.

Filipcev, B., Simurina, O., Bodroza-Solarov, M., Vujakovic, M. 2011a. Evaluation of physical, textural and microstructural properties of dough and honey biscuits enriched with buckwheat and rye. *Chem Ind Chem Eng Q* 17(3): 291–298.

Filipcev, B., Simurina, O., Sakac, M., Sedej, I., Jovanov, P., Pestoric, M., Bodroza-Solarov, M. 2011b. Feasibility of use of buckwheat flour as an ingredient in ginger nut biscuit formulation. *Food Chem* 125(1): 164–170.

Ge, R.H., Wang, H. 2020. Nutrient components and bioactive compounds in Tartary buckwheat bran and flour as affected by thermal processing. *Int J Food Prop* 23(1): 127–137.

Gimenez-Bastida, J., Piskuła, M., Zielinski, H. 2015. Recent advances in processing and development of buckwheat derived bakery and non-bakery products-a review. *Pol J Food Nutr Sci* 65(1): 9–20.

González-Sarrías, A., Larrosa, M., García-Conesa, M.T., Tomás-Barberán, F.A., Espín, J.C. 2013. Nutraceuticals for older people: Facts, fictions and gaps in knowledge. *Maturitas* 75(570): 313–334.

Hromadkova, Z., Stavova, A., Ebringerova, A., Hirsch, J. 2007. Effect of buckwheat hull hemicelluloses addition of bread-making quality of wheat flour. *J Food Nutr Res* 46: 158–166.

Huda, M.N., Lu, S., Jahan, T., Ding, M., Jha, R., Zhang, K., Zhou, M. 2021. Treasure from garden: Bioactive compounds of buckwheat. *Food Chem* 335: 127653.

Kaur, P., Dhull, S.B., Sandhu, K.S., Salar, R.K., Purewal, S.S. 2018. Tulsi (*Ocimum tenuiflorum*) seeds: In vitro DNA damage protection, bioactive compounds and antioxidant potential. *J Food Meas Char* 12(3): 1530–1538.

Kidwai, M.K., Singh, A., Malik, T., Dhull, S.B., Punia, S. 2020. Essential fatty acid bioavailability: A dietary perspective. In: *Essential Fatty Acids: Sources, Processing Effects, and Health Benefits* by Dhull, S.B., Punia, S., Sandhu, CRC Press, 129–156.

Kim, S.J., Maeda, T., Sarker, M.Z., Takigawa, S., Matsuura-Endo, C., Yamauchi, H., Mukasa, Y., Saito, K., Hashimoto, N., Noda, T., Saito, T., Suzuki, T. 2007. Identification of anthocyanins in the sprouts of buckwheat. *J Agric Food Chem* 55(15): 6314–6318.

Kalinova, J., & Vrchotova, N. (2009). Level of catechin, myricetin, quercetin and isoquercitrin in buckwheat (*Fagopyrum esculentum* Moench), changes of their levels during vegetation and their effect on the growth of selected weeds. *J Agri Food Chem* 57: 2719–2725.

Kim, S. L., Kim, S. K., & Park, C. H. (2004). Introduction and nutritional evaluation of buckwheat sprouts as a new vegetable. *Food Res Int* 37(4): 319–327.

Kim, J., & Hwang, K. T. (2020). Fagopyrins in different parts of common buckwheat (*Fagopyrum esculentum*) and Tartary buckwheat (*Fagopyrum tataricum*) during growth. *J Food Comp Analysis* 86: 103354.

Klepacka, J., Najda, A., & Klimek, K. (2020). Effect of buckwheat groats processing on the content and bioaccessibility of selected minerals. *Food* 9(6): 832.

Kowalska, H., Marzec, A., Mucha, M. 2012. Ocena sensoryczna wybranych rodzajow pieczywa funkcjonalnego oraz preferencje pieczywa wsrod konsumentow. *Zesz Problemowe Postepow Nauk Rolniczych* 571: 67–78.

Krkošková, B., Mrázová, Z. 2005. Prophylactic components of buckwheat. *Food Res Int* 38(5): 561–568.

Kwon, S.J., Roy, S.K., Choi, J., Park, J., Cho, S., Sarker, K., Woo, S.H. 2018. Recent research updates on functional components in Buckwheat. *J Agric Sci* 34: 1–8.

Li, S.Q., Zhang, Q.H. 2001. Advances in the development of functional foods from buckwheat. *Crit Rev Food Sci Nutr* 41(6): 451–464.

Li, F., Zhang, X., Zheng, S., Lu, K., Zhao, G., & Ming, J. (2016). The composition, antioxidant and antiproliferative capacities of phenolic compounds extracted from Tartary buckwheat bran (*Fagopyrum tartaricum* (L.) Gareth). *J Func Foods* 22: 145–155.

Lin, L.- Y., Liu, H.- M., Yu, Y.- W., Lin, S.- D., Mau, J.- L. 2009b. Quality and antioxidant property of buckwheat enhanced wheat bread. *Food Chem* 112(4): 987–991.

Liu, C.L., Chen, Y.S., Yang, J.H., Chiang, B.H., Hsu, C.K. 2007. Trace element water improves the antioxidant activity of buckwheat (*Fagopyrum esculentum* Moench) sprouts. *J Agric Food Chem* 55(22): 8934–8940.

Loredana, I.M., Petru, R.B., Daniela, S., Ioan, T.T., Monica, N. 2015. Sensory evaluation of some sweet gluten-free bakery products based on rice and buckwheat flour. *J Biotech* 208: 5–120.

Luitel, D.R., Siwakoti, M., Jha, P.K., Jha,A.K., Krakauer, N. 2017. An overview: Distribution, production, and diversity of local landraces of buckwheat in Nepal. *Adv Agri* 2017: 2738045. https://doi.org/10.1155/2017/2738045

Lv, L., Xia, Y., Zou, D., Han, H., Wang, Y., Fang, H., Li, M. 2017. *Fagopyrum tataricum* (L.) Gaertn.: A review on its traditional uses, phytochemical and pharmacology. *Food Sci Technol Res* 23(1): 1–7.

Ma, Y.J., Guo, X.D., Liu, H., Xu, B.N., Wang, M. 2013. Cooking, textural, sensorial, and antioxidant properties of common and Tartary buckwheat noodles. *Food Sci Biotechnol* 22(1): 153–159.

Malik, M., Sindhu, R., Dhull, S.B., Bou-Mitri, C., Singh, Y., Panwar, S., Khatkar, B.S. 2023. Nutritional composition, functionality, and processing technologies for amaranth. *J Food Process Preserv* 2023: 1753029. https://doi.org/10.1155/2023/1753029.

Martínez-Villaluenga, C., Peñas, E., Hernández-Ledesma, B. 2020. Pseudocereal grains: Nutritional value, health benefits and current applications for the development of gluten-free foods. *Food Chem Toxicol* 137: 111178.

Mir, N. A., Riar, C. S., & Singh, S. (2018). Nutritional constituents of pseudo cereals and their potential use in food systems: A review. *Trends Food Sci Tech* 75: 170–180.

Nakamura, K., Naramoto, K., Koyama, M. 2013. Blood-pressure-lowering effect of fermented buckwheat sprouts in spontaneously hypertensive rats. *J Funct Foods* 5(1): 406–404.

Nic Phiaris, B.P., Mauch, A., Schehl, B.D., Zarnkow, M., Gastl, M., Herrmann, M., Zaninni, E., Arendt, E.K. 2010. Processing of top fermented beer brewed from 100% buckwheat malt with sensory and analytical characterization. *J Inst Brew* 116(3): 265–274.

Orožen, L., Vogel Mikuš, K., Likar, M., Nečemer, M., Kump, P., & Regvar, M. (2012). Elemental composition of wheat, common buckwheat, and Tartary buckwheat grains under conventional production. *Acta biologica slovenica* 55(2): 13–24.

Park, C. H., Kim, Y. B., Choi, Y. S., Heo, K., Kim, S. L., Lee, K. C., Chang, K. J., & Lee, H. B. (2008). Rutin content in food products processed from groats, leaves, and flowers of buckwheat. *Fagopyrum* 17: 63–66.

Petr, J., Michalik, I., Tlaskalova, H., Capouchova, I., Famera, O., Urminska, D., Tuckova, L., & Knoblochova, H. (2003). Extention of the spectra of plant products for the diet in coeliac disease. *Czech J Food Sci* 21: 59–70.

Phiarais, B.P.N., Wijngaard, H.H., Arendt, E.K. 2005. The impact of kilning on enzymatic activity of buckwheat malt. *J Inst Brew* 111(3): 290–298.

Podeszwa, T. 2013. The use of pseudocereals for the production of gluten-free beer. *Eng Sci Technol Wydawnictwo Uniwersytetu Ekonomicznego we Wrocławiu, Wrocław* 3(10): 92–1.

Podolska, G., Gujska, E., Klepacka, J., Aleksandrowicz, E. 2021. Bioactive compounds in different buckwheat species. *Plants (Basel)* 10(5): 961.

Punia, S., Dhull, S.B. 2019. Chia seed (*Salvia hispanica* L.) mucilage (a heteropolysaccharide): Functional, thermal, rheological behaviour and its utilization. *Int J Biol Macromol* 140: 1084–1090.

Punia, S., Dhull, S.B., Siroha, A.K., Sandhu, K.S., Chaudhary, V. 2020. Mechanism of action of essential fatty acids. In *Essential Fatty Acids: Sources, Processing Effects, and Health Benefits* by Dhull, S.B., Punia, S., Sandhu, K.S., CRC Press, 89–100.

Punia, S., Sandhu, K.S., Siroha, A.K., Dhull, S.B. 2019. Omega 3-metabolism, absorption, bioavailability and health benefits-A review. *PharmaNutrition* 10: 100162. https://doi.org/10.1016/j.phanu.2019.100162

Qin, P., Wang, Q., Shan, F., Hou, Z., Ren, G. 2010. Nutritional composition and flavonoids content of flour from different buckwheat cultivars. *Int J Food Sci Technol* 45(5): 951–958.

Raguindin, P.F., Itodo, O.A., Stoyanov, J., Dejanovic, G.M., Gamba, M., Asllanaj, E., Kern, H. 2021. A systematic review of phytochemicals in oat and buckwheat. *Food Chem* 338: 127982.

Ruan, J., Zhou, Y., Yan, J., Zhou, M., Woo, S. H., Weng, W., Cheng, J., & Zhang, K. (2020). Tartary buckwheat: An under-utilized edible and medicinal herb for food and nutritional security. *Food Rev Int* 38: 1–15.

Sanchez, A., Schuster, T., Burke, J., Kron, K. 2011. Taxonomy of Polygonoideae (Polygonaceae): 555 A new tribal classification. *Taxon* 60(1): 151–160. 556.

Selimovic, A., Milicevic, D., Jasic, M., Selimovic, A., Ackar, Đ., Pesic, T. 2014. The effect of baking temperature and buckwheat flour addition on the selected properties of wheat bread. *Croat J Food Sci Technol* 6: 43–50.

Sidorova YuS., Petrov, N.A., Shipelin, V.A., Mazo, V.K. 2020. Spinach and quinoa - promising food sources of biologically active substances. *Problems of Nutr* 89(2): 100–106.

Skrabanja, V., Kreft, I., Golob, T., Modic, M., Ikeda, S., Ikeda, K., Kosmelj, K. 2004. Nutrient content in buckwheat milling fractions. *Cereal Chem* 81(2): 172–176.

Sofi, S.A., Ahmed, N., Farooq, A., Rafiq, S., Zargar, S.M., Kamran, F., Dar, T.A., Mir, S.A., Dar, B.N., Mousavi Khaneghah, A. 2023. Nutritional and bioactive characteristics of buckwheat, and its potential for developing gluten-free products: An updated overview. *Food Sci Nutr* 11(5): 2256–2276.

Stember, R.H. 2006. Buckwheat allergy. *Allergy Asthma Proc* 27(4): 393–395.

Sturza, A., Păucean, A., Chiş, M. S., Mureşan, V., Vodnar, D. C., Man, S. M., & Muste, S. (2020). Influence of buckwheat and buckwheat sprouts flours on the nutritional and textural parameters of wheat buns. *Applied Sciences* 10(22): 7969.

Sun Lim, K. I. M. I., Young Koo, S., Jong Jin HWANGl, S. K. K., Han Sun, H. U. R., & PARK, C. H. (2001). Development and utilization of buckwheat sprouts as functional vegetables. *Fagopyrum* 18(49): 4.

Suzuki, T., Morishita, T., Noda, T., Ishiguro, K., Otsuka, S., Katsu, K. 2021. Breeding of buckwheat to reduce bitterness and Rutin hydrolysis. *Plants (Basel)* 10(4): 791.

Taguchi, G. 2016. *Flavonoid Biosynthesis in Buckwheat. Molecular Breeding and Nutritional Aspects of Buckwheat*. Elsevier Inc.

Taylor, J., Awika, J. eds. 2017. *Gluten-Free Ancient Grains: Cereals, Pseudocereals, and Legumes: Sustainable, Nutritious, and Health-Promoting Foods for the 21st Century*. Woodhead Publishing, pp. 1–342.

Watanabe, M. 1998. Catechins as antioxidants from buckwheat (*Fagopyrum esculentum* Moench.) groats. *J Agri Food Chem* 46(3): 839–845.

Wei, Y.M., Hu, X.Z., Zhang, G.Q., Ouyang, S.H. 2003. Studies on the amino acid and mineral content of buckwheat protein fractions. *Nahrung Food* 47(2): 114–116.

Wijngaard, H.H., Renzetti, S., Arendt, E.K. 2007. Microstructure of buckwheat and barley during malting observed by confocal scanning laser microscopy and scanning electron microscopy. *J Inst Brew* 113(1): 34–41.

Wijngaard, H.H., Ulmer, H.M., Arendt, E.K. 2005. The effect of germination temperature on malt quality of buckwheat. *J Am Soc Brew Chem* 63(1): 31–36.

Wijngaard, H.H., Ulmer, H.M., Neumann, M., Arendt, E.K. 2005. The effect of steeping time on the final malt quality of buckwheat. *J Inst Brew* 111(3): 275–281.

Woo, S.H., Kamal, A.H.M., Tatsuro, S., Campbell, C.G., Adachi, T., Yun, S.H., Chung, K.Y., Choi, J.S. 2010. Buckwheat (*Fagopyrum esculentum* Moench.): Concepts, prospects and potential. *Eur J Plant Sci Biotech* 4: 1–16.

Woo, S.H., Roy, S.K., Kwon, S.J., Cho, S.W., Sarker, K., Lee, M.S., Chung, K.Y., Kim, H.H. 2016. *Concepts, Prospects, and Potentiality in Buckwheat (Fagopyrum esculentum Moench.): A Research Perspective. Molecular Breeding and Nutritional Aspects of Buckwheat*. Academic Press, 21–49.

Wronkowska, M., Troszynska, A., Soral-Smietana, M., Wolejszo, A. 2008. Effects of buck-wheat flour (*Fagopyrum esculentum* Moench.) on the quality of gluten-free bread. *Pol J Food Nutr Sci* 58: 211–216.

Wronkowska, M., Zielinska, D., Szawara-Nowak, D., Troszynska, A., Soral- Smietana, M. 2012. Antioxidative and reducing capacity, macroelements content and sensorial properties of buckwheat-enhanced gluten-free bread. *Int J Food Sci Technol* 45(10): 1993–2000.

Yilmaz, H.Ö., Ayhan, N.Y., Meriç, Ç.S. 2020. Buckwheat: A useful food and its effects on human health. *Curr Nutr Food Sci* 16(1): 29–34.

Zhang, Z., Zhou, M., Tang, Y., Li, F., Tang, Y., Shao, J., Xue, W., Wu, Y. 2012. Bioactive compounds in functional buckwheat food. *Food Res Int* 49(1): 389–395.

Zheng, S., Cheng-hua, H.A.N., Huan, K.F. 2011. Research on Se content of different Tartary buckwheat genotypes. *Agric Sci Technol* 12: 102–104.

Zhou, Q., Wintersteen, C.L., Cadwallader, K.R. 2002. Identification and quantification of aroma-active components that contribute to the distinct malty flavor of buckwheat honey. *J Agric Food Chem* 50(7): 2016–2021.

Živković, A., Polak, T., Cigić, B., Požrl, T. 2021. Germinated buckwheat: Effects of dehu-lling on phenolics profile and antioxidant activity of buckwheat seeds. *Foods* 10(4): 740.

Zhu, F. (2016). Chemical composition and health effects of Tartary buckwheat. *Food Chem* 203: 231–245.

Zou, L., Wu, D., Ren, G., Hu, Y., Peng, L., Zhao, J., Xiao, J. 2021. Bioactive compounds, health benefits, and industrial applications of Tartary buckwheat (*Fagopyrum tatari-cum*). *Crit Rev Food Sci Nutr* 63(5): 657–673.

Chapter 6

Chemistry and Application of Starches and Gums of Pseudocereals

Madhu Sharma, Harshvardhan Patel,
Aarti Bains, Prince Chawla

6.1 Introduction

Pseudocereals, a class of non-grass plants, have gained popularity in recent years due to their nutritional value and versatility in a variety of industries. The term "pseudocereals" refers to plants that do not belong to the grass family (Poaceae) but have seeds and fruits that are used as flour for staple foods like bread (Das 2016). The top three pseudocereals known are quinoa (*Chenopodium quinoa* ssp. *quinoa*; family Amaranthaceae) amaranth (*Amaranthus caudatus, Amaranthus cruentus, Amaranthus hypochondriacus*; family Amaranthaceae), and buckwheat (*Fagopyrum esculentum*; family Polygonaceae). Unlike many true cereals, such as wheat, rice, and barley, which are monocotyledonous, these pseudocereals are dicotyledonous. However, their seed's function and composition are similar to those of true cereals, which is why they are referred to as pseudocereals (Mir et al. 2018; Malik et al. 2023).

Starch is the main component found in pseudocereal grains, playing critical roles in determining their functional properties and applications. This overview covers the chemistry and applications of pseudocereal starches and gums, emphasising

DOI: 10.1201/9781003325277-6

their potential as sustainable alternatives in the food and pharmaceutical industries (Alonso-Miravalles and O'Mahony 2018). Pseudocereals, which include amaranth, buckwheat, and quinoa, are broadleaf plants that produce grains or seeds that look and taste like true cereals. Pseudocereals, despite belonging to different botanical families, have been cultivated for centuries by various Indigenous cultures and are highly valued for their exceptional nutritional composition and adaptability to diverse environmental conditions (Wang et al. 2021). Starch, a complex polysaccharide made up of glucose units, is an important carbohydrate reserve in plants (Chandak et al. 2022a, 2022b; Dhull et al. 2022a). It is an important source of energy and contributes significantly to food texture and stability (Punia et al. 2020a; Kaur et al. 2020; Dhull et al. 2020a). Pseudocereal starches have distinct physicochemical properties that set them apart from other starch sources. For example, they frequently have a high amylose content, small granule size, and excellent gel-forming ability, all of which contribute to their potential applications in a variety of food products (Agama-Acevedo et al. 2019).

The chemistry and applications of pseudocereal starches and gums have attracted the interest of researchers. Because of their unique properties, pseudocereal starches have a wide range of potential applications (Martínez-Villaluenga et al. 2020). They can be used in the food industry to improve the texture, mouthfeel, and stability of various products. Pseudocereal starches, for example, can be used to improve the sensory qualities and shelf life of bakery goods, confectionery, dairy products, and meat and seafood products (Hosseini et al. 2019). Furthermore, pseudocereal gums have shown significant potential in a variety of industries. The rheological properties of gums and mucilages make them useful in the formulation and stabilisation of a wide range of food and beverage products (Punia and Dhull 2019; Dhull et al. 2020b). Understanding the physicochemical properties of pseudocereal starches and gums is critical for their effective use (McClements et al. 2019). Thermal properties, rheological behaviour, structural properties, water-holding capacity, gelation, retrogradation, emulsification, and foaming properties all have an impact on the functionality of starches and gums in various applications (Chawla et al. 2019, 2020; Punia et al. 2019a, 2019b). Investigating these properties can aid in optimising their performance and developing new formulations (Asaithambi et al. 2022).

Despite numerous advances in the field, there are still challenges and future prospects. Ongoing research focuses on optimising processing methods, improving functional properties, addressing sustainability and environmental concerns, and exploring new applications (Bibri and Krogstie 2017). Pseudocereal starches are promising alternatives that can contribute to a more sustainable and diverse ingredient landscape as demand for sustainable and plant-based ingredients grows. Overall, pseudocereal starches and gums have distinct properties and functions, making them valuable ingredients in the food, pharmaceutical, and cosmetic industries (Boukid et al. 2023). Their chemistry and applications have been thoroughly studied, with ongoing research aimed at better understanding their properties and exploring new applications. Pseudocereal starches, with their potential as sustainable alternatives, have the potential to contribute to the development of innovative and environmentally friendly products.

6.2 Chemical Composition of Starch of Pseudocereals

6.2.1 Quinoa Starch

Carbohydrates make up the majority of the nutritional content of pseudocereal grains, accounting for 60 to 80% of seed dry weight. Quinoa (*Chenopodium quinoa*) starch is a valuable component found in quinoa seeds (Li and Zhu 2018a). It's widely extracted and used in the food industry for a variety of purposes. Quinoa, as a pseudocereal, has a unique composition that sets it apart from traditional grains. Starch, the main biopolymeric storage constituent of plants (grains, tubers, and seeds), is typically found in the form of granules of various shapes and sizes (Li et al. 2023). Starchy compounds can be found in the seeds' perisperm as simple units or as spherical aggregates. Starch is the most important carbohydrate in quinoa, accounting for approximately 58.1–64.2% of the total dry matter. It is primarily made up of two types of polysaccharides: amylose and amylopectin (Brenda and Torres-Vargas 2019). Amylose is a linear polymer of glucose molecules, whereas amylopectin is a branched polymer of glucose molecules. Quinoa is composed of carbohydrates, which range from 67% to 74% of the dry matter, and the amylose content is approximately 11–12%, which is lower than in cereals (Tao et al. 2019). This feature contributes to its distinct functional properties, which include increased water-holding capacity and gel formation ability. Insoluble polysaccharides (78% of total dietary fibre content) in quinoa include homogalacturonans and rhamnogalacturonan-I with arabinan sidechains (55–60%), as well as highly branched xyloglucans (30%) and cellulose (López-Fernández et al. 2021). Soluble polysaccharides composed of xyloglucans (40–60% of the soluble fibre fraction) and arabinose-rich pectic polysaccharides (34–55% of the soluble fibre fraction) account for 22% of total dietary fibre in quinoa (Thakur et al. 2019). Its starches form stable gels with good textural properties, making them suitable for use in a variety of food products. The physicochemical and pasting properties of quinoa and discovered that the diameter of starch granules was smaller than that of maize (range 1–2 μm) or wheat (2–40 μm) (Mir et al. 2018). It has also been found that starches with small granules have a higher gelatinisation temperature (Jan et al. 2017). The temperature range for quinoa is 57–64 °C. Other carbohydrates include monosaccharides (2%) and disaccharides (2.3%), crude fibre (2.5–3.9%), and pentosans (2.9–3.6%). Quinoa isolated starch may be suitable for applications requiring improved binding and reduced breakability (Mir et al. 2018). Because of its small granule diameter, quinoa starch can meet these requirements. Furthermore, they found that the isolated starch's swelling power was associated with highly restricted swelling, which is desirable for products such as noodles and composite blends (Yadav et al. 2014). Quinoa starch has a lower gelatinisation temperature than other common starches, which means it requires less energy to reach its gelatinisation point. This property is advantageous in food processing because it allows for faster gel formation and increased manufacturing efficiency (Li and Zhu 2018b). Quinoa starch is a preferred ingredient for people who are gluten intolerant or have coeliac disease because it is gluten free. It is a great alternative to wheat-based starches because it provides similar functionalities while being gluten free. Quinoa starch has nutritional benefits in addition to its functional properties (Xing et al. 2021). It contains essential amino acids, dietary fibre, vitamins, and minerals, which improve the overall nutritional profile of quinoa-fortified foods. Overall, quinoa starch, derived from quinoa plant seeds, is a versatile ingredient with distinct functional and nutritional properties (Rayner et al. 2012). Its high amylose content, lower gelatinisation temperature, and gluten-free status make it a valuable component in the food industry, helping to develop more diverse and healthier food products.

6.2.2 Amaranth Starch

The primary constituents of amaranth starch are carbohydrates, specifically amylose and amylopectin. These carbohydrates are made up of glucose units that are joined together by glycosidic bonds. Amaranth seeds contain 65–75% starch as well as non-starchy polysaccharide components (Skendi and Papageorgiou 2019). Amaranth starch granules are exceptionally small, ranging from 0.8 to 2.5 μm in size, when compared with starch granules from other grains such as wheat (3–34 μm), rice (3–8 μm), and maize (5–25 μm) (Thakur et al. 2021). Compared with other cereals, amaranth starch granules have a higher swelling capacity, a better water binding capacity, a lower gelatinization temperature, and greater resistance to amylases. Amaaranth starch has a relatively high amylose content (7.8–34.3% of total starch) (Repo-Carrasco-Valencia and Arana 2017). Amylose is a linear polymer of D-glucopyranose units linked by 1,4-glycosidic bonds, and is one of the main components of starch (Punia et al. 2020b, 2020c; Dhull et al. 2021b). Insoluble polysaccharides in amaranth include homogalacturonans and rhamnogalacturonan-I with arabinan side chains (55–60%), as well as highly branched xyloglucans (30%) and cellulose (78% of total dietary fibre content) (Chandla et al. 2017). Soluble polysaccharides made up of xyloglucans (40–60% of the soluble fibre fraction) and arabinose-rich pectic polysaccharides (34–55% of the soluble fibre fraction) account for 22% of amaranth total dietary fibre. With varying degrees of polymerisation, it forms a relatively straight chain structure. The functional properties of starches, such as gelatinisation, retrogradation, and texture formation, are influenced by the length of the amylose chains (Bangar et al. 2021; Punia et al. 2021). Amaranth starch, on the other hand, typically contains 15% to 30% amylose and the remaining percentage is amylopectin (Everth et al. 2018). Because of these factors, amaranth grain starch may have good gelatinisation properties as well as good thaw/freeze stability for the food industry. Amylopectin is the most important component of amaranth starch, accounting for the majority of its structure (Sindhu et al. 2021). It has a highly branched structure with both 1,4- and 1,6-glycosidic linkages. The 1,4- linkages, as in amylose, form linear chains, whereas the 1,6- linkages create branching points within the amylopectin molecule (Espinosa-Solis et al. 2020). The branching points occur at regular intervals, resulting in a complex and compact starch molecule organisation. The degree and distribution of branching influence the physicochemical properties of starches, including viscosity, gel strength, and digestion characteristics (Singla et al. 2020; Dhull et al. 2019a, 2021a). Amylopectin, due to its branched structure, increases the swelling and solubility of amaranth starch, contributing to its thickening and binding capabilities in a variety of applications (Condés et al. 2015).

6.2.3 Buckwheat Starch

Starch is a major storage component of buckwheat grains, and it is distributed as energy material in the endosperm, powering seedling growth. The starch content of whole buckwheat grain varies from 59% to 70% by dry weight due to climatic fluctuations and variable cultivation techniques (Gao et al. 2022). The degree of polymerisation and amylose content of starch granules range from 15% to 52% and 12–45 units of glucose, respectively. Buckwheat starch has a relatively higher amylose content (18.3–47% of total starch) (Zhu 2016). Buckwheat starch has unique structural properties, which contribute to its unique properties and functionality. Buckwheat starch granules are typically between 2 and 10 µm in diameter. They have a rounded or irregular shape that varies depending on the variety of buckwheat (Liu et al. 2015). The surface of the granule may have irregularities and slight roughness, which contribute to its textural characteristics. Buckwheat starch is made up of both amylose and amylopectin components. Amylopectin makes up the majority of the starch structure and is responsible for the starch's distinct functional properties (Liu et al. 2016). Buckwheat starch has intriguing linkage patterns. Amylose is composed of linear chains of D-glucopyranose units linked by 1,4- glycosidic bonds in buckwheat starch. Amylopectin, on the other hand, has a branched structure with both 1,4- and 1,6- glycosidic linkages (Goel et al. 2020). The presence of branching points at regular intervals contributes to the starch molecule's complex and compact organisation. Buckwheat grains have more fibre (17.8%) than other pseudocereals but have a lower soluble-to-insoluble dietary fibre ratio (0.5–0.28). Buckwheat dietary fibre contains 6.8% lignin, 2.2% hemicellulose, and 10.6% cellulose (Zhou et al. 2021). Mono- and disaccharides are minor carbohydrates found in pseudocereal grains. The main monosaccharides found in pseudocereals are glucose, fructose, arabinose, and xylose, while sucrose and maltose are the most prominent disaccharides (Torbica et al. 2022). Buckwheat starch has a semi-crystalline structure with regions of ordered and disordered glucose chain arrangements. The alignment of amylose and amylopectin chains forms the crystalline regions, whereas the amorphous regions lack this ordered arrangement (Akhila et al. 2022). During processing and cooking, buckwheat starch has unique gelatinisation properties. It gelatinises at low temperatures and swells quickly, enhancing its thickening and stabilising properties in a variety of food applications. Buckwheat starch undergoes retrogradation, or the realignment of starch molecules into more rigid structures, upon cooling but at a slower rate than other starches (Ma et al. 2022).The structural properties of buckwheat starch, such as granule morphology, amylose and amylopectin content, linkage patterns, and crystalline structure, are important in determining its functional behaviour. These properties influence the texture, viscosity, and stability of buckwheat starch-containing foods and products, making it a valuable ingredient with distinct properties.

6.3 Physicochemical Properties of Pseudocereal Starch

6.3.1 Gelatinisation and Retrogradation

When pseudocereal starches are exposed to heat and water during cooking or food processing, a crucial process called gelatinisation takes place. The formation of a gel-like matrix is caused by the swelling and structural changes that occur as the starch granules absorb heat and water (Fang et al. 2020). Due to its impact on the final product's texture, viscosity, and general quality, the gelatinisation process is crucial in a variety of food applications. When compared to conventional cereal starches, pseudocereal starches show different gelatinisation temperatures and peak viscosities (Deriu et al. 2022). For instance, the gelatinisation temperature of quinoa starch is higher than that of maize or wheat starch. Because of this quality, pseudocereal starches are appropriate for particular food processing methods that call for higher temperatures or longer cooking times. By carefully controlling the gelatinisation process, food manufacturers can achieve the desired texture and consistency in products like sauces, puddings, and fillings (Geyang and Morris 2017). After gelatinisation, the gelatinised starch cools and goes through recrystallisation, which is the process known as retrogradation. The starch molecules reorganise during retrogradation to create a more organised structure (Li et al. 2021).

Retrogradation can help improve texture and moisture absorption. Pseudocereal starches can prolong the shelf life of baked goods by reducing staling thanks to their special retrogradation properties (Šmídová and Rysová 2022). For instance, due to their capacity to postpone retrogradation, adding pseudocereal starches like quinoa or amaranth starch to bread formulations can lead to a softer crumb structure and prolonged freshness. This is especially helpful when making gluten-free baked goods because it can be difficult to get the best texture and moisture retention (Tsatsaragkou et al. 2016). However, excessive retrogradation can cause undesirable textural changes in some food products, such as increased firmness or syneresis (water release). In applications where prolonged shelf life and consistent quality are important considerations, such as starch-based desserts, sauces, and ready-to-eat meals, managing retrogradation is essential (Wang and Copeland 2015). To achieve the desired texture, stability, and sensory qualities in a variety of food products, food technologists and manufacturers carefully manipulate the gelatinisation and retrogradation properties of pseudocereal starches. Understanding and regulating these characteristics enables the development of unique, high-quality pseudocereal-based foods that satisfy dietary needs and consumer preferences (Bender and Schönlechner 2021). Overall, the functionality and application of pseudocereal starches in the food industry depend critically on their gelatinisation and retrogradation properties. A variety of products, from sauces and baked goods to gluten-free substitutes and convenience foods, can benefit from improved texture, moisture retention, and shelf life thanks to the distinct properties of these starches.

6.3.2 *Thickening and Binding Properties*

Pseudocereal starches, such as those derived from quinoa, amaranth, and buckwheat, possess exceptional thickening and binding properties, making them highly sought-after ingredients in the food industry. When these starches are heated in the presence of water, they undergo gelatinisation, a process in which the starch granules absorb water and swell, resulting in the formation of a thickened and viscous consistency. This unique property allows pseudocereal starches to serve as natural thickeners in a wide variety of food applications (Goff and Guo 2019). Pseudocereal starches are particularly useful in the creation of sauces, gravies, soups, and fillings as a thickening agent. By incorporating pseudocereal starches into these formulations, food manufacturers can achieve the desired texture, viscosity, and mouthfeel (Mohammed et al. 2021). The gelatinised starch acts as a thickening agent, improving the overall sensory experience and ensuring the proper cling and coating of the sauce or filling on various dishes. Moreover, pseudocereal starches offer a superior alternative to conventional thickeners, such as wheat flour or cornstarch, especially for individuals following gluten-free diets (McClements 2019).In addition to their thickening properties, pseudocereal starches exhibit excellent binding capabilities, making them invaluable in various food formulations. When used as binders, these starches create a cohesive matrix that helps hold ingredients together and enhance the structure of different products. For instance, in the production of meat analogues and vegetarian patties, pseudocereal starches play a critical role in maintaining the shape, firmness, and integrity of the final product (Mouritsen and Styrbæk 2017). They prevent crumbling during cooking or processing, providing a satisfying texture to plant-based meat alternatives. Similarly, in gluten-free bakery items like bread, cookies, and cakes, pseudocereal starches contribute to achieving the desired texture and structure that gluten would typically provide in traditional recipes (Singh et al. 2021).

Furthermore, pseudocereal starches excel in moisture retention, which is another valuable attribute. These starches can hold on to moisture, preventing baked goods from becoming dry or stale. In gluten-free baking, where the absence of gluten can result in products that are prone to dryness, pseudocereal starches help to maintain moisture levels, ensuring longer shelf life and improved eating experience (Yang 2016). The compatibility of pseudocereal starches with a wide range of ingredients is yet another advantage. They can be easily incorporated into both hot and cold food preparations without compromising their thickening and binding capabilities. This versatility allows for their application in diverse culinary creations, including sauces, dressings, chilled desserts, and fillings, where they contribute to the desired texture, mouthfeel, and overall quality of the final products (Egharevba 2019).Overall, pseudocereal starches offer outstanding thickening and binding properties, making them valuable in the food industry. Their ability to gelatinise, thicken, bind, and retain moisture contributes to the texture, structure, and sensory experience of various food formulations, including sauces, gravies, soups, meat analogues, vegetarian patties, and gluten-free bakery items. These starches serve as natural alternatives, providing functional and versatile solutions to achieve desired characteristics in a wide array of culinary creations

6.3.3 Film-Forming Properties

Pseudocereal starches, like that found in quinoa, amaranth, and buckwheat, are very good at forming films. These starches go through a process called gelatinisation when heated in a liquid medium, which causes a thin, transparent film to form on the surface. This film serves as a barrier and offers the food industry several advantages (Rai et al. 2018). Food packaging is a common use for pseudocereal starch films. These starch films can be used to coat a variety of food items, including fruits, vegetables, confections, and other foods, in an edible or biodegradable layer. The food's shelf life is increased by the film's ability to hold onto moisture, preventing the food from drying out. The film also serves as a barrier against microbial contamination and oxygen, protecting the packaged food's quality and freshness Menegalli 2016. Pseudocereal starches' ability to form films is also used in the cosmetics sector. They can be used in cosmetic formulations as all-natural, environmentally friendly substitutes for synthetic film-forming agents. Pseudocereal starch films can act as a barrier for the skin or hair, retaining moisture and improving the effectiveness of cosmetics (Henning et al. 2022).

6.3.4 Emulsifying Properties

Pseudocereal starches are useful ingredients for stabilising oil-in-water emulsions because they have emulsifying properties. Emulsions are mixtures of immiscible liquids held together by an emulsifier, such as oil and water (Yupeng et al. 2018). By lowering the surface tension between the oil and water phases, pseudocereal starches serve as emulsifiers, facilitating more even mixing and preventing phase separation (Lorenzo et al. 2018).Pseudocereal starches are used to formulate a variety of products in the food industry, where their emulsifying properties are particularly advantageous. To produce stable and homogeneous emulsions, pseudo-cereal starches are frequently used in dairy products, mayonnaise, sauces, and salad dressings. Pseudocereal starches enhance the texture, appearance, and mouthfeel of these food products by preventing the separation of oil and water, giving consumers an excellent sensory experience (Arslan et al. 2019). Additionally, the stability and shelf life of food products is aided by the emulsifying abilities of pseudocereal starches. The risk of oil or water separation is reduced by maintaining a consistent emulsion, ensuring that the product maintains its visual appeal and flavour over time. Overall, the film-forming and emulsifying properties of pseudocereal starches have important applications in the food and cosmetic industries (Silva et al. 2020). Their capacity to create barrier films on food surfaces prolongs shelf life and maintains the calibre of the final product. Their stabilising effects on oil-in-water emulsions improve the texture of various food and cosmetic formulations while preventing phase separation.

6.4 Applications of Pseudocereal Starch

6.4.1 Food Industry

Pseudocereals, which include quinoa, amaranth, and buckwheat, are non-grass plants that have gained popularity as alternative grains. Despite not being true cereals, some pseudocereals have superior nutritional profiles, with higher quantities of protein, essential amino acids, dietary fiber, and other beneficial components. This chapter investigates their uses in many industries, emphasizing their versatility and value (Ugural and Akyol 2022).

6.4.1.1 Bakery Products

Pseudocereal starches are widely used in the bread and confectionery industries. These starches increase numerous products' texture, stability, and sensory qualities, resulting in higher product quality and consumer satisfaction. In the baking industry, pseudocereal starches play an important role in improving the properties of bread, cakes, pastries, and cookies (Bender and Schönlechner 2021). One of the primary roles of pseudocereal starches is to thicken. They improve dough handling features such as greater elasticity, decreased stickiness, and improved viscosity control. Pseudocereal starches improve bakery product shaping and moulding by optimizing dough rheology (Padalino et al. 2016). Furthermore, pseudocereal starches retain moisture well, resulting in a longer shelf life and enhanced suppleness. These starches can absorb and hold moisture during baking, preventing items from becoming dry or stale. This property is especially useful in gluten-free baking items, as gluten-free flours have a decreased water-holding capacity (Petrova and Petrov 2020). Pseudocereal starches compensate for this problem in the absence of gluten by increasing the overall water absorption and moisture retention of the dough or batter, resulting in higher quality and longer freshness of bakery items. Furthermore, pseudocereal starches can be used as viable substitutes for regular wheat-based starches, making bread items suitable for people who are gluten intolerant or have coeliac disease. Gluten-free pseudocereal starches provide a viable alternative for gluten-free bread items without sacrificing texture, flavour, or functionality (Lorenzo et al. 2018). They aid in the preservation of the structure and volume of gluten-free products, offering a pleasurable eating experience for consumers with special dietary needs. Pseudocereal starches have nutritional benefits in addition to their functional features in baked products. Pseudocereals have a high protein content and high levels of essential amino acids, vitamins, and minerals. (Sasthri et al., Muthugopal, and Krishnakumar 2020). These starches add to the overall nutritional profile of the finished goods when utilized in baking formulations. They add important nutrients to bread, cakes, and cookies, making them a healthier alternative for consumers (Goldfein and Slavin 2015). Moreover, pseudocereal starches are used to make a variety of sweet delicacies. They are used to thicken fillings, icings, creams, glazes, and other confectionary applications. Pseudocereal starches contribute to these goods' smoothness, stability, and texture, ensuring ideal sensory qualities and long shelf life. Pseudocereal starches offer the proper viscosity in fillings, preventing excessive liquid separation or "syneresis" (May and Henry 2020). They produce a smooth and creamy texture, improving the overall mouthfeel of confectionary foods. Pseudocereal starches also help to stabilise icings, creams, and glazes, allowing them to keep their texture, form, and glossy appearance over time. Pseudocereal starches in confectionary items, like those used in bread products, can provide gluten-free alternatives. This is essential because many traditional confectionary items rely on wheat-based starches, which can be problematic for people who are gluten intolerant (Schoenlechner 2017). Pseudocereal starches enable the creation of gluten-free confectionaries without sacrificing quality or sensory appeal. Pseudocereal starches also help to increase the nutritional content of confectionary items. They contain dietary fibre, minerals, and vitamins, making these delectable delights healthy. Manufacturers can improve the nutritional profile of confectionary foods and their appeal to health-conscious consumers by using pseudocereal starches (Moradi et al. 2021). Overall, pseudocereal starches are widely used in the bread and confectionary industries, where they contribute to the texture, stability, and sensory qualities of numerous goods. These starches work as thickeners in bread items, improving dough handling and viscosity control while increasing moisture retention and shelf life. They also offer gluten-free options for those with special dietary requirements. Pseudocereal starches are crucial thickening agents in confectionary goods, ensuring smoothness, stability, and enhanced texture (Žuljević and Oručević 2021). They also add to the nutrient content of these decadent delicacies. Overall, pseudocereal starches provide useful alternatives for producing high-quality, functional, and nutritional baking and confectionary goods that satisfy today's consumer demands.

6.4.1.2 Dairy and Frozen Desserts

Dairy and frozen desserts encompass a diverse range of goods that delight people of all ages. These delicacies have become food industry classics, ranging from creamy ice creams to silky yogurts and delicious puddings. Manufacturers have turned to pseudocereal starches derived from quinoa, amaranth, and buckwheat to improve the texture, mouthfeel, and stability of these products (Morgeson et al. 2019). These starches have specific qualities that make them valuable components in the dairy and frozen dessert industries, ensuring product quality and consumer satisfaction. The generation of ice crystals is one of the most difficult tasks in the production of frozen desserts. When ice crystals form, they can hurt the texture and creaminess of the product, making it less appealing to eat. Pseudocereal starches are efficient at preventing ice crystal formation, ensuring that frozen sweets remain smooth and creamy throughout their shelf life (Bender and Schönlechner 2021). The starches serve as stabilisers, binding water molecules and forming a network that prevents ice crystal formation and movement. Manufacturers can supply consumers with frozen sweets that have a wonderful texture and mouthfeel by integrating pseudocereal starches into their recipes (Mujinga 2020). Furthermore, pseudocereal starches are widely used in a variety of dairy products, including yogurts, puddings, and dairy-based desserts. These starches serve as stabilisers in dairy formulas, reducing phase separation and enhancing overall product uniformity (Mccarroll 2015). Pseudocereal starches, for example, aid in the homogeneity of the yogurt matrix, avoiding whey separation and maintaining a smooth and creamy texture. Furthermore, pseudocereal starches help to reduce syneresis, the undesirable process in which liquid separates from the solid element of a product (Jeske 2018). This not only improves the overall sensory experience but also extends the product's shelf life. The potential of pseudocereal starches to improve the quality of dairy and frozen desserts demonstrates their versatility. These starches provide a variety of functional benefits, such as viscosity control, water binding capacity, and texture enhancement (Pua et al. 2022). Manufacturers can get the desired texture and stability for their unique products by altering the concentration and processing procedures. Pseudocereal starches are often included during the formulation step of the manufacturing process of dairy and frozen desserts. The starches are disseminated in liquid before being heated to activate their thickening capabilities (Nwachukwu and Aluko 2021). This procedure ensures that the starches are spread consistently throughout the product, resulting in consistent results. Furthermore, in addition to their functionality, pseudocereal starches have other benefits. They are naturally gluten-free, making them acceptable for anyone who is gluten-intolerant or has coeliac disease. Manufacturers may now cater to a broader consumer base, capitalising on the increased demand for gluten-free products (Aristotelous 2018).

Furthermore, pseudocereal starches are thought to have a higher nutritional value than typical cereal starches. They contain essential amino acids, nutritional fibre, and other bioactive elements, making dairy and frozen dessert products healthier. As customer preferences grow, so does the desire for cleaner labelling and natural ingredients (Moreira et al. 2018). Because they are produced from whole grains and undergo little processing, pseudocereal starches fit in well with these trends. This allows businesses to match consumer expectations for wholesome, natural products (Klerks et al. 2019). To make the best use of pseudocereal starches, manufacturers must undergo extensive testing and formulation optimisation. This includes calculating the optimum usage levels, comprehending the precise interactions with other additives, and evaluating the impact on product properties such as texture, viscosity, and stability. Manufacturers can attain the necessary qualities while ensuring overall product quality by fine-tuning the composition (Szakály and Kiss 2023). At last, pseudocereal starches play an important part in the dairy and frozen dessert industries. They have a wide range of applications, such as reducing ice crystal formation in frozen desserts and functioning as stabilisers in dairy products. Pseudocereal starches help these goods achieve the desired texture, mouthfeel, and stability, boosting the whole sensory experience for customers (Fiorentini et al. 2020). Furthermore, their gluten-free status, nutritional value, and alignment with clean-label trends make them an appealing option for manufacturers looking to fulfil customers' changing demands. Pseudocereal starches are anticipated to remain a valuable element in the manufacture of tasty and high-quality dairy and frozen dessert items as the food industry continues to innovate (Mende et al. 2016).

6.4.1.3 Sauces, Dressings, and Condiments

Sauces, dressings, and condiments enhance the flavour, texture, and aesthetic appeal of many dishes. Pseudocereal starches, obtained from grains such as quinoa, amaranth, and buckwheat, serve as a thickening and stabilising factor in these culinary concoctions (Woomer and Adedeji 2021). They improve the overall quality and sensory experience of sauces, dressings, and condiments by improving viscosity, preventing ingredient separation, maintaining uniform dispersion, providing clinginess, and improving appearance (Bender and Schönlechner 2021).

The capacity of pseudocereal starches to thicken sauces, dressings, and condiments is one of their primary roles. When these starches are heated or combined with liquids, a process known as gelatinisation occurs. Gelatinisation is the process by which starch granules swell and hydrate, resulting in the creation of a gel-like structure that thickens the mixture (Sethi et al. 2022). Manufacturers can get the necessary viscosity in sauces such as gravies, pasta sauces, and curry sauces by adding pseudocereal starches, resulting in a rich and silky texture that enriches the entire sensory experience of the food. Pseudocereal starches, in addition to thickening, act as stabilisers in sauces, dressings, and condiments. They are critical in preventing component separation, particularly in emulsified dressings with oil and water phases

(Obadi et al. 2021). Pseudocereal starches contribute to the formation of a stable emulsion by forming a network of links that hold the oil and water phases together. This keeps the dressing homogeneous and well-blended, resulting in a constant flavour and texture. Furthermore, these starches help to keep solid particles or herbs suspended in condiments like mustard or salsa, avoiding settling and guaranteeing a uniform distribution of components throughout the product (Quiles et al. 2022). Another key characteristic that pseudocereal starches provide to sauces, dressings, and condiments is clinginess. This involves creating a coating effect that improves the sauce's capacity to attach to food surfaces, resulting in better coverage and flavour distribution (Woomer and Adedeji 2021). When a barbecue sauce containing pseudocereal starch is smeared upon grilled meats, for example, it clings to the surface, creating a tasty and visually pleasing glaze. Similarly, pseudocereal starches can add to salad dressings' creamy texture and clinginess, ensuring that the leaves are equally coated and stick nicely to the salad toppings (Petrova and Petrov 2020). Pseudocereal starches contribute to the overall appearance of sauces, dressings, and condiments in addition to their functional capabilities. These starches contribute to a smooth and glossy texture, which improves the product's visual attractiveness. The glossy appearance of dressings and condiments is especially desirable since it relates to freshness and quality (Gómez et al. 2020). A visually appealing sauce or dressing can influence consumer perception, enhancing their willingness to try the product and possibly leading to repeat sales. Pseudocereal starches have some advantages over typical cereal starches such as corn or wheat starch. Because of their increased protein content and superior amino acid profiles, they are frequently seen as healthier options (Amini Sarteshnizi et al. 2015). Pseudocereal starches also have a lower glycaemic index, which is advantageous for people who need to control their blood sugar levels. These nutritional benefits add to the popularity of pseudocereal-based sauces, dressings, and condiments among health-conscious customers (Pietrasik and Soladoye 2021).Additionally, pseudocereal starches are extremely useful in the manufacturing of sauces, dressings, and condiments. Their thickening and stabilising qualities ensure that the desired viscosity is achieved, that component separation is avoided, and that uniform dispersion is maintained. Furthermore, these starches add clinginess to the sauce, boosting its capacity to attach to food surfaces and guaranteeing even flavour distribution (Balestra et al. 2019). Again, pseudocereal starches improve the appearance of sauces, dressings, and condiments by generating a smooth and glossy texture that improves the product's aesthetic appeal. Pseudocereal starches, with their functional and nutritional benefits, continue to play an essential role in the food business, improving the quality and sensory experience of numerous culinary creations.

6.4.1.4 Meat and Seafood Industry

Pseudocereal starches have achieved widespread attention and application in the meat and seafood industries. Their distinct qualities make them valuable ingredients as binders, fillers, and stabilisers, boosting the quality, texture, and shelf life of processed meat and seafood products (Younis et al. 2022).

6.4.1.4.1 Meat Products

Due to their practical qualities, pseudocereal starches, like quinoa or amaranth starch, are now widely used in the processing of meat. One of their main functions is to serve as binders, which improve water and fat retention in the meat matrix (Pathiraje et al. 2023). As a result, processed meat products, like sausages, meatballs, and nuggets, have better juiciness, softness, and overall sensory qualities. Pseudocereal starches work to bind the fluids and fat, preventing dryness and preserving succulence even after cooking or reheating (Santos et al. 2021). Consumers have a more tasty and gratifying eating experience as a result. Additionally, pseudocereal starches help emulsified meat products stay stable. For instance, emulsified sausages are vulnerable to the separation of fat during processing or cooking. Pseudocereal starches work as stabilisers to maintain the end product's desired texture and appearance by inhibiting fat exudation (Bavaro et al. 2021). The starches maintain a uniform and constant meat matrix, improving the quality of the finished product.

6.4.1.4.2 Seafood Products

Pseudocereal starches are helpful in the fish business as well, especially for enhancing the flavour and texture of seafood products. These starches aid in the retention of moisture, minimizing drip loss throughout preparation and cooking (Martin et al. 2022). This moisture retention keeps seafood products from drying out and maintains their succulence and juiciness. Pseudocereal starches have a critical role in maintaining the proper sensory qualities, whether they are used in fish cakes, goods made with surimi, or breaded seafood items (Haros and Sanz-Penella 2017). Additionally, pseudocereal starches help keep frozen fish products from losing their texture. Processes involving freezing and thawing frequently result in textural changes, such as a loss of hardness or an unfavourable texture. Pseudocereal starches serve as texture modifiers, assisting seafood items in maintaining their ideal texture and quality even after freezing and subsequent thawing. This guarantees that the items, when consumed, have a pleasant mouthfeel and a fresh-like quality (Bangar et al. 2022).

6.4.1.4.3 Coating and Breading

The coating and breading of meat and fish products use pseudocereal starches. Due to the adhesive qualities of these starches, the coating can successfully cling to the product's surface. This produces a crisp, golden exterior, boosting the final product's texture and aesthetic appeal. Pseudocereal starches offer a pleasing textural contrast between the coating and the underlying product, whether it be a crunchy breaded fish fillet or a crispy chicken nugget, adding to the entire sensory experience (Alencar and de Carvalho Oliveira 2019).

6.4.1.4.4 Reduced-Fat Formulations

Pseudocereal starches have also been used in the production of low-fat meat and seafood items in response to the rising demand for healthier dietary options. These starches can mimic the sensation and texture of fat by mimicking its increased fat content. Manufacturers can make products with lower fat content while still keeping a pleasant sensory experience by integrating pseudocereal starches into formulas. This enables customers to choose healthier options without sacrificing flavour or texture (Robin et al. 2015).

6.4.1.5 Snacks and Extruded Products

Pseudocereal starches, made from grains like quinoa, amaranth, and buckwheat, have become popular components in the creation of these snacks because they offer crucial functions that support texture, structure, and stability. In this subsection, we will delve further into the uses of pseudocereal starches in snacks and extruded goods, examining how they contribute to satisfying consumer expectations for appealing sensory experiences (Paucar-Menacho et al. 2022). Pseudocereal starches serve as binders in snacks, ensuring proper adhesion and shape retention in extruded snacks like chips, crisps, and pellets. Pseudocereal starches transform into a gelatinised matrix during the extrusion process, serving as a binder to bind the ingredients of the snack and give the product structural integrity. This keeps the snacks from crumbling or breaking while they are being made, transported, or consumed, ensuring that they keep their desired shape and presentation (Kringel et al. 2020). In the production of extruded snacks, where maintaining structural integrity is essential for delivering a satisfying eating experience, pseudocereal starches' binding abilities are essential. Pseudocereal starches' contribution to texture and mouthfeel in extruded snacks is another significant factor (Wheate and Limantoro 2016). The starches are heated, pressed, and sheared during the extrusion process, which causes gelatinisation and expansion. The unique gelatinisation properties of pseudocereal starches cause a gel network to form during extrusion. The expansion and crispiness of extruded snacks are greatly influenced by this gel network, which gives them a light and crunchy texture that consumers adore. The pseudocereal starches' ability to gelatinise and form a gel influences the expansion of the snacks, making them more palatable to eat. In addition, pseudocereal starches are used in a variety of snack products in addition to extruded snacks (Suárez-González et al. 2021). Pseudocereal starches aid in the texture, binding, and moisture management of granola bars, energy bars, and other similar snacks (Alencar and de Carvalho Oliveira 2019). To ensure that the ingredients stay together and do not crumble, these starches aid in giving structure and stability to the bar. The pseudocereal starches serve as binders, keeping the bars' different parts together and preventing ingredient separation. The bars' desired texture, whether chewy or crispy, is maintained by them as well by assisting in moisture control. Pseudocereal starches are present in these snack bars, which adds to their overall sensory appeal and patron satisfaction (Robin et al. 2015). Additionally, compared with conventional cereal starches, pseudocereal starches have better nutritional qualities, making them a desirable option for consumers who are concerned about their health. The higher protein and dietary fibre content of pseudocereals is evident in their starches. These starches, which offer more protein and fibre than conventional cereal starches, can help to improve the overall nutritional value of snack foods. Pseudocereal-based snacks provide a nutrient-rich alternative as consumers prioritise healthier snacking options.

6.4.2 Other Applications

Pseudocereal starches have become valuable ingredients in the pharmaceutical sector, with a variety of uses and advantages. This subsection explains in detail how pseudocereal starches are used in the pharmaceutical industry. Topics covered include tablet binding and formulation, controlled drug release systems, their use as excipients in oral dosage forms, and their capacity to stabilise liquid and semi-solid formulations. Pseudocereal starches, like quinoa, amaranth, and buckwheat starch, are frequently utilised as binders in tablet formulations (Paucar-Menacho et al. 2022). These starches' binding abilities are essential for tablet compression because they ensure the cohesion of powdered ingredients and the creation of tablets with desirable physical properties. Pseudocereal starches enhance the integrity, hardness, and dissolution properties of tablets. As binders, they make it easier for the active pharmaceutical ingredient to be evenly distributed throughout the tablet, which is necessary for reliable drug release and therapeutic effectiveness. Moreover, the potential for creating controlled drug release systems is one of the key benefits of pseudocereal starches in pharmaceutical applications (Kringel et al. 2020). These starches have special gelatinisation and retrogradation properties that enable the prolonged controlled release of drugs. Pseudocereal starches can be altered to create matrices that control the rate of drug release through processes like cross-linking or blending with other polymers. This controlled-release mechanism lessens the frequency of dosing, increases patient compliance, and improves the therapeutic effects. Pseudocereal starches are still being researched and developed by the pharmaceutical industry for a variety of applications (Wheate and Limantoro 2016). To improve the functionality and adaptability of these starches in drug delivery systems, researchers are looking into new processing methods, modifications, and combinations with other excipients. Pseudocereal starches are used in pharmaceutical applications, which present opportunities for the creation of novel and patient-friendly formulations (Rathi et al. 2022). In conclusion, pseudocereal starches have proven to have enormous potential and versatility in the pharmaceutical sector. Their importance is highlighted by their uses as stabilisers in liquid and semi-solid formulations, excipients in oral dosage forms, binders in tablet formulations, and contributors to controlled drug release systems (Shahrizan et al. 2022).

6.5 Conclusion

Pseudocereal starch and gum chemistry and applications present both exciting opportunities and challenges. These adaptable ingredients have distinct properties that can be used in a variety of applications in the food, pharmaceutical, and other industrial sectors. However, realising their full potential necessitates addressing issues related to processing optimisation, functional property improvement, sustainability, and the exploration of new applications. The chemistry and applications of pseudocereal starch and gums will continue to evolve as a result of ongoing research, innovation, and collaboration among academia, industry, and policymakers. Their application can help to develop more sustainable, healthier, and functional products that can meet the demands of a rapidly changing market. With careful attention to processing techniques, functionalisation strategies, and sustainable practices, pseudocereal starch and gums have the potential to make a significant impact on the food and pharmaceutical industries, benefitting both consumers and the environment.

References

Agama-Acevedo, Edith, Pamela Celeste Flores-Silva, and Luis Arturo Bello-Perez. 2019. "Cereal Starch Production for Food Applications." In *Starches for Food Application*, 71–102. Elsevier. https://doi.org/10.1016/b978-0-12-809440-2.00003-4.

Akhila, Plachikkattu Parambil, Kappat Valiyapeediyekkal Sunooj, Ganesh Revathi, Basheer Aaliya, Muhammed Navaf, Cherakkathodi Sudheesh, Sarasan Sabu, et al. 2022. "Incrementing Effect on Cold Water Solubility, Structural and Functional Properties of Alcohol-Alkali Treated *Plectranthus rotundifolius* Starch by Organic Acids." *Applied Food Research* 2(2): 100237. https://doi.org/10.1016/j.afres.2022.100237.

Alencar, Natalia Manzatti Machado, and Ludmilla de Carvalho Oliveira. 2019. "Advances in Pseudocereals: Crop Cultivation, Food Application, and Consumer Perception." In *Bioactive Molecules in Food*, 1695–1713. Springer International Publishing. https://doi.org/10.1007/978-3-319-78030-6_63.

Alonso-Miravalles, Loreto, and James O'Mahony. 2018. "Composition, Protein Profile and Rheological Properties of Pseudocereal-Based Protein-Rich Ingredients." *Foods (Basel, Switzerland)* 7(5): 73. https://doi.org/10.3390/foods7050073.

Amini Sarteshnizi, R., H. Hosseini, A. Mousavi Khaneghah, and N. Karimi. 2015. "A Review on Application of Hydrocolloids in Meat and Poultry Products." *International Food Research Journal* 22(3), 872–887.

Aristotelous, Aimee. 2018. *The Whole Pregnancy: A Complete Nutrition Plan for Gluten-Free Moms to Be*. Simon and Schuster.

Arslan, Muhammad, Allah Rakha, Zou Xiaobo, and Muhammad Arsalan Mahmood. 2019. "Complimenting Gluten Free Bakery Products with Dietary Fiber: Opportunities and Constraints." *Trends in Food Science and Technology* 83(January): 194–202. https://doi.org/10.1016/j.tifs.2018.11.011.

Asaithambi, Niveditha, Poonam Singha, and Sushil Kumar Singh. 2022. "Recent Application of Protein Hydrolysates in Food Texture Modification." *Critical Reviews in Food Science and Nutrition*, June, 1–32. https://doi.org/10.1080/10408398.2022.2081665.

Balestra, Federica, Maurizio Bianchi, and Massimiliano Petracci. 2019. "Applications in Meat Products." In *Dietary Fiber: Properties, Recovery, and Applications*, 313–344. Elsevier. https://doi.org/10.1016/b978-0-12-816495-2.00010-1.

Bangar, Punia, Nitya Sneh, Arashdeep Sharma, Yuthana Singh, and Charles S. Phimolsiripol. 2022. "Glycaemic Response of Pseudocereal-Based Gluten-Free Food Products: A Review." *International Journal of Food Science and Technology* 57(8): 4936–4944.

Bangar, S. P., A. O. Ashogbon, S. B. Dhull, R. Thirumdas, M. Kumar, M. Hasan, V. Chaudhary, and S. Pathem. 2021. "Proso-Millet Starch: Properties, Functionality, and Applications." *International Journal of Biological Macromolecules* 190: 960–968.

Bavaro, Anna Rita, Mariaelena Di Biase, Amalia Conte, Stella Lisa Lonigro, Leonardo Caputo, Annamaria Cedola, Matteo Alessandro Del Nobile, Antonio Francesco Logrieco, Paola Lavermicocca, and Francesca Valerio. 2021. "*Weissella cibaria* Short-Fermented Liquid Sourdoughs Based on Quinoa or Amaranth Flours as Fat Replacer in Focaccia Bread Formulation." *International Journal of Food Science and Technology* 56(7): 3197–3208. https://doi.org/10.1111/ijfs.14874.

Bender, D., and R. Schönlechner. 2021. "Recent Developments and Knowledge in Pseudocereals Including Technological Aspects." *Acta Alimentaria* 50(4): 583–609. https://doi.org/10.1556/066.2021.00136.

Bibri, Simon Elias, and John Krogstie. 2017. "Smart Sustainable Cities of the Future: An Extensive Interdisciplinary Literature Review." *Sustainable Cities and Society* 31(May): 183–212. https://doi.org/10.1016/j.scs.2017.02.016.

Boukid, Fatma, Abdo Hassoun, Ahmed Zouari, Mehmet Çağlar Tülbek, Marina Mefleh, Abderrahmane Aït-Kaddour, and Massimo Castellari. 2023. "Fermentation for Designing Innovative Plant-Based Meat and Dairy Alternatives." *Foods (Basel, Switzerland)* 12(5). https://doi.org/10.3390/foods12051005.

Brenda, Olga L., and Mario E. Torres-Vargas. 2019. "Physicochemical Characterization of Quinoa (*Chenopodium quinoa*) Flour and Isolated Starch." *Food Chemistry* 298,124982.

Chandak, A., S. B. Dhull, P. Chawla, M. Fogarasi, and S. Fogarasi. 2022. "Effect of Single and Dual Modifications on Properties of Lotus Rhizome Starch Modified by Microwave and γ-Irradiation: A Comparative Study." *Foods* 11(19): 2969.

Chandak, A., S. B. Dhull, S. Punia Bangar, and A. V. Rusu. 2022. "Effects of Cross-Linking on Physicochemical and Film Properties of Lotus (*Nelumbo nucifera* G.) Seed Starch." *Foods* 11(19): 3069.

Chandla, Narender K., C. Dharmesh, and Sukhcharn Saxena. 2017. "Amaranth (*Amaranthus* spp.) Starch Isolation, Characterization, and Utilization in Development of Clear Edible Films." *Journal of Food Processing and Preservation* 41(6),e13217.

Chawla, P., N. Kumar, A. Bains, S. B. Dhull, M. Kumar, R. Kaushik, and S. Punia. 2020. "Gum Arabic Capped Copper Nanoparticles: Synthesis, Characterization, and Applications." *International Journal of Biological Macromolecules* 146: 232–242.

Chawla, P., N. Kumar, R. Kaushik, and S. B. Dhull. 2019. "Synthesis, Characterization and Cellular Mineral Absorption of Nanoemulsions of *Rhododendron arboreum* Flower Extracts Stabilized with Gum Arabic." *Journal of Food Science and Technology* 56(12): 5194–5203.

Condés, Maria Cecilia, María Cristina Añón, Adriana Noemi Mauri, and Alain Dufresne. 2015. "Amaranth Protein Films Reinforced with Maize Starch Nanocrystals." *Food Hydrocolloids* 47(May): 146–157. https://doi.org/10.1016/j.foodhyd.2015.01.026.

Das, Saubhik. 2016. "Pseudocereals: An Efficient Food Supplement." In *Amaranthus: A Promising Crop of Future*, 5–11. Springer Singapore. https://doi.org/10.1007/978-981-10-1469-7_2.

Deriu, Aloisa G., J. Antonio, and Felicidad Vela. 2022. "Techno-Functional and Gelling Properties of Acha (Fonio) (*Digitaria exilis* Stapf) Flour: A Study of Its Potential as a New Gluten-Free Starch Source in Industrial Applications." *Foods* 11(2),183.

Dhull, S. B., M. Anju, S. Punia, R. Kaushik, and P. Chawla. 2019a. "Application of Gum Arabic in Nanoemulsion for Safe Conveyance of Bioactive Components." In Prasad, R., Kumar, V., Kumar, M.,Choudhary, D. (eds) *Nanobiotechnology in Bioformulations*, 85–98. Springer.

Dhull, S. B., S. P. Bangar, R. Deswal, P. Dhandhi, M. Kumar, M. Trif, and A. Rusu. 2021. "Development and Characterization of Active Native and Cross-Linked Pearl Millet Starch-Based Film Loaded with Fenugreek Oil." *Foods* 10(12): 3097.

Dhull, S. B., A. Chandak, M. N. Collins, S. P. Bangar, P. Chawla, and A. Singh. 2022. "Lotus Seed Starch: A Novel Functional Ingredient with Promising Properties and Applications in Food—A Review." *Starch-Stärke* 74(9–10): 2200064.

Dhull, S. B., T. Malik, R. Kaur, P. Kumar, N. Kaushal, and A. Singh. 2021. "Banana Starch: Properties Illustration and Food Applications—A Review." *Starch-Stärke* 73(1–2): 2000085.

Dhull, S. B., S. Punia, M. Kumar, S. Singh, and P. Singh. 2020a. "Effect of Different Modifications (Physical and Chemical) on Morphological, Pasting, and Rheological Properties of Black Rice (*Oryza sativa* L. *indica*) Starch: A Comparative Study." *Starch-Stärke* 73(1–2): 2000098.

Dhull, S. B., K. S. Sandhu, S. Punia, M. Kaur, P. Chawla, and A. Malik. 2020b. "Functional, Thermal and Rheological Behavior of Fenugreek (*Trigonella foenum–graecum* L.) Gums from Different Cultivars: A Comparative Study." *International Journal of Biological Macromolecules* 159: 406–414.

Egharevba, Henry. 2019. "Chemical Properties of Starch and Its Application in the Food Industry." In *Chemical Properties of Starch* 9, IntechOpen eBooks,

Espinosa-Solis, Vicente, Yunia Verónica García-Tejeda, Everth Jimena Leal-Castañeda, and Víctor Barrera-Figueroa. 2020. "Effect of the Degree of Substitution on the Hydrophobicity, Crystallinity, and Thermal Properties of Lauroylated Amaranth Starch." *Polymers* 12(11): 2548. https://doi.org/10.3390/polym12112548.

Everth, J., Yunia García-Tejeda, Humberto Hernández-Sánchez, Liliana Alamilla-Beltrán, I. Darío, Georgina Téllez-Medina, Hugo S. Calderón-Domínguez, and Gustavo F. García. 2018. "Pickering Emulsions Stabilized with Native and Lauroylated Amaranth Starch." *Food Hydrocolloids* 80: 177–185.

Fang, Chenlu, Junrong Huang, Qi Yang, Huayin Pu, Shuxing Liu, and Zhenbao Zhu. 2020. "Adsorption Capacity and Cold-Water Solubility of Honeycomb-Like Potato Starch Granule." *International Journal of Biological Macromolecules* 147(March): 741–749. https://doi.org/10.1016/j.ijbiomac.2020.01.224.

Fiorentini, Martina, Amanda J. Kinchla, and Alissa A. Nolden. 2020. "Role of Sensory Evaluation in Consumer Acceptance of Plant-Based Meat Analogs and Meat Extenders: A Scoping Review." *Foods (Basel, Switzerland)* 9(9): 1334. https://doi.org/10.3390/foods9091334.

Gao, Licheng, Chenxi Wan, Jiale Wang, Pengke Wang, Xiaoli Gao, Mia Eeckhout, and Jinfeng Gao. 2022. "Relationship between Nitrogen Fertilizer and Structural, Pasting and Rheological Properties on Common Buckwheat Starch." *Food Chemistry* 389(132664): 132664. https://doi.org/10.1016/j.foodchem.2022.132664.

Geyang, Craig F., and Kevin M. Morris. 2017. "Quinoa Starch Characteristics and Their Correlations with the Texture Profile Analysis (TPA) of Cooked Quinoa." *Journal of Food Science* 82(10): 2387–2395.

Goel, Charu, Anil Dutt Semwal, Khan Ayub, Sunny Kumar, and Gopal Kumar Sharma. 2020. "Physical Modification of Starch: Changes in Glycemic Index, Starch Fractions, Physicochemical and Functional Properties of Heat-Moisture Treated Buckwheat Starch." *Journal of Food Science and Technology* 57(8): 2941–2948. https://doi.org/10.1007/s13197-020-04326-4.

Goff, H. Douglas, and Qingbin Guo. 2019. "The Role of Hydrocolloids in the Development of Food Structure." In *Handbook of Food Structure Development*, 1–28. The Royal Society of Chemistry. https://doi.org/10.1039/9781788016155-00001.

Goldfein, Kara R., and Joanne L. Slavin. 2015. "Why Sugar Is Added to Food: Food Science 101." *Comprehensive Reviews in Food Science and Food Safety* 14(5): 644–656. https://doi.org/10.1111/1541-4337.12151.

Gómez, Inmaculada, Rasmi Janardhanan, Francisco C. Ibañez, and María José Beriain. 2020. "The Effects of Processing and Preservation Technologies on Meat Quality: Sensory and Nutritional Aspects." *Foods (Basel, Switzerland)* 9(10): 1416. https://doi.org/10.3390/foods9101416.

Haros, Claudia Monika, and Juan Mario Sanz-Penella. 2017. "Food Uses of Whole Pseudocereals." In *Pseudocereals*, 163–192. John Wiley & Sons, Ltd. https://doi.org/10.1002/9781118938256.ch8.

Henning, Fernanda Gabriela, Vivian Cristina Ito, Ivo Mottin Demiate, and Luiz Gustavo Lacerda. 2022. "Non-conventional Starches for Biodegradable Films: A Review Focussing on Characterisation and Recent Applications in Food Packaging." *Carbohydrate Polymer Technologies and Applications* 4(100157): 100157. https://doi.org/10.1016/j.carpta.2021.100157.

Hosseini, Hamed, Zhila Tajiani, and Seid Mahdi Jafari. 2019. "Improving the Shelf-Life of Food Products by Nano/Micro-encapsulated Ingredients." In *Food Quality and Shelf Life*, 159–200. Elsevier. https://doi.org/10.1016/b978-0-12-817190-5.00005-7.

Jan, Khan Nadiya, P. S. Panesar, J. C. Rana, and Sukhcharn Singh. 2017. "Structural, Thermal and Rheological Properties of Starches Isolated from Indian Quinoa Varieties." *International Journal of Biological Macromolecules* 102(September): 315–322. https://doi.org/10.1016/j.ijbiomac.2017.04.027.

Jeske, Stephanie. 2018. *Evaluation and Improvement of Technological and Nutritional Properties of Plant-Based Milk Substitutes*. University College Cork.

Kaur, L., S. B. Dhull, P. Kumar, and A. Singh. 2020. "Banana Starch: Properties, Description, and Modified Variations-A Review." *International Journal of Biological Macromolecules* 165(Part B): 2096–2102.

Klerks, Michelle, Maria Jose Bernal, Sergio Roman, Stefan Bodenstab, Angel Gil, and Luis Manuel Sanchez-Siles. 2019. "Infant Cereals: Current Status, Challenges, and Future Opportunities for Whole Grains." *Nutrients* 11(2): 473. https://doi.org/10.3390/nu11020473.

Kringel, Dianini Hüttner, Shanise Lisie Mello El Halal, Elessandra da Rosa Zavareze, and Alvaro Renato Guerra Dias. 2020. "Methods for the Extraction of Roots, Tubers, Pulses, Pseudocereals, and Other Unconventional Starches Sources: A Review." *Die Starke* 72(11–12): 1900234. https://doi.org/10.1002/star.201900234.

Li, Guantian, and Fan Zhu. 2018a. "Quinoa Starch: Structure, Properties, and Applications." *Carbohydrate Polymers* 181(February): 851–861. https://doi.org/10.1016/j.carbpol.2017.11.067.

———. 2018b. "Rheological Properties in Relation to Molecular Structure of Quinoa Starch." *International Journal of Biological Macromolecules* 114(July): 767–775. https://doi.org/10.1016/j.ijbiomac.2018.03.039.

Li, Hua, Fengyan Zhai, Jianfeng Li, Xuanxuan Zhu, Yanyan Guo, Beibei Zhao, and Baocheng Xu. 2021. "Physicochemical Properties and Structure of Modified Potato Starch Granules and Their Complex with Tea Polyphenols." *International Journal of Biological Macromolecules* 166(January): 521–528. https://doi.org/10.1016/j.ijbiomac.2020.10.209.

Li, Jing, Mengting Wang, Guangxin Liu, Wei Wang, Aijun Hu, and Jie Zheng. 2023. "Effects of Microwave and Conventional Heating on Physicochemical, Digestive, and Structural Properties of Debranched Quinoa Starch-Oleic Acid Complexes with Different Water Addition." *Journal of the Science of Food and Agriculture* 103(4): 2146–2154. https://doi.org/10.1002/jsfa.12415.

Liu, Hang, Xudan Guo, Wuxia Li, Xiaofang Wang, Manman Lv, Qiang Peng, and Min Wang. 2015. "Changes in Physicochemical Properties and *In Vitro* Digestibility of Common Buckwheat Starch by Heat-Moisture Treatment and Annealing." *Carbohydrate Polymers* 132(November): 237–244. https://doi.org/10.1016/j.carbpol.2015.06.071.

Liu, Hang, Lijing Wang, Rong Cao, Huanhuan Fan, and Min Wang. 2016. "In Vitro Digestibility and Changes in Physicochemical and Structural Properties of Common Buckwheat Starch Affected by High Hydrostatic Pressure." *Carbohydrate Polymers* 144(June): 1–8. https://doi.org/10.1016/j.carbpol.2016.02.028.

López-Fernández, María Paula, Silvio David Rodríguez, Leonardo Cristian Favre, Verónica María Busch, and María del Pilar Buera. 2021. "Physicochemical, Thermal and Rheological Properties of Isolated Argentina Quinoa Starch." *Lebensmittel-Wissenschaft Und Technologie [Food Science and Technology]* 135(110113): 110113. https://doi.org/10.1016/j.lwt.2020.110113.

Lorenzo, Gabriel, Meli Sosa, and Alicia Califano. 2018. "Alternative Proteins and Pseudocereals in the Development of Gluten-Free Pasta." In *Alternative and Replacement Foods*, 433–458. Elsevier. https://doi.org/10.1016/b978-0-12-811446-9.00015-0.

Ma, Hao, Mei Liu, Ying Liang, Xueling Zheng, Le Sun, Wenqian Dang, Jie Li, Limin Li, and Chong Liu. 2022. "Research Progress on Properties of Pre-gelatinized Starch and Its Application in Wheat Flour Products." *Grain & Oil Science and Technology* 5(2): 87–97. https://doi.org/10.1016/j.gaost.2022.01.001.

Malik, M., R. Sindhu, S. B. Dhull, C. Bou-Mitri, Y. Singh, S. Panwar, and B. S. Khatkar. 2023. "Nutritional Composition, Functionality, and Processing Technologies for Amaranth." *Journal of Food Processing and Preservation* 2023: 1753029. https://doi.org/10.1155/2023/1753029.

Martin, Anna, Verena Schmidt, Raffael Osen, Jürgen Bez, Eva Ortner, and Stephanie Mittermaier. 2022. "Texture, Sensory Properties and Functionality of Extruded Snacks from Pulses and Pseudocereal Proteins." *Journal of the Science of Food and Agriculture* 102(12): 5011. https://doi.org/10.1002/jsfa.11041.

Martínez-Villaluenga, Cristina, Elena Peñas, and Blanca Hernández-Ledesma. 2020. "Pseudocereal Grains: Nutritional Value, Health Benefits and Current Applications for the Development of Gluten-Free Foods." *Food and Chemical Toxicology: An International Journal Published for the British Industrial Biological Research Association* 137(111178): 111178. https://doi.org/10.1016/j.fct.2020.111178.

May, S. M., and Christiani Jeyakumar Henry. 2020. "Reducing the Glycemic Impact of Carbohydrates on Foods and Meals: Strategies for the Food Industry and Consumers with Special Focus on Asia." *Comprehensive Reviews in Food Science and Food Safety* 19(2): 670–702.

Mccarroll, Leigh. 2015. *Evaluation of Mageu-Based Gluten-Free Bread in South Africa*, University of Johannesburg.

McClements, David Julian. 2019. "Food Architecture: Building Better Foods." In *Future Foods*, 27–60. Springer International Publishing. https://doi.org/10.1007/978-3-030-12995-8_2.

McClements, David Julian, Emily Newman, and Isobelle Farrell McClements. 2019. "Plant-Based Milks: A Review of the Science Underpinning Their Design, Fabrication, and Performance." *Comprehensive Reviews in Food Science and Food Safety* 18(6): 2047–2067. https://doi.org/10.1111/1541-4337.12505.

Mende, Susann, Harald Rohm, and Doris Jaros. 2016. "Influence of Exopolysaccharides on the Structure, Texture, Stability and Sensory Properties of Yoghurt and Related Products." *International Dairy Journal* 52(January): 57–71. https://doi.org/10.1016/j.idairyj.2015.08.002.

Menegalli, Florencia. 2016. "Films and Coatings from Starch and Gums." In *Edible Films and Coatings*, 143–160. CRC Press.

Mir, Nisar Ahmad, Charanjit Singh Riar, and Sukhcharn Singh. 2018. "Nutritional Constituents of Pseudo Cereals and Their Potential Use in Food Systems: A Review." *Trends in Food Science and Technology* 75(May): 170–180. https://doi.org/10.1016/j.tifs.2018.03.016.

Mohammed, Abdulrahman A. B. A., A. Abdoulhdi, Borhana Omran, Zaimah Hasan, R. A. Ilyas, and S. M. Sapuan. 2021. "Wheat Biocomposite Extraction, Structure, Properties and Characterization: A Review." *Polymers* 13(21): 3624. https://doi.org/10.3390/polym13213624.

Moradi, Mahdis, Marzieh Bolandi, Majid Arabameri, Mahdi Karimi, Homa Baghaei, Fariborz Nahidi, and Mohadeseh Eslami Kanafi. 2021. "Semi-volume Gluten-Free Bread: Effect of Guar Gum, Sodium Caseinate and Transglutaminase Enzyme on the Quality Parameters." *Journal of Food Measurement and Characterization* 15(3): 2344–2351. https://doi.org/10.1007/s11694-021-00823-y.

Moreira, Ana, Joana Jacob Martins, and Maria Carolina Magalhães Fechas Momade. 2018. "Connect to Success Consulting Program-Business Plan." PhD Diss.

Morgeson, Frederick P., T. Michael, and Edward L. Brannick. 2019. *Job and Work Analysis: Methods, Research, and Applications for Human Resource Management*. Sage Publications.

Mouritsen, Ole G., and Klavs Styrbæk. 2017. "Playing Around with Mouthfeel." In *Mouthfeel*, 5, 113–206. Columbia University Press.

Mujinga, Nyembwe. 2020. *Utilisation of Marama Bean [*Tylosema eEsculentum *(Burchell) A. Schreiber] Flour in Gluten-Free Bread Making*.

Nwachukwu, Ifeanyi D., and Rotimi E. Aluko. 2021. "CHAPTER 1. Food Protein Structures, Functionality and Product Development." In *Food Chemistry, Function and Analysis*, 1–33. Royal Society of Chemistry. https://doi.org/10.1039/9781839163425-00001.

Obadi, Mohammed, Yajing Qi, and Bin Xu. 2021. "Highland Barley Starch (Qingke): Structures, Properties, Modifications, and Applications." *International Journal of Biological Macromolecules* 185(August): 725–738. https://doi.org/10.1016/j.ijbiomac.2021.06.204.

Padalino, Lucia, Amalia Conte, and Matteo Alessandro Del Nobile. 2016. "Overview on the General Approaches to Improve Gluten-Free Pasta and Bread." *Foods (Basel, Switzerland)* 5(4): 87. https://doi.org/10.3390/foods5040087.

Pathiraje, Darshika, Janelle Carlin, Tanya Der, Janitha P. D. Wanasundara, and Phyllis J. Shand. 2023. "Generating Multi-functional Pulse Ingredients for Processed Meat Products-Scientific Evaluation of Infrared-Treated Lentils." *Foods (Basel, Switzerland)* 12(8). https://doi.org/10.3390/foods12081722.

Paucar-Menacho, Luz María, Williams Esteward Castillo-Martínez, Wilson Daniel Simpalo-Lopez, Anggie Verona-Ruiz, Alicia Lavado-Cruz, Cristina Martínez-Villaluenga, Elena Peñas, Juana Frias, and Marcio Schmiele. 2022. "Performance of Thermoplastic Extrusion, Germination, Fermentation, and Hydrolysis Techniques on Phenolic Compounds in Cereals and Pseudocereals." *Foods (Basel, Switzerland)* 11(13): 1957. https://doi.org/10.3390/foods11131957.

Petrova, Penka, and Kaloyan Petrov. 2020. "Lactic Acid Fermentation of Cereals and Pseudocereals: Ancient Nutritional Biotechnologies with Modern Applications." *Nutrients* 12(4): 1118. https://doi.org/10.3390/nu12041118.

Pietrasik, Zeb, and Olugbenga P. Soladoye. 2021. "Functionality and Consumer Acceptability of Low-Fat Breakfast Sausages Processed with Non-meat Ingredients of Pulse Derivatives." *Journal of the Science of Food and Agriculture* 101(11): 4464–4472. https://doi.org/10.1002/jsfa.11084.

Pua, Aileen, Vivien Chia Yen Tang, Rui Min Vivian Goh, Jingcan Sun, Benjamin Lassabliere, and Shao Quan Liu. 2022. "Ingredients, Processing, and Fermentation: Addressing the Organoleptic Boundaries of Plant-Based Dairy Analogues." *Foods (Basel, Switzerland)* 11(6): 875. https://doi.org/10.3390/foods11060875.

Punia, S., and S. B. Dhull. 2019. "Chia Seed (*Salvia hispanica* L.) Mucilage (a Heteropolysaccharide): Functional, Thermal, Rheological Behaviour and Its Utilization." *International Journal of Biological Macromolecules* 140: 1084–1090.

Punia, S., S. B. Dhull, P. Kunner, and S. Rohilla. 2020a. "Effect of γ-Radiation on Physico-chemical, Morphological and Thermal Characteristics of Lotus Seed (*Nelumbo nucifera*) Starch." *International Journal of Biological Macromolecules* 157: 584.

Punia, S., S. B. Dhull, K. S. Sandhu, and M. Kaur. 2019a. "Faba Bean (*Vicia faba*) Starch: Structure, Properties, and In Vitro Digestibility—A Review." *Legume Science* 1(1): e18.

Punia, S., S. B. Dhull, K. S. Sandhu, M. Kaur, and S. S. Purewal. 2020b. "Kidney Bean (*Phaseolus vulgaris*) Starch: A Review." *Legume Science* 2(3): e52.

Punia, S., M. Kumar, A. K. Siroha, J. F. Kennedy, S. B. Dhull, and W. S. Whiteside. 2021. "Pearl Millet Grain as an Emerging Source of Starch: A Review on Its Structure, Physicochemical Properties, Functionalization, and Industrial Applications." *Carbohydrate Polymers* 260: 117776.

Punia, S., K. S. Sandhu, S. B. Dhull, and M. Kaur. 2019b. "Dynamic, Shear and Pasting Behaviour of Native and Octenyl Succinic Anhydride (OSA) Modified Wheat Starch and Their Utilization in Preparation of Edible Films." *International Journal of Biological Macromolecules* 133: 110–116.

Punia, S., K. S. Sandhu, S. B. Dhull, A. K. Siroha, S. S. Purewal, M. Kaur, and M. K. Kidwai. 2020c. "Oat Starch: Physico-chemical, Morphological, Rheological Characteristics and Its Application-A Review." *International Journal of Biological Macromolecules* 154: 493–498.

Quiles, Amparo, Empar Llorca, Gemma Moraga, and Isabel Hernando. 2022. "Clean Label Foods with Reduced Fat Content." In *The Age of Clean Label Foods*, 103–133. Springer International Publishing. https://doi.org/10.1007/978-3-030-96698-0_4.

Rai, Sweta, Amarjeet Kaur, and C. S. Chopra. 2018. "Gluten-Free Products for Celiac Susceptible People." *Frontiers in Nutrition* 5(December): 116. https://doi.org/10.3389/fnut.2018.00116.

Rathi, Ritu, Sanshita, Alpesh Kumar, Vivekanand Vishvakarma, Kampanart Huanbutta, Inderbir Singh, and Tanikan Sangnim. 2022. "Advancements in Rectal Drug Delivery Systems: Clinical Trials, and Patents Perspective." *Pharmaceutics* 14(10). https://doi.org/10.3390/pharmaceutics14102210.

Rayner, Marilyn, Anna Timgren, Malin Sjöö, and Petr Dejmek. 2012. "Quinoa Starch Granules: A Candidate for Stabilising Food-Grade Pickering Emulsions." *Journal of the Science of Food and Agriculture* 92(9): 1841–1847. https://doi.org/10.1002/jsfa.5610.

Repo-Carrasco-Valencia, Ritva, and Jenny Valdez Arana. 2017. "Carbohydrates of Kernels." In *Pseudocereals*, 49–70. John Wiley & Sons, Ltd. https://doi.org/10.1002/9781118938256.ch3.

Robin, Frédéric, Christine Théoduloz, and Sathaporn Srichuwong. 2015. "Properties of Extruded Whole Grain Cereals and Pseudocereals Flours." *International Journal of Food Science and Technology* 50(10): 2152–2159. https://doi.org/10.1111/ijfs.12893.

Santos, João Marcos Dos, Eduardo Oliveira Ignácio, Camila Vespúcio Bis-Souza, and Andrea Carla da Silva-Barretto. 2021. "Performance of Reduced Fat-Reduced Salt Fermented Sausage with Added Microcrystalline Cellulose, Resistant Starch and Oat Fiber Using the Simplex Design." *Meat Science* 175(108433): 108433. https://doi.org/10.1016/j.meatsci.2021.108433.

Sasthri, Vijaykrishnaraj, Nivedha Muthugopal, and Pichan Krishnakumar. 2020. "Advances in Conventional Cereal and Pseudocereal Processing." In *Innovative Processing Technologies for Healthy Grains*: 61–81.

Schoenlechner, Regine. 2017. "Quinoa: Its Unique Nutritional and Health-Promoting Attributes." In *Gluten-Free Ancient Grains*, 105–129. Elsevier. https://doi.org/10.1016/b978-0-08-100866-9.00005-4.

Sethi, Swati, Poonam Choudhary, Prerna Nath, and O. P. Chauhan. 2022. "Starch Gelatinization and Modification." In *Advances in Food Chemistry*, 65–116. Springer Nature Singapore. https://doi.org/10.1007/978-981-19-4796-4_3.

Shahrizan, Muhammad, Zul Hadif Abd Syakir Mohd, and Haliza Aziz. 2022. "Fluid Gels: A Systematic Review Towards Their Application in Pharmaceutical Dosage Forms and Drug Delivery Systems." *Journal of Drug Delivery Science and Technology* 67: 102947.

Silva, Aline R. A., Marselle M. N. Silva, and Bernardo D. Ribeiro. 2020. "Health Issues and Technological Aspects of Plant-Based Alternative Milk." *Food Research International (Ottawa, Ont.)* 131(108972): 108972. https://doi.org/10.1016/j.foodres.2019.108972.

Sindhu, Ritu, Amita Devi, and B. S. Khatkar. 2021. "Morphology, Structure and Functionality of Acetylated, Oxidized and Heat Moisture Treated Amaranth Starches." *Food Hydrocolloids* 118(106800): 106800. https://doi.org/10.1016/j.foodhyd.2021.106800.

Singh, Meenakshi, Nitin Trivedi, Manoj Kumar Enamala, Chandrasekhar Kuppam, Punita Parikh, Maria P. Nikolova, and Murthy Chavali. 2021. "Plant-Based Meat Analogue (PBMA) as a Sustainable Food: A Concise Review." *European Food Research and Technology* 247(10): 2499–2526. https://doi.org/10.1007/s00217-021-03810-1.

Singla, D., A. Singh, S. B. Dhull, P. Kumar, T. Malik, and P. Kumar. 2020. "Taro Starch: Isolation, Morphology, Modification and Novel Applications Concern - A Review." *International Journal of Biological Macromolecules* 163: 1283–1290.

Skendi, Adriana, and Maria Papageorgiou. 2019. "Low Glycemic Index Ingredients and Modified Starches in Food Products." In *The Role of Alternative and Innovative Food Ingredients and Products in Consumer Wellness*, 167–195. Elsevier. https://doi.org/10.1016/b978-0-12-816453-2.00006-1.

Šmídová, Zuzana, and Jana Rysová. 2022. "Gluten-Free Bread and Bakery Products Technology." *Foods (Basel, Switzerland)* 11(3). https://doi.org/10.3390/foods11030480.

Suárez-González, Javier, María Magariños-Triviño, Eduardo Díaz-Torres, Amor R. Cáceres-Pérez, Ana Santoveña-Estévez, and José B. Fariña. 2021. "Individualized Orodispersible Pediatric Dosage Forms Obtained by Molding and Semi-solid Extrusion by 3D Printing: A Comparative Study for Hydrochlorothiazide." *Journal of Drug Delivery Science and Technology* 66(102884): 102884. https://doi.org/10.1016/j.jddst.2021.102884.

Szakály, Zoltán, and Marietta Kiss. n.d. "Consumer Acceptance of Different Cereal-Based 'Healthy Foods'." In *Developing Sustainable and Health Promoting Cereals and Pseudocereals*, 467–488. Elsevier.

Tao, Keyu, Cheng Li, Yu Wenwen, Robert G. Gilbert, and Enpeng Li. 2019. "How Amylose Molecular Fine Structure of Rice Starch Affects Functional Properties." *Carbohydrate Polymers* 204(January): 24–31. https://doi.org/10.1016/j.carbpol.2018.09.078.

Thakur, Priyanka, Krishan Kumar, and Harcharan Singh Dhaliwal. 2021. "Nutritional Facts, Bio-active Components and Processing Aspects of Pseudocereals: A Comprehensive Review." *Food Bioscience* 42(101170): 101170. https://doi.org/10.1016/j.fbio.2021.101170.

Thakur, Rahul, Penta Pristijono, Christopher J. Scarlett, Michael Bowyer, S. P. Singh, and Quan V. Vuong. 2019. "Starch-Based Films: Major Factors Affecting Their Properties." *International Journal of Biological Macromolecules* 132(July): 1079–1089. https://doi.org/10.1016/j.ijbiomac.2019.03.190.

Torbica, Aleksandra, Miloš Radosavljević, Miona Belović, T. Tamilselvan, and Pichan Prabhasankar. 2022. "Biotechnological Tools for Cereal and Pseudocereal Dietary Fibre Modification in the Bakery Products Creation-Advantages, Disadvantages and Challenges." *Trends in Food Science and Technology* 129: 194–209.

Tsatsaragkou, Kleopatra, Styliani Protonotariou, and Ioanna Mandala. 2016. "Structural Role of Fibre Addition to Increase Knowledge of Non-gluten Bread." *Journal of Cereal Science* 67(January): 58–67. https://doi.org/10.1016/j.jcs.2015.10.003.

Ugural, Aysegul, and Aslı Akyol. 2022. "Can Pseudocereals Modulate Microbiota by Functioning as Probiotics or Prebiotics?." *Critical Reviews in Food Science and Nutrition* 62(7): 1725–1739. https://doi.org/10.1080/10408398.2020.1846493.

Wang, Shujun, and Les Copeland. 2015. "Effect of Acid Hydrolysis on Starch Structure and Functionality: A Review." *Critical Reviews in Food Science and Nutrition* 55(8): 1081–1097. https://doi.org/10.1080/10408398.2012.684551.

Wang, Yaqin, Ndegwa Henry Maina, Rossana Coda, and Kati Katina. 2021. "Challenges and Opportunities for Wheat Alternative Grains in Breadmaking: Ex-Situ-versus In-Situ-Produced Dextran." *Trends in Food Science and Technology* 113: 232–244.

Wheate, Nial J., and Christina Limantoro. 2016. "Cucurbit [n] Urils as Excipients in Pharmaceutical Dosage Forms." *Supramolecular Chemistry* 28(9): 849–856.

Woomer, Joseph S., and Akinbode A. Adedeji. 2021. "Current Applications of Gluten-Free Grains-A Review." *Critical Reviews in Food Science and Nutrition* 61(1): 14–24.

Xing, Bao, Cong Teng, Menghan Sun, Qinping Zhang, Bangwei Zhou, Hongliang Cui, Guixing Ren, Xiushi Yang, and Peiyou Qin. 2021. "Effect of Germination Treatment on the Structural and Physicochemical Properties of Quinoa Starch." *Food Hydrocolloids* 115(106604): 106604. https://doi.org/10.1016/j.foodhyd.2021.106604.

Yadav, Baljeet S., Ritika B. Yadav, Manisha Kumari, and Bhupender S. Khatkar. 2014. "Studies on Suitability of Wheat Flour Blends with Sweet Potato, Colocasia and Water Chestnut Flours for Noodle Making." *Lebensmittel-Wissenschaft Und Technologie [Food Science and Technology]* 57(1): 352–358. https://doi.org/10.1016/j.lwt.2013.12.042.

Yang, Tianyi. 2016. *Development of Gluten-Free Wrap Bread: A Thesis Submitted in Partial Fulfilment of the Requirements for the Degree of Master of Food Technology*, Massey University, Albany, New Zealand. https://mro.massey.ac.nz/handle/10179/10726

Younis, Kaiser, Owais Yousuf, Ovais Shafiq Qadri, Kausar Jahan, Khwaja Osama, and Rayees Ul Islam. 2022. "Incorporation of Soluble Dietary Fiber in Comminuted Meat Products: Special Emphasis on Changes in Textural Properties." *Bioactive Carbohydrates and Dietary Fibre* 27(100288): 100288. https://doi.org/10.1016/j.bcdf.2021.100288.

Yupeng, Marlene E., Busarawan Janes, Margaret A. Chaiya, Charles S. Brennan, and Witoon Brennan. 2018. "Gluten-Free Bakery and Pasta Products: Prevalence and Quality Improvement." *International Journal of Food Science and Technology* 53(1): 19–32.

Zhou, Yiming, Qingyi Jiang, Sijia Ma, and Xiaoli Zhou. 2021. "Effect of Quercetin on the In Vitro Tartary Buckwheat Starch Digestibility." *International Journal of Biological Macromolecules* 183(July): 818–830. https://doi.org/10.1016/j.ijbiomac.2021.05.013.

Zhu, Fan. 2016. "Buckwheat Starch: Structures, Properties, and Applications." *Trends in Food Science and Technology* 49(March): 121–135. https://doi.org/10.1016/j.tifs.2015.12.002.

Žuljević, Sanja, and Asima Oručević. 2021. "Flour-Based Confectionery as Functional Food." *Functional Foods-Phytochemicals and Health Promoting Potential* 351–377.

Chapter 7

Properties and Applications of Proteins from Pseudocereals

Indumathi Mullaiselvan, Puja Nelluri,
Baghya Nisha, Debapam Saha

7.1 Introduction

Pseudocereals are a type of grain that are edible seeds originating from dicotyledonous plants. They earned the name "pseudocereals" because they resemble true cereals in their physical appearance and possess a high starch content, similar to the grains from the Poaceae family, which are monocotyledonous. The proteins derived from seeds like quinoa, amaranth, chia, and teff have gained recognition as promising sources of plant-based protein for human consumption. Pseudocereals are often referred to as "sub-exploited or underutilized crops" (Alvarez-Jubete et al., 2010; Malik et al., 2023). Pseudocereals and their proteins can be utilized for numerous food applications which include the development of packaging materials for various food products, a replacement for animal-based proteins, and a source of edible and biodegradable packages (Ciudad-Mulero et al., 2019). Since the biological values of proteins in the pseudocereals are very high, it could be a possible alternative for a nutritious supplement for countries with less valuable protein sources. This chapter reviews and explains the properties and applications of pseudocereal proteins.

DOI: 10.1201/9781003325277-7

7.2 Proteins from Pseudocereals

7.2.1 Quinoa Protein

Quinoa (*Chenopodium quinoa*) is an annual plant of the Amaranthaceae family that originally hails from regions in Ecuador but is now cultivated in many parts of the world, particularly in Asia and Europe. Its significant nutritional properties have made it a popular choice among farmers and consumers alike.Quinoa has gained recognition as a valuable plant-based protein source, as it contains all the essential amino acids and peptides required by the body (Nascimento et al., 2014; Aluko & Monu, 2003). Over time, the cultivation of quinoa has seen a considerable increase, and its consumption has surged tenfold, driven by numerous studies that have highlighted its nutritional and functional benefits (Gonzalez et al., 2015). It is reported that quinoa contains higher fractions of albumin (29–50%) and globulins (7–39%). The 11S-type of quinoa-specific globulin called chenopodin is present in quinoa seeds at concentrations up to 37%. The lowest-frequency fraction found in quinoa protein is prolamins (2%).

7.2.2 Amaranth Protein

Amaranth (*Amaranthus* spp.) are annual plants primarily cultivated in South Africa and India. It belongs to the Amaranthaceae family and is known to possess comparable nutritional qualities to quinoa and buckwheat (Day, 2013). Amaranth is reported to be adaptable to both tropical and subtropical climates. The grains and leaves of the amaranth plant possess high nutritional and functional value. Amaranth is reported to have high concentrations of bioactive compounds which confer various health benefits. Amaranth grains are highly protein-rich, constituting 60% of their total seed composition. The grains contain a combination of albumins and globulins, offering a well-balanced amino acid profile. Notably, amaranth is particularly rich in essential amino acids like methionine and lysine, which are often limited in many other true cereal grains. Because the protein percentage is very high in amaranth, it could be a potential alternative source for animal feed.

7.2.3 Chia Protein

Chia (*Salvia hispanica*) originated from regions of South America and belongs to the Lamiaceae family. The added value of chia seeds to drinks and desserts is becoming more common among people throughout the world (Punia & Dhull, 2019). Chia seeds are greatly favoured by health-conscious individuals due to the abundance of omega-3 fatty acids and high-quality protein. These seeds are classified as oilseeds and have a significant lipid content, primarily composed of omega-3 fatty acids (up to 60%) (Dhull & Punia, 2020a, 2020b; Dhull et al., 2020). Additionally, chia seeds boast high levels of protein (15–24%), dietary fiber (18–30%), and carbohydrates (26–41%). The predominant protein component in chia seeds is globulin, which accounts for up to 52% of the total seed protein and is composed of 11S and 7S protein fractions. Chia seed proteins also have an exceptional balance of amino acids, with high concentrations of methionine, cysteine, and lysine, all of which are 10-fold higher than in the true cereals.

7.3 Protein Extraction and Isolation

Pseudocereals, unlike monocotyledonous cereals, are dicotyledonous plants. They derive their name from the resemblance of their grains to those of cereals. The protein content of pseudocereals ranges between 10 and 20%. Protein extraction from pseudocereals is becoming increasingly essential as they are employed in many food formulations. Proteins are primarily isolated by dissolving the protein source in a high-pH environment, typically away from its isoelectric point, and subsequently precipitating the solubilized proteins by adjusting the pH toward their isoelectric point. In the case of pseudocereals, the separation of proteins has mostly been achieved by utilizing alkalis like NaOH, which efficiently break down the proteins present in the seeds.

7.3.1 Methods of Protein Isolation

Proteins in the pseudocereals can be either intracellular or extracellular, and they can be retrieved from various sites such as the membrane, cytoplasm, and nucleus. In order to extract the proteins enclosed within these parts of the cells, various methods can be employed. These include treating the cells with chemical agents or enzymes, or subjecting them to external forces that induce changes, ultimately leading to the rupture of the cell membrane or the formation of tiny pores in the cell wall. These alterations enable the release of the cytosol, containing proteins and other biomolecules, from within the cells.

7.3.1.1 Primary Isolation

This step involves the separation of proteins through different methods:

1. Concentration
2. Cell disruption
3. Refolding

7.3.1.1.1 Concentration (Extracellular Proteins)

For extracellular protein concentration, various techniques can be used. Extracellular proteins from cells are typically concentrated via ultrafiltration. Hydrostatic pressure is used to force the protein solution across a selectively permeable membrane; ultrafiltration is a type of membrane filtration. While molecules with a high molecular weight are confined in the membrane, water and low-molecular-weight solutes pass through. The surface of the membrane contains adsorbed protein molecules, which are separated and purified.

7.3.1.1.2 Cell Disruption

The cell disruption techniques for extracting proteins from pseudocereals can be divided into two categories: mechanical methods and non-mechanical methods (Figure 7.1).

7.3.1.2 Non-mechanical Methods

Non-mechanical cell disruption methods, also referred to as "chemical methods", involve the application of acids, alkalis, or organic solvents to treat the cells. These methods utilize different mechanisms, such as enzyme activity, osmotic pressure, or protein interference and precipitation within the cell wall.

7.3.1.2.1 Chemical Methods

In this method, chemicals or solvents are used for protein extraction. Cells or tissues are immersed in an alkaline solution with a pH ranging from 10 to 12. Within a short period of time, this alkaline solution effectively breaks down the cell membrane, enabling the release of the cytosol from the cells (Figure 7.2).

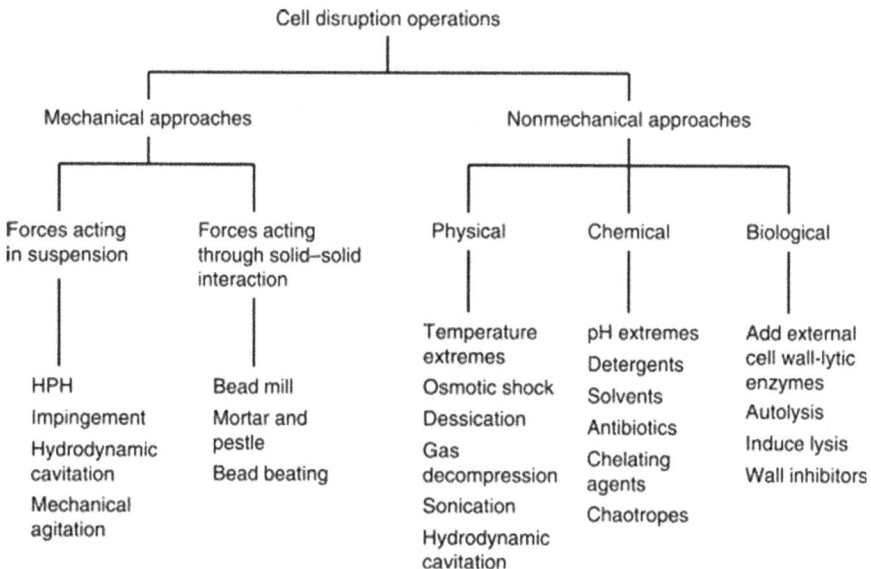

Figure 7.1 Different methods for extracting proteins.

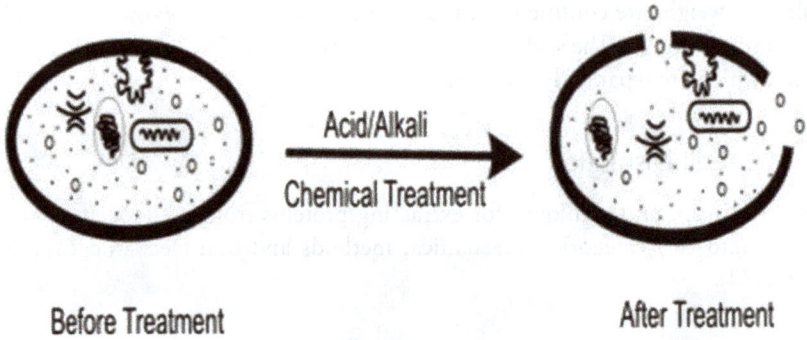

Figure 7.2 Chemical treatment (Wang et al., 2021).

Various organic solvents, including alcohols, dimethyl sulfoxide, methyl ethyl ketone, and toluene, can be utilized for this purpose. These solvents also exert an effect on the outer membrane of the cell, causing it to dissolve and facilitating the release of the cytosol and other cellular components from within the cell.

7.3.1.2.2 Treatment with Detergent

Detergents are used in affinity purification, immunoassay, and electrophoresis as well as to solubilize membrane proteins and lipids, release soluble proteins, prevent protein crystallization, and reduce nonspecific binding. Examples include sodium dodecyl sulfate (SDS), Triton X-100, ethyltrimethylammonium bromide, etc. (Figure 7.3).

7.3.1.2.3 Treatment with Enzymes

Another method for achieving cell lysis is to use digestive enzymes that dissolve the cell wall. Different enzymes are required for disruption of different cell types and

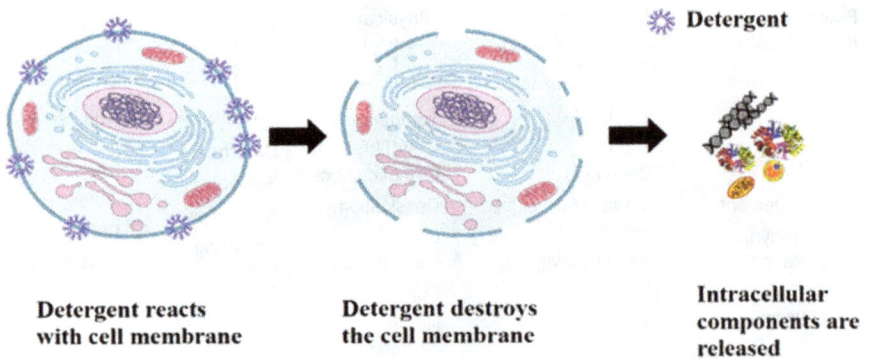

Detergent reacts with cell membrane

Detergent destroys the cell membrane

Intracellular components are released

Figure 7.3 Detergent treatment (Wang et al., 2021).

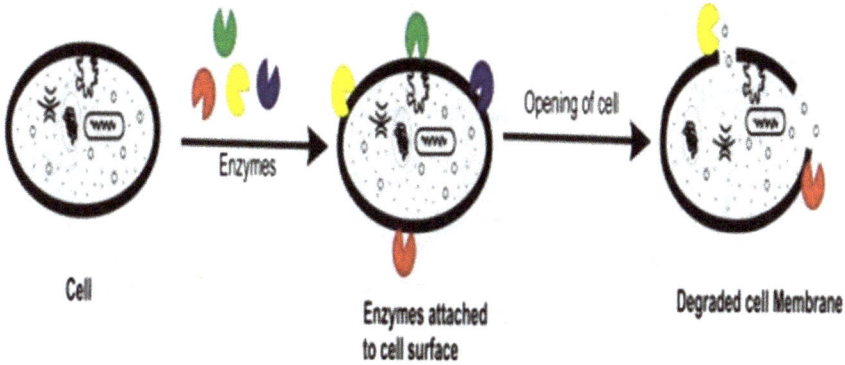

Figure 7.4   Enzyme treatment (Wang et al., 2021).

membranes. Enzymes, including xylanases, pectinases, and chitinases, are commonly employed to disrupt the cell walls of fungi and yeast cells. In addition, lysozyme is utilized to break down proteins in microbial cell walls. Cellulases are well-known for their ability to break down cellulose present in plant cell walls. These enzymes act on the surface of the plant cell, effectively activating their enzymatic activity (Figure 7.4).

After the appropriate incubation period, the plant's cell wall is broken down, allowing various biomolecules, including proteins, to exit the plant cell along with internal subcellular organelles.

7.3.1.2.4. Osmotic Shock

The "osmosis" (or the diffusion of water across a selectively permeable membrane) theory is the foundation of this technique. When a cell is immersed in a hypertonic solution, water will exit the cell, causing it to shrink or contract (a process known as exo-osmosis). This occurs because water naturally moves from an area of lower solute concentration to an area of higher solute concentration, following the concentration gradient. As a result, pressure inside the cell increases and the cell explodes (Figure 7.5).

7.3.1.3 Mechanical Methods

The two types of mechanical extraction methods, solid shear and liquid shear, can be generally classified. Using either solid particles or liquid layers, these two techniques produce external stress and movement. Cells are subjected to high pressures in either types of cell disruption technique, which are produced by physical forces like rapid bead agitation or external forces like ultrasonic waves.

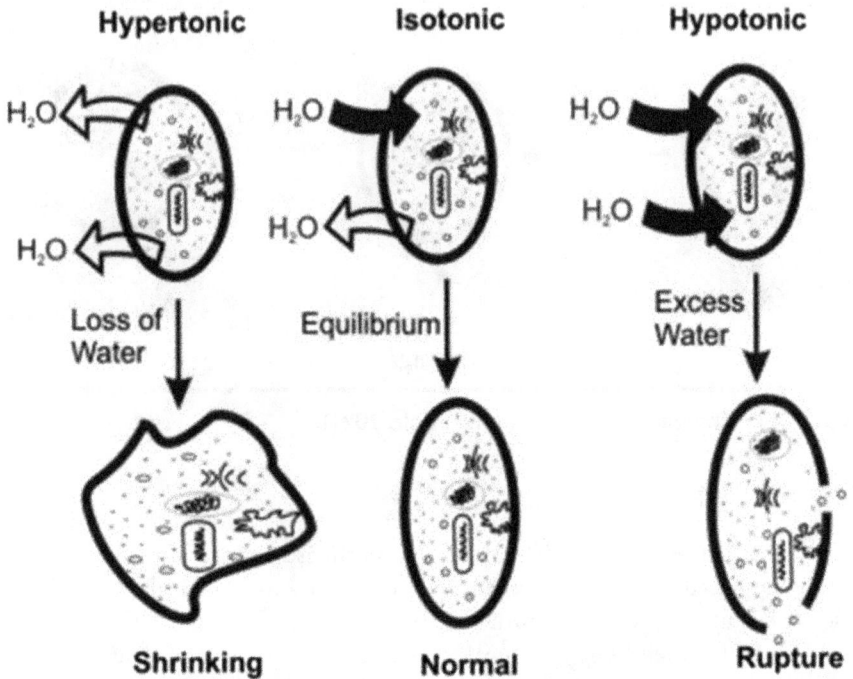

Figure 7.5   Osmotic shock treatment (Wang et al., 2021).

7.3.1.3.1 Solid Disruption Methods

7.3.1.3.1.1 Bead Mill Method   A small revolving cylinder is half filled with tiny beads in the device. These beads are often made of zirconia, a strong and durable zirconium-based material. When rotated clockwise, the cylinder will convert kinetic energy into mechanical energy in the beads it holds. The beads and cells inside the rolling cylinder collide as a result of mechanical energy, producing a shear force and causing cell disruption and protein elution (Figure 7.6).

7.3.1.3.1.2 French Press   This type of disruption is driven by the pressure that the piston produces. Cells will explode if there is a sudden drop in pressure inside the steel cylinder or if the pressure inside the cylinder changes. The pressure used in this procedure can reach 1500 bar. The piston applies pressure, and when it suddenly releases pressure, the cells are under some stress, which causes them to burst. With the outlet tube open, the cell extract will elute.

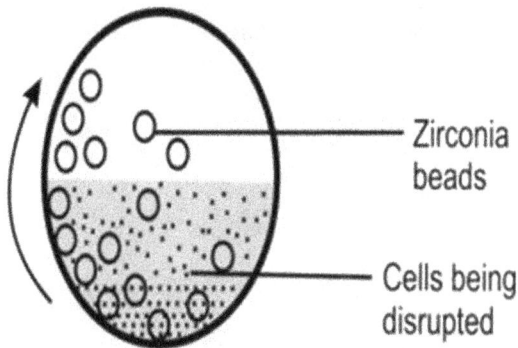

Figure 7.6   Disruption using bead mill (Lopez et al., 2018).

7.3.1.3.2 Liquid Disruption Methods

7.3.1.3.2.1 Homogenization   In various industries, this is one of the extensively employed mechanical cell disruption methods that involves applying external pressure at approximately 1200 PSI. This pressure generates shear forces as cells pass through narrow tunnels. Upon reaching the opposite end of the tunnel, the cells experience expansion, resulting in a phenomenon known as cavitation, where cavities form inside the cells. This cavitation event further leads to the disruption of the cells.

7.3.1.3.2.2 Ultrasonication   High-frequency sound waves are used in ultrasonic cell lysis and extraction to damage cells and extract their contents. Cavitation is the process by which sound waves cause tiny bubbles to grow and burst in the liquid that they are in contact with. These bubbles produce highly concentrated mechanical forces that are localized, which have the power to rupture cells and release their contents into the surrounding solution (Figure 7.7).

7.3.1.3.2.3 Freeze–Thaw   The described process is referred to as temperature-induced lysis, as rapid temperature fluctuations lead to cell disintegration. To carry out this procedure, cells are first immersed in a suitable cell storage solution or buffer. Subsequently, they are placed in specifically prepared vials (small containers or Eppendorf tubes) and exposed to extremely low temperatures using dry ice, ethanol, or deep freezers. These vials are removed from the freezer and gradually thawed (defrosted) at room temperature. By doing so, the temperature will rise and the cells will be disrupted by the sudden change in temperature. Continuous thawing, on the other hand, is not recommended because it raises the temperature and denatures heat-sensitive components such as enzymes, proteins, and nucleic acids (Figure 7.8).

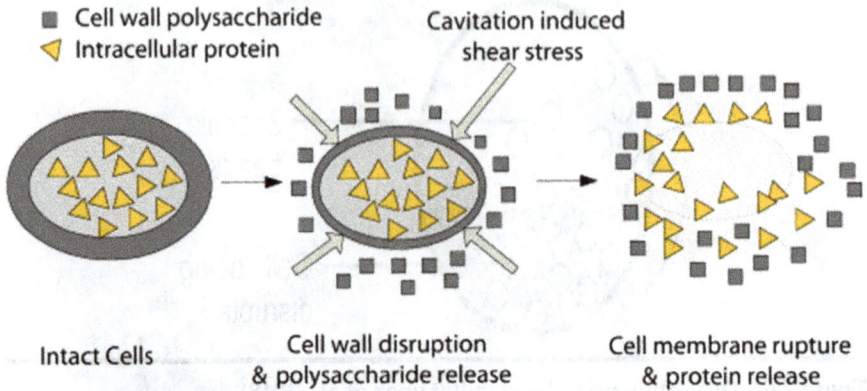

Figure 7.7 **Disruption using ultrasonication (Lopez et al., 2018).**

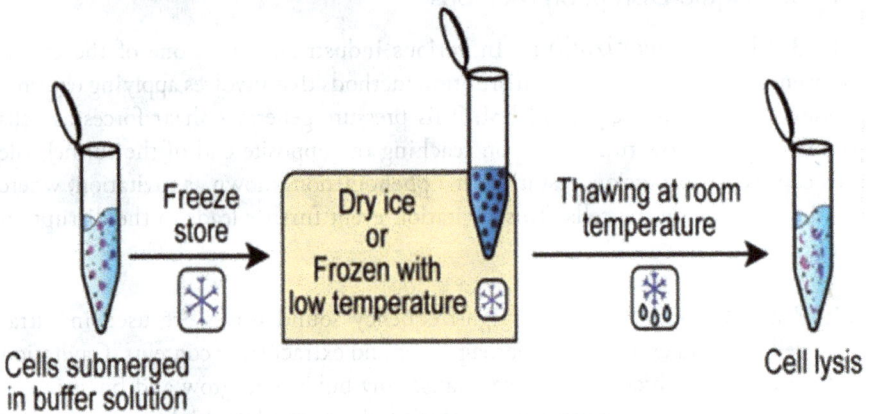

Figure 7.8 **Disruption using freeze-thaw (Lopez et al., 2018).**

7.3.1.3.2.4 Ultracentrifugation   When particles of various sizes or densities are floating in a liquid, centrifugation is a technique used to separate them, using the force of spinning. Each particle experiences an outward force proportional to its mass when a protein mixture is swiftly spun in a tube or bottle. This force may lead a particle to attempt to pass through a liquid. The fluid, though, resists the particle and pushes it away. As a result, small, dense, and heavy particles flow outward more quickly than less dense, lighter, or particles that "drag" more in the liquid, which are also known as less dense or heavier particles.

7.3.1.1.3 Refolding

The initial stage of the refolding process involves dissolving the inclusion bodies, obtained through concentration, in a robust chaotropic solution containing 6 M urea and 2 M thiourea. Chaotropic agents act as denaturing agents, disrupting the intramolecular forces between water molecules and facilitating the dissolution of proteins and other macromolecules. Subsequently, the denatured protein is renatured by removing the chaotropic agent through methods such as dilution, dialysis, or chromatographic separation.

7.4 Protein Purification

Protein purification is a difficult and constantly changing process. Each separation is diverse due to the variety of molecular structures, characteristics, and amounts of both valuable components and their impurities. Protein purification protocols have a wide range of sizes. Plant tissues contain a variety of proteins, a robust cell wall comprised of cellulosic material, and phenolic chemicals that can precipitate proteins, all of which prevent the production of a pure plant protein through a single-step purification procedure. Purification of a protein depends on the following factors:

- The selection of the specific source material and the location of the target protein (whether intracellular or extracellular)
- The quantity of protein needed determines the amount of raw material that needs to be processed
- Physical, chemical, and biological properties of the protein

7.4.1 Chromatography for Protein Purification

In most protein purification methods, one or more chromatographic stages are employed. The initial step involves passing the protein solution through a column that contains diverse components. Different proteins exhibit specific interactions with the material present in the column. This means that the time it takes for a protein to transit through the column, as well as the conditions it requires to exit the column, can be utilized to separate the proteins. Their absorbance at 280 nm is commonly employed to detect proteins as they exit the column.

7.4.1.1 Size-Exclusion Chromatography

Gel filtration, also known as size-exclusion chromatography, is a widely used method for protein separation based on size or molecular weight. The column consists of a densely packed porous matrix, which leads to a delay in the elution of proteins. As the matrix contains small pores, smaller proteins can pass through and elute later, while proteins with higher molecular weight elute first. By using porous gels, chromatography can be used to separate proteins in solution or under conditions that cause them to break down.

7.4.1.2 Ion-exchange Chromatography

Depending on the electric property of the proteins, cations or anions are coupled to resin beads in ion-exchange chromatography. In the case of a negatively charged protein, positively charged resin beads are utilized in the column. The resin beads are densely packed within the column. Upon introducing the sample onto the column, the desired negatively charged protein adheres to the resin beads, while positively charged proteins elute first. By adjusting the pH of the wash buffer or using a high salt solution during the washing process, the desired negatively charged protein is subsequently eluted, resulting in its partial purification and collection.

When the sample is introduced onto the resin, the solute molecules exchange places with the buffer ions in an attempt to bind to the resin. The amount of time each solute is retained is determined by the strength of its charge. During ion-exchange chromatography, compounds with the lowest charge are the first to elute, followed by those with higher charges. The separation process is influenced by factors such as pH, buffer type, buffer concentration, and temperature. These parameters play crucial roles in controlling the separation process. Ion-exchange chromatography is an effective method for removing contaminants from proteins and is commonly employed for both analytical and practical purposes in compound separation.

7.4.1.3 Affinity Chromatography

Affinity chromatography involves introducing a substance with a specific affinity for the target protein onto the resin, such as coating the resin with an antibody against the desired protein. The resin, along with the bound antibody, is packed into a column. When a protein mixture is applied to the column, only the proteins that possess a special affinity for the resin (due to their interaction with the antibody) adhere to the column, while the remaining proteins are washed away. Specifically, the unwanted proteins are eluted, leaving behind the desired protein bound to the column. To release the target protein from the column, the ionic strength of the elution solution can be adjusted to disrupt the binding between the desired protein and the resin. Alternatively, a specific chemical can be included in the elution solution to alter the equilibrium state and facilitate the elution of the protein of interest.

Affinity chromatography is a method of separating products based on the geometry of molecules. Only specific resins are frequently utilized. These resins have ligands on their surfaces that are unique to the substances that need to be separated. Most of the time, these ligands interact in the same way as antibodies and antigens do. Many membrane proteins are glycoproteins, and they can be purified using lectin affinity chromatography.

7.4.1.4 High-performance Liquid Chromatography

High-performance liquid chromatography (HPLC), also referred to as high-pressure liquid chromatography, utilizes elevated pressure to enhance the movement of solutes through the column, resulting in faster elution. Since there are fewer instances of spreading as a result, the resolution is improved. The most common HPLC method is "reversed phase" because the substance in the column dislikes water. To extract the proteins, a gradient of progressively higher concentrations of an organic solvent, such as acetonitrile, is utilized. Depending on their solubility in water, the proteins will emerge. The protein is in a solution containing only volatile components following HPLC purification. HPLC purification frequently results in the proteins losing their form, hence it cannot be utilised on proteins that do not refold on lyophilization.

7.4.1.5 Hydrophobic Chromatography

Hydrophobic interaction chromatography is employed in the purification of proteins by utilizing their surface hydrophobicity. Proteins possess dispersed hydrophobic residue groups on their surface, contributing to their unique characteristics. In an aqueous solution, these hydrophobic groups are shielded by a layer of water. However, the addition of salt exposes the hydrophobic groups, leading to their interaction. Hydrophobic interaction chromatography involves filling the column with hydrophobic beads, such as those containing phenyl or acetyl groups. As the sample is introduced into the column, the hydrophobic proteins interact with the hydrophobic matrix. The elution of proteins is achieved by employing a salting-out chemical, such as ammonium sulfate, at decreasing concentrations, resulting in the elution of proteins with lower hydrophobicity first. Non-ionic detergents, such as Tween-20, Triton X-100, or others, are utilized to facilitate elution.

7.5 Protein Characterization

There are several methods of protein characterization.

7.5.1 Electrophoresis

SDS-PAGE (sodium dodecyl sulfate-polyacrylamide gel electrophoresis) is a widely utilized technique for separating charged particles in the presence of an electric field. It is particularly valuable for analyzing protein mixtures and monitoring protein purification processes. SDS-PAGE separates proteins based on their molecular weight. The procedure involves the creation of a polyacrylamide gel with a well. The protein sample is treated by boiling it for 5 minutes in the presence of β-mercaptoethanol and sodium dodecyl sulfate (SDS). Proteins are denatured during the boiling process. Each SDS molecule interacts with two denatured protein amino acid molecules. Due to its highly negative charge, the addition of SDS molecules to the protein renders the protein negatively charged as well. During electrophoresis, the negatively charged protein migrates towards the positively charged anode. Smaller proteins migrate quicker, while larger proteins move slower, resulting in various bands. By dyeing the gel with Coomassie brilliant blue (CBB), the protein bands can be seen.

7.5.2 Peptide Sequencing

This process known as Edman degradation, developed by Pehr Edman, involves the sequential determination of amino acid residues in a polypeptide chain. Under mild alkaline conditions, the polypeptide reacts with phenylisothiocyanate, resulting in the transformation of the amino terminus of the peptide to a phenylthiocarbamoyl (PTC) derivative. The PTC derivatives are then washed with an organic solvent, such as benzene, and dried. Subsequently, the dried PTC derivative is treated with anhydrous acid, such as heptafluorobutyric acid, which cleaves the PTC-polypeptide near the PTC substitution. This cleavage releases the N-terminal amino acid as a thioazoline derivative. The thioazoline derivative is stable and further transformed into a thiohydantoin derivative containing the specific amino acid. The identification and quantification of the amino acid are achieved using HPLC.

7.5.3 Tryptic Mapping

To determine the amino acid sequence from a protein's N-terminus, the Edman degradation method needs a free amino group. However, after post-translational modification, formyl, acetyl, or acryl groups block the N-terminus of 50–70% of proteins. Since sequencing such a protein is not conceivable, endopeptidases have been used to break the protein down into a peptide, which is then sequenced. An example of an endopeptidase that cleaves the C-terminal of lysine and arginine is the trypsin enzyme. Other endopeptidases also have unique, constrained cleavage sites. The small fragment that is produced is sequenced using the Edman method and then the amino acids are identified using HPLC.

7.5.4 Analytical Ultracentrifugation

This technique analyzes the homogeneity of the collected material as well as a number of features of the protein sample, such as molecular weight and interactions with other molecules.

7.5.5 Spectroscopy

It is used to analyze a wide variety of samples. Spectroscopy can be used to evaluate proteins that contain metals (co-factors). The electromagnetic spectrum varies depending on the co-factor.

7.5.6 Biosensors

■ A biosensor is an instrument employed for the detection of specific proteins within cells.
■ The device is equipped with specific antibodies that specifically bind to the desired protein.
■ When a sample is added, the specific protein binds to the antibodies, resulting in a detectable signal on the device.

7.5.7 Mass Spectrometry

■ Mass spectrometry is an analytical technique that offers valuable insights into the molecular structure of both organic and inorganic compounds.
■ The mass spectrometry technique enables the determination of the mass of a specific protein.
■ It is capable of detecting post-translational modifications or any structural variations in proteins.

7.6 Properties of Pseudocereal Proteins

Proteins from pseudocereals possess many functional properties owing to their native characteristics like size, tertiary structure, surface charge, and amino acid sequence and composition (Timilsena et al., 2016). The functional properties also depend upon many factors like pH, method of extraction, solvent or salt, temperature, and physical, chemical, or enzymatic modification methods (Klupsaite & Juodeikiene, 2015). The proteins have key functional qualities such as solubility, emulsification, foaming, gelling, rheological properties, and absorption. Owing to the versatile functionality of proteins, they have numerous applications in the food, cosmetic, and pharmaceutical industries (Agarwal et al., 2020; Santhakumar et al., 2022). The different properties of pseudocereal proteins will be discussed briefly in the following subsections.

7.6.1 Physical Properties

7.6.1.1 Color

Color plays a crucial role in the food industry as it influences the acceptability and visual appeal of the final product, making it an important property closely tied to the raw materials used. The color of the proteins can be measured with a colorimeter and can be quantified in terms of L*, a* and b*. Cerdán-Leal et al. (2020) found that L*, a*, and b* of isolated quinoa proteins varied from 62.42 to 69.18, 4.44 to 5.20, and 21.18 to 28.34, respectively. They found that freeze-dried isolated quinoa proteins gave a better color when compared to other methods. Isolated amaranth proteins were found to possess more lightness index (82 to 87.22) and less yellowish than quinoa (Shevkani et al., 2014), whereas, in comparison with amaranth and quinoa, isolated chia seed protein exhibited a higher degree of redness (Timilsena et al., 2016). Canary seed (*P. canariensis*) protein was reported to have L*, a*, and b* values of 84.40, 0.23 and 12.94, respectively (Perera et al., 2022). Isolated buckwheat protein was found to be yellowish in color than canary seed protein (Jin et al., 2022). The color of pseudocereals is also influenced by factors such as the specific cultivar, variety, and source of the grains (Lopez et al., 2019).

7.6.1.2 Particle Size

Particle size greatly affects the functionality of the pseudocereal proteins. It can be analyzed by using different techniques, like laser diffraction particle size analyzer, dynamic light scattering technique, and many others (He et al., 2022; Constantino et al., 2020). The hydrodynamic diameters of quinoa, amaranth, chia and buckwheat protein were 7.37 to 49.69 nm, 414.40 to 478.67 d nm, 29.5 to 32 μm and 427.7 to 2663.3 nm, respectively. Isolation method and amino acid composition were found to significantly affect the particle size distribution (Shen et al., 2021; Jin et al., 2022).

7.6.1.3 Zeta Potential

The zeta potential of proteins usually depends on the surface charge present on the protein molecules. It is usually measured by suspending the proteins in solutions followed by pH adjustment and then analysis using a Zetasizer. Protein solubility and other functional properties are significantly influenced by their zeta potential. In a study conducted by Shevkani et al. (2014), the zeta potential of isolated amaranth proteins was observed to range between –40.1 and –43.7 mV. Similarly, Jin et al. (2022) reported that the zeta potential of isolated buckwheat protein fell within the range of –48.7 to –58.7 mV. Another study by Lopez et al. (2018a) indicated that isolated chia protein exhibited a zeta potential of approximately –54 mV.

7.6.2 Functional Properties

7.6.2.1 Foaming Properties

Foaming is an essential functional property of proteins which is associated with their solubility, film formation, and hydrogen bonding (Klupsaite & Juodeikiene, 2015). Foaming capacity refers to the protein's capability to create a stable and resilient foam when mechanically dispersed in a solution. Protein isolates of amaranth exhibited the highest foaming capacity of 250% (pH 7) followed by chia (130% at pH 11), with the lowest foaming capacity being shown by quinoa (50% at pH 8), whereas the highest foam stability was exhibited by chia (100 % at pH 11), followed by amaranth (95% at pH 6) and lowest by quinoa (68% at pH 4) (Lopez et al., 2019). Canary seed protein isolate exhibited a foaming capacity of 28.33% and foam stability of 74.67% (Perera et al., 2022).

7.6.2.2 Emulsifying Properties

Emulsifying activity (EAI) and stability index (ESI) are the major parameters which indicate the emulsifying behavior of a protein. They quantify the amount of emulsion formed per total volume and its stability after a prescribed time. The EAI of quinoa, amaranth, chia, canary seed, buckwheat protein isolates were 61.01% (at pH 8), 80% (at pH 2), 55% (at pH 8), 257.29 m^2/g (at pH 8) and 550 m^2/g (at pH 13), respectively. The ESI of quinoa, amaranth, chia, canary seed, and buckwheat proteins (all at pH 8) were 57.98%, 100%, 98.2%, 88.67%, and 300 min, respectively (Lopez et al., 2019; Perera et al., 2022; Wu et al., 2021).

7.6.2.3 Gelling Properties

Gel formation is also an essential functional property of proteins. Generally, proteins form different kinds of gels in response to heat treatment, which usually depends on water content, temperature, pH, type of protein and its concentration (Klupsaite & Juodeikiene, 2015). The least gelation concentration is a parameter that measures the minimum protein concentration needed to form a gel upon heating. According to Ruiz et al. (2016), freeze-dried quinoa protein exhibited a more robust gel compared with spray-dried quinoa protein. Proteins with higher solubility exhibited stronger gels (Shen et al., 2021). Avanza et al. (2005) reported that the least gelation concentration of amaranth protein isolate was 7% (w/v) at a temperature of 70°C. Chia seed protein isolate was found to have a least gelation concentration of 20% (w/v) (Oilvos-Lugo et al., 2010).

7.6.2.4 Absorption Properties

The water/oil absorption capacity of proteins refers to their ability to retain or bind water or oil. These properties play a significant role in determining the flavor and texture of the end products. The presence of polar and non-polar amino groups on the protein molecule's surface largely influences these properties (Lopez et al., 2018b). The water absorption capacity (WAC) of quinoa, amaranth, chia, canary seed and teff protein isolates was found to be 2.76, 3.3, 6.0, 2.52, 1.853 g of water/g of sample, respectively. The oil absorption capacity (OAC) of quinoa, amaranth, chia, canary seed and teff protein isolates was found to be 3.19, 6.4, 7.1, 3.59 g of oil/g of sample, 375% (Wang et al., 2022; Perera et al., 2022; Lopez et al., 2018b; Shen et al., 2021 and Argundade (2006)). WAC and OAC are also affected by the temperature, cultivar, method of extraction, and other factors.

7.6.2.5 Solubility

The solubility of proteins is a crucial property determined by the distribution of hydrophobic and hydrophilic sites on the protein molecule's surface. The functionality of a protein is closely linked to its solubility, which is influenced by factors such as pH, ionic strength, and processing conditions. Protein solubility of amaranth protein isolate was found to be highest (83.9%) at pH 2 and lowest (5.3%) at its isoelectric point, pH 5.0 (Shevkani et al., 2014). Wang et al. (2021) found that quinoa protein isolate exhibited high solubility (74.82%) at pH greater than 7.0 but low solubility at pH 3 and 5. Canary seed protein isolate demonstrated comparatively low solubility (49.3%) in comparison with other pseudocereal proteins, such as quinoa and amaranth (Perera et al., 2022). Buckwheat protein isolates exhibited excellent solubility of 88% at pH values ranging from 7 to 10 (Wu et al., 2021). Urbizo-Reyes et al. (2019) observed that the solubility of chia seed protein isolate (77.17%) increased with an increase in pH; solubility of chia seed protein was found to be highest at pH levels above or below the isoelectric point, primarily due to electrostatic and hydrophobic interactions.

7.6.2.6 Thermal Properties

Thermal properties are usually measured by differential scanning calorimetry. It presents endothermic peaks, which, in turn, gives insights into glass transition temperatures, denaturation temperatures, and others. The denaturation temperature of quinoa protein isolate was found to be in the range of 79.81 to 83.92°C (Mir et al., 2021). Shevkani et al. (2014) found that amaranth protein isolate had two denaturation temperatures, namely 69.2 and 96.6°C. Jin et al. (2022) found that the denaturation temperature of buckwheat protein isolate was 100.4°C. Lopez et al. (2018a) found that chia seed protein isolate has two endothermic peaks at 58.3 and 104.9°C.

7.7 Challenges and Future Perspectives of Pseudocereals

Different cultures have historically employed pseudocereals, and they have played a significant role in their diet and medicine. These grains contain an extensive amount of minerals and health-improving phytochemicals, including polyphenols, phytosterols, squalene, and saponins, making them extremely nutritious and therapeutic, with positive benefits on health. They are rich in essential and other important amino acids and contain high biological value proteins. Pseudocereals, in contrast to traditional cereals, are recognized as a rich source of essential micro and macronutrients. Their exceptional nutritional profile has earned them the reputation of being the "superfood for the future" due to their significant health benefits and superior nutritional value. Also, pseudocereals show a great deal of promise for use as gluten-free ingredients in the creation of gluten-free goods for people suffering with celiac disease or gluten intolerance. With regard to addressing hidden poverty and offering options for income generation, these underutilized crops have great promise for the functional food industry. Different proteins from pseudocereals are of great quality, but they are underutilized due to a lack of information; hence, further investigations on seed protein extraction, isolation, and purification are required.

Furthermore, research studies play a vital role in advancing our understanding of the compositional, structural, and functional properties of pseudocereal proteins. This valuable knowledge is essential for effectively utilizing these proteins in the development of stable and high-quality food formulations, both in terms of fundamental understanding and practical applications.

7.8 Conclusion

In conclusion, despite the nutritional advantages offered by specialty cereals, pseudocereals, and legumes, their utilization in food products remains limited. However, research has demonstrated the potential of these species in creating value-added food products. It is crucial to focus on developing food items with sensory appeal to encourage wider consumption of these products. With their high nutritional value, pseudocereals should be included in the daily diet of the global population. As we gain a better understanding of the nutritional and bioactive properties of these plants, the interest in these foods continues to grow. Pseudocereals, such as quinoa, amaranth, chia, and buckwheat, hold significant potential as natural sources of biologically active compounds, including peptides and protein hydrolysates that offer health benefits. Pseudocereal phytocompounds exhibit antioxidant and antihypertensive activities, and products based on pseudocereals can play a crucial role in improving the health and quality of life for individuals with celiac disease.

Bibliography

Abugoch, L., Castro, E., Tapia, C., Añón, M. C., Gajardo, P., & Villarroel, A. (2009). Stability of quinoa flour proteins (Chenopodium quinoa Willd.) during storage. *International journal of food science & technology*, 44(10), 2013–2020.

Agarwal, A., Pathera, A. K., Kaushik, R., Kumar, N., Dhull, S. B., Arora, S., & Chawla, P. (2020). Succinylation of milk proteins: Influence on micronutrient binding and functional indices. *Trends in Food Science and Technology*, 97, 254–264.

Agboola, S. O., & Aluko, R. E. (2009). Isolation and structural properties of the major protein fraction in Australian wattleseed (*Acacia victoriae* Bentham). *Food Chemistry*, 115(4), 1187–1193.

Alonso-Miravalles, L., & O'Mahony, J. A. (2018). Composition, protein profile and rheo-logical properties of pseudocereal-based protein-rich ingredients. *Foods, 7*(5), 73.

Aluko, R. E., & Monu, E. (2003). Functional and bioactive properties of quinoa seed pro-tein hydrolysates. *Journal of Food Science, 68*(4), 1254-1258.

Alvarez-Jubete, L., Arendt, E. K., & Gallagher, E. (2010). Nutritive value of pseudocereals and their increasing use as functional gluten-free ingredients. *Trends in Food Science and Technology, 21*(2), 106–113.

Arogundade, L. A. (2006). Functional characterization of tef (*Eragostics tef*) protein con-centrate: Influence of altered chemical environment on its gelation, foaming, and water hydration properties. *Food Hydrocolloids, 20*(6), 831–838.

Avanza, M. V., Puppo, M. C., & Añon, M. C. (2005). Rheological characterization of amaranth protein gels. *Food Hydrocolloids, 19*(5), 889–898.

Cerdán-Leal, M. A., López-Alarcón, C. A., Ortiz-Basurto, R. I., Luna-Solano, G., & Jiménez-Fernández, M. (2020). Influence of heat denaturation and freezing–lyophili-zation on physicochemical and functional properties of quinoa protein isolate. *Cereal Chemistry, 97*(2), 373–381.

Chalamaiah, M., Hemalatha, R., Jyothirmayi, T. (2012). Fish protein hydrolysates: Prox-imate composition, amino acid composition, antioxidant activities and applications: A review. *Food Chemistry, 135*(4), 3020–3038.

Ciudad-Mulero, M., Fernández-Ruiz, V., Matallana-González, M. C., & Morales, P. (2019). Dietary fiber sources and human benefits: The case study of cereal and pseu-docereals. In Advances in food and nutrition research, Academic Press, (Vol. 90, pp. 83–134).

Comai, S., Bertazzo, A., Bailoni, L., Zancato, M., Costa, C. V., & Allegri, G. (2007). The content of proteic and nonproteic (free and protein-bound) tryptophan in quinoa and cereal flours. *Food Chemistry, 100*(4), 1350–1355.

Constantino, A. B. T., & Garcia-Rojas, E. E. (2020). Modifications of physicochemical and functional properties of amaranth protein isolate (*Amaranthus cruentus* BRS Alegria) treated with high-intensity ultrasound. *Journal of Cereal Science, 95*, 103076.

Day, L. (2013). Proteins from land plants—Potential resources for human nutrition and food security. *Trends in Food Science and Technology, 32*(1), 25–42.

Dhull, S. B., & Punia, S. (2020a). Essential fatty acids: Introduction. In *Essential Fatty Acids: Sources, Processing Effects, and Health Benefits* by Dhull, S.B., Punia, S., Sandhu, CRC Press, 1–18.

Dhull, S. B., & Punia, S. (2020b). Sources: Plants, animals and microbial Essential Fatty Acids. In *Essential Fatty Acids: Sources, Processing Effects, and Health Benefits* by Dhull, S.B., Punia, S., Sandhu, CRC Press, 19–56.

Dhull, S. B., Punia, S., & Sandhu, K. S. (Eds.). (2020c). *Essential Fatty Acids: Sources, Processing Effects, and Health Benefits*. CRC Press.

Elsohaimy, S., Refaay, T., &Zaytoun, M. (2015). Physicochemical and functional proper-ties of quinoa protein isolate. *Annals of Agricultural Sciences, 60*(2), 297–305.

González-Martín, I., Álvarez-García, N., & González-Cabrera, J. M. (2006). Near-infrared spectroscopy (NIRS) with a fibre-optic probe for the prediction of the amino acid composition in animal feeds. *Talanta, 69*(3), 706–710.

González, J. A., Eisa, S. S., Hussin, S. A., & Prado, F. E. (2015). Quinoa: an Incan crop to face global changes in agriculture. *Quinoa: Improvement and sustainable production*, 1–18.

Gross, R., Koch, F., Malaga, I., De Miranda, A. F., Schoeneberger, H., &Trugo, L. C. (1989). Chemical composition and protein quality of some local Andean food sources. *Food Chemistry, 34*(1), 25–34. International Journal of Food Science & Technology, *44*(10), 2013–2020.

Hatti-Kaul, R., &Mattiasson, B. (Eds.). (2003). *Isolation and Purification of Proteins.* CRC Press.

He, X., Wang, B., Zhao, B., Meng, Y., Chen, J., & Yang, F. (2022). Effect of hydrothermal treatment on the structure and functional properties of quinoa protein isolate. *Foods, 11*(19), 2954.

Jin, J., Okagu, O. D., &Udenigwe, C. C. (2022). Differential influence of microwave and conventional thermal treatments on digestibility and molecular structure of buckwheat protein isolates. *Food Biophysics, 17*(2), 198–208.

Kakko, N., Ivanona, N., & Rantasalo, A. (2016). Cell disruption methods. Available online: https://www.mlsu.ac.in/econtents/404_Unit%204-%20Physical%20and%20Chemical%20Cell%20disruption%20methods.pdf (accessed on 18 July 2021).

Karaca, A. C. (2022). Proteins from pseudocereal grains. *Pseudocereals,* 35.

Kareem, M. Abdul. (2021). Proteins isolation and separation. Indira Gandhi National Open University, New Delhi. http://egyankosh.ac.in//handle/123456789/71692.

Klose, C. ,Schehl, B. D., & Arendt, E. K. (2009). Fundamental study on protein changes taking place during malting of oats. *Journal of Cereal Science, 49*(1), 83–91 .

Klupšaitė, D., & Juodeikienė, G. (2015). Legume: Composition, protein extraction and functional properties. A review. *Chemical Technology, 66*(1), 5–12.

Kozioł, M. (1992). *Chemical composition and nutritional evaluation of quinoa (Chenopodium quinoa* Willd.). *Journal of Food Composition and Analysis, 5*(1), 35–68.

López, D. N., Galante, M., Raimundo, G., Spelzini, D., & Boeris, V. (2019). Functional properties of amaranth, quinoa and chia proteins and the biological activities of their hydrolyzates. *Food Research International, 116,* 419–429.

López, D. N., Ingrassia, R., Busti, P., Bonino, J., Delgado, J. F., Wagner, J., ... & Spelzini, D. (2018). Structural characterization of protein isolates obtained from chia (*Salvia hispanica* L.) seeds. *LWT, 90,* 396–402.

López, D. N., Ingrassia, R., Busti, P., Wagner, J., Boeris, V., & Spelzini, D. (2018). Effects of extraction pH of chia protein isolates on functional properties. *LWT, 97,* 523–529.

Mahoney, A. W. , Lopez, J. G., & Hendricks, D. G. (1975). Evaluation of the protein quality of quinoa. *Journal of Agricultural and Food Chemistry, 23*(2), 190–193 .

Malik, A. M., & Singh, A. (2022). Pseudocereals proteins-A comprehensive review on its isolation, composition and quality evaluation techniques. *Food Chemistry Advances, 1,* 100001.

Malik, M., Sindhu, R., Dhull, S. B., Bou-Mitri, C., Singh, Y., Panwar, S., & Khatkar, B. S. (2023). Nutritional composition, functionality, and processing technologies for amaranth. *Journal of Food Processing and Preservation, 2023,* 1753029. https://doi.org/10.1155/2023/1753029.

Mir, N. A., Riar, C. S., & Singh, S. (2021). Improvement in the functional properties of quinoa (*Chenopodium quinoa*) protein isolates after the application of controlled heat-treatment: Effect on structural properties. *Food Structure, 28,* 100189.

Morales, D., Miguel, M., & Garcés-Rimón, M. (2021). Pseudocereals: A novel source of biologically active peptides. *Critical Reviews in Food Science and Nutrition, 61*(9), 1537–1544.

Mota, C., Santos, M., Mauro, R., Samman, N., Matos, A. S., Torres, D., & Castanheira, I. (2016). Protein content and amino acids profile of pseudocereals. *Food Chemistry*, *193*, 55–61.

Nascimento, A. C., Mota, C., Coelho, I., Gueifão, S., Santos, M., Matos, A. S., ... & Castanheira, I. (2014). Characterisation of nutrient profile of quinoa (Chenopodium quinoa), amaranth (Amaranthus caudatus), and purple corn (Zea mays L.) consumed in the North of Argentina: Proximates, minerals and trace elements. Food chemistry, 148, 420–426.

Nirmala, K., Vig, L., & Dhamija, N. Extraction and purification of enzymes. EgPathashala. 16:9. Microsoft Word - CHE_P16_M9_etext.docx (duhslibrary.ac.in) Dt. 16.04.23.

Olivos-Lugo, B., Valdivia-López, M., &Tecante, A. (2010). Thermal and physicochemical properties and nutritional value of the protein fraction of Mexican chia seed (*Salvia hispanica* L). *Food Science and Technology International*, *16*(1), 89–96.

Perera, S. P., Konieczny, D., Ding, K., Hucl, P., L'Hocine, L., & Nickerson, M. T. (2022). Techno-functional and nutritional properties of full-bran and low-bran canaryseed flour, and the effect of solvent-de-oiling on the proteins of low-bran flour and isolates. *Cereal Chemistry*, *99*(4), 762–785.

Pérez-Álvarez, E. P., Garde-Cerdán, T., García-Escudero, E., & Martínez-Vidaurre, J. M. (2014). Amino acid content in Tempranillo must from three soil types.

Písa ř íková, B., Kráčmar, S., & Herzig, I. (2005). Amino acid contents and biological value of protein in various amaranth species. *Czech Journal of Animal Science*, *50*(4), 169–174.

Pomeranz, Y. ,& Robbins, G. S. (1972). Amino acid composition of buckwheat. *Journal of Agricultural and Food Chemistry*, *20*(2), 270–274.

Protein Purification and characterization - Online Biology notes. Dt. 19.08.2020.

Protein purification methods, MN Editors. Protein Purification Methods (microbiology-note.com) Dt. (13.9.2022).

Punia, S., & Dhull, S. B. (2019). Chia seed (*Salvia hispanica* L.) mucilage (a heteropolysaccharide): Functional, thermal, rheological behaviour and its utilization. *International Journal of Biological Macromolecules*, *140*, 1084–1090.

Ruiz, G. A., Xiao, W., van Boekel, M., Minor, M., &Stieger, M. (2016). Effect of extraction pH on heat-induced aggregation, gelation and microstructure of protein isolate from quinoa (*Chenopodium quinoa* Willd). *Food Chemistry*, *209*, 203–210.

Shanthakumar, P., Klepacka, J., Bains, A., Chawla, P., Dhull, S. B., & Najda, A. (2022). The current situation of pea protein and its application in the food industry. *Molecules*, *27*(16), 5354.

Shehadul Islam, M., Aryasomayajula, A., & Selvaganapathy, P. R. (2017). A review on macroscale and microscale cell lysis methods. *Micromachines*, *8*(3), 83.

Shen, Y., Tang, X., & Li, Y. (2021). Drying methods affect physicochemical and functional properties of quinoa protein isolate. *Food Chemistry*, *339*, 127823.

Shevkani, K., Singh, N., Rana, J. C., & Kaur, A. (2014). Relationship between physicochemical and functional properties of amaranth (Amaranthus hypochondriacus) protein isolates. *International Journal of Food Science and Technology*, *49*(2), 541–550.

Stikic, R., Glamoclija, D., Demin, M., Vucelic-Radovic, B., & Jovanovic, Z., Milo-jkovic-Opsenica, D., ... Milovanovic, M. (2012). Agronomical and nutritional eval- uation of quinoa seeds (*Chenopodium quinoa* Willd.) as an ingredient in bread formulations. *Journal of Cereal Science*, *55*(2), 132–138 .

Timilsena, Y. P., Adhikari, R., Barrow, C. J., & Adhikari, B. (2016). Physicochemical and functional properties of protein isolate produced from Australian chia seeds. *Food Chemistry, 212,* 648–656.

Toapanta, A., Carpio, C., Vilcacundo, R., & Carrillo, W. (2016). Analysis of protein isolate from quinoa (Chenopodium quinoa Willd). *Asian J. Pharm. Clin. Res, 9*(2), 332–334.

Urbizo-Reyes, U., San Martin-González, M. F., Garcia-Bravo, J., Vigil, A. L. M., & Liceaga, A. M. (2019). Physicochemical characteristics of chia seed (*Salvia hispanica*) protein hydrolysates produced using ultrasonication followed by microwave-assisted hydrolysis. *Food Hydrocolloids, 97,* 105187.

Wang, L., Dong, J. L., Zhu, Y. Y., Shen, R. L., Wu, L. G., & Zhang, K. Y. (2021). Effects of microwave heating, steaming, boiling and baking on the structure and functional properties of quinoa (*Chenopodium quinoa* Willd.) protein isolates. *International Journal of Food Science and Technology, 56*(2), 709–720.

Wright, K. H. , Pike, O. A. , Fairbanks, D. J. , & Huber, C. S. (2002). Composition of Atriplex hortensis, sweet and bitter *Chenopodium quinoa* seeds. *Journal of Food Science, 67*(4), 1383–1385.

Wu, L., Li, J., Wu, W., Wang, L., Qin, F., &Xie, W. (2021). Effect of extraction pH on functional properties, structural properties, and *in vitro* gastrointestinal digestion of Tartary buckwheat protein isolates. *Journal of Cereal Science, 101,* 103314.

Chapter 8

Different Minerals and Their Bioavailability in Pseudocereals

Vidhi Gupta, Atul Anand Mishra

8.1 Introduction

Due to globalization, people are moving toward Western lifestyles and dietary habits. As a result, there has been a drastic increase in the demand for wheat and wheat-based food products. It is estimated that wheat consumption will rise by 12% by 2030 and therefore the global production of wheat must increase from 87 Mt to 840 Mt to ensure global food security and meet future demands (Shewry & Hey, 2015). The COVID-19 pandemic has resulted in long-lasting challenges in wheat production and disruptions in the supply chain. Consequently, this has added more pressure on the food systems, especially in vulnerable regions of the world. Moreover, climate changes and fluctuations in crop yield and wheat prices have led to uncertainty about wheat availability in the future (Shiferaw et al., 2013). Although grains such as wheat, rice, and maize contribute to 80% of global consumption and are an essential part of the human diet, they are deficient in micronutrients, like vitamins and minerals. It is estimated that approximately 2 billion people worldwide suffer from micronutrient deficiency diseases such as anemia and iodine deficiency disorder (IDD) (Pirzadah & Malik, 2020). Hence, the cereal grains have been biofortified to increase the levels of minerals. But considering the food crisis, overdependence on cereal grains, and disruptions in the supply chain, it is the need of the hour to make use of underutilized crops as a source of nutritious food to support the

 DOI: 10.1201/9781003325277-8

world's food basket. Pseudocereals are one such underutilized and neglected group of crops that have immense potential to combat hunger and nutritional crisis (Malik et al., 2023; Punia & Dhull, 2019). Pseudocereals, such as quinoa, amaranth, and buckwheat, have a rich nutritional profile in terms of protein, essential amino acids, fatty acids, phytochemicals, and minerals. Therefore, their incorporation into our regular diet plan can be beneficial to elevate levels of minerals and combat the global prevalence of micronutrient deficiencies. The nutritional composition of different pseudocereals and their comparison with commonly consumed true cereal grains is depicted in Table 8.1.

In addition to being a rich source of minerals, pseudocereals also contain certain anti-nutritional substances, like saponins, as well as mineral inhibitors (phytates and polyphenols) in their grains (Reguera & Haros, 2016). Phytates, the primary mineral inhibitors, bind with minerals and form a stable and insoluble complex that becomes indigestible by the body (Dhull et al., 2020a, 2020b). This prevents the body from using and metabolizing these essential minerals leading to a reduction in the bioavailability of iron, zinc, and calcium (Schlemmer et al., 2009). Mineral bioavailability can be enhanced by lowering the phytate level of pseudocereal food before consumption (Chawla et al., 2017). Various processing methods, like cooking, soaking, fermentation, roasting, and thermal treatments, have been employed to reduce the level of anti-nutritional compounds and improve bioavailability (Dhull et al., 2022a, 2022b, 2023). Over the past few decades, extensive research has been carried out to fully explore the mineral profile of different pseudocereals, understand their health benefits, investigate their bioavailability, the factors affecting bioavailability, and the processing methods that could be applied to improve mineral bioavailability, which is discussed later in this chapter.

8.2 Minerals Available in Pseudocereals

Most of the mineral compounds found in pseudocereal grains are present in the bran, making pseudocereal wholegrains an excellent source of minerals. The mineral profile of different pseudocereals is highlighted in Table 8.2. The highest concentration of minerals is found in amaranth, followed by quinoa and buckwheat (Joshi et al., 2019). In general, with a few exceptions, pseudocereals tend to have higher mineral content (such as calcium, potassium, magnesium, phosphorus, iron, and zinc) than their more commonly consumed staple food counterparts (wheat and rice). Gluten-free diets recommended for people with celiac disease or gluten intolerance often lack minerals like iron, magnesium, and calcium. Pseudocereals like buckwheat, amaranth, and quinoa are excellent sources of these minerals as well as other minerals. Amaranth seeds have a high calcium concentration, more than five times that found in wheat and even much more than in quinoa (Alvarez-Jubete et al., 2010). The higher

Table 8.1 Nutritional Profile of Pseudocereals and Its Comparison with Cereal Grains

Pseudocereal/Cereal	Energy (kcal)	Carbohydrates (%)	Protein (%)	Fat (%)	Total dietary fiber (%)	References
Buckwheat	355	59–75	5.7–18.9	1–6.9	17.8–27.4	Martínez-Villaluenga et al., 2020; Pirzadah & Malik, 2020; Thakur, Kumar, & Dhaliwal, 2021
Amaranth	371	63.1–69	13.5–17.6	5.6–10.6	3.1–17.3	
Quinoa	368	54.1–77	10–16.7	4.5–12.4	3.8–26.5	
Wheat	346	71.18	11.8–15	1.6–2.5	2.5–12.5	
Rice	345	75.61	2–7	1.5–6.8	2.1–4.5	
Maize	365	75.39	7.1–9.4	3.5–4.7	2–7.5	

Table 8.2 Important Minerals Present in Different Pseudocereals and Cereals (mg/100 g)

Mineral	Buckwheat	Amaranth	Quinoa	Rice	Wheat	Maize	References
Calcium (Ca)	110	159	47	3.35	60.02	12.95	Bekkering& Tian, 2019; Nagar & Rajput, 2022
Iron (Fe)	4	7.61	4.57	59.33	67.22	58.35	
Magnesium (Mg)	390	248	197	23.67	140.73	77.62	
Phosphorus (P)	330	557	457	-	-	-	
Manganese (Mn)	3.4	3.33	2.03	-	-	-	
Zinc (Zn)	0.8	2.87	3.1	9.27	11.73	9.45	
Potassium (K)	450	508	563	183.33	416.67	300	
Sodium (Na)	-	4	5	126.67	383.33	333.33	

calcium concentration found in amaranth seeds holds significant importance for individuals with celiac disease, as there is a well-documented prevalence of osteopenia and osteoporosis among them. Therefore, consumption of amaranth could be beneficial for the management of celiac disease and help to balance the calcium intake of celiac patients. Other minerals, like phosphorus, potassium, and sodium, are found in reasonable levels, whereas only moderate amounts of zinc, copper, and manganese have been reported in amaranth. Another mineral that is found in higher concentrations in amaranth is iron (Fe), ranging from 4.6 to 10.7 mg/100 g. This is approximately two to three times higher than the iron content found in wheat. The high iron concentration in amaranth might be of great relevance to pregnant women to provide health benefits, considering that they have a higher iron requirement and are more prone to becoming anemic. In addition to this, essential trace elements, such as molybdenum (Mo) at 59.6 µg/100 g and chromium (Cr) ranging from 6.8 to 14.4 µg/100 g, have been identified in amaranth in significant concentrations that are essential for human nutritional requirements (Zhu, 2016). In a nutshell, amaranth exhibits a rich and well-balanced mineral profile. As a result, amaranth can be utilized as a potential solution for addressing global nutritional deficiencies.

The mineral profile of buckwheat is similar to that of amaranth but higher than in quinoa. The macroelements, like sodium, potassium, calcium, and magnesium, and microelements, such as iron, zinc, copper, and manganese, are found in remarkably higher levels in buckwheat. Buckwheat demonstrates an increased bioavailability of zinc, copper, and potassium. A 100 g serving of buckwheat flour can supply 13–89% of the recommended dietary allowance (RDA) for magnesium, zinc, copper, and manganese. Tartary buckwheat (bitter buckwheat) seeds contain higher concentrations of minerals like iron, zinc, selenium, copper, and nickel than do common buckwheat seeds, with a majority of the minerals being found in the cotyledons while calcium is present in the pericarp (Zhu, 2016). Minerals like K, Mg, P, Zn, and Co are mainly stored in the form of phytate within protein bodies located in the aleurone layer and embryo tissues. During germination, phytic acid gets hydrolyzed and the minerals get liberated, thereby elevating their level. Minerals such as copper (Cu), iron (Fe), manganese (Mn), zinc (Zn), molybdenum (Mo), aluminum (Al), and nickel (Ni) are predominantly found in both the hull and seed coat of the grain (Wijngaard & Arendt, 2006). Apart from amaranth and buckwheat, quinoa is also an excellent source of essential minerals. The mineral content of quinoa is almost twice that present in true cereal grains like wheat and rice. Overall, the total mineral profile of quinoa (approximately 2.74 g/100 g) surpasses that of wheat (1.81 g/100 g), maize (1.34 g/100 g), and rice (0.72 g/100 g) (Tanwar et al., 2019). The main minerals present in quinoa include iron, zinc, calcium, and magnesium. Potassium and magnesium are primarily found in the embryo of the grain, whereas calcium and phosphorus are associated with pectic compounds present in the cell wall of the pericarp (Vega-Gálvez et al., 2010).

8.3 Health Benefits Associated with Minerals Present in Pseudocereals

Diseases caused by deficiencies of different minerals are becoming a serious health problem throughout the world. Despite efforts undertaken to address this problem through biofortification, supplementation, and other strategies, a large segment of the world population still suffers from these deficiencies. The burden of mineral deficiency diseases exists worldwide and is most prevalent in children and pregnant women, which may lead to life-threatening issues. Among these deficiencies, iron, zinc, and iodine deficiencies are the most prevalent ones. According to WHO, globally 2 billion people suffer from iodine deficiency disorders (IDD) and more than 40% of children and pregnant women have inadequate iron deposits and are victims of anemia. According to data from the National Family Health Survey-4 (NFHS-4), India carries the highest global burden of anemia. In 2016, the prevalence of anemia was recorded to be highest among children (58.6%), followed by non-pregnant (53.2%) and pregnant women (50.4%) (Han et al., 2022). On the other hand, 19% of Indian pre-school children and 32% of adolescents suffered from zinc deficiency in 2019 as stated by the Comprehensive National Nutrition Survey of Children (CNNSC) (Venkatesh et al., 2021).

One possible strategy to combat these deficiencies could be the utilization of pseudocereals. As discussed above, these pseudocereals have high mineral concentrations, which could prove to be an efficient strategy to overcome these challenges and lower the public health risks associated with mineral deficiencies. Several research works have demonstrated that the minerals found in amaranth, buckwheat, and quinoa exert notable health benefits. For example, inadequate intake of calcium is associated with poor bone health, leading to osteopenia and osteoporosis in severe cases. A high-calcium diet helps in metabolizing fat and keeps the body weight in check. Pseudocereals contain high amounts of calcium compared with wheat and rice. Their consumption can be helpful for good bone, teeth, and heart health (Bonjour et al., 2009).

Furthermore, it has been reported that gluten-free diets are often deficient in minerals such as zinc, calcium, and magnesium, and zinc deficiency can influence the synthesis of proteins in the body and result in growth retardation. Magnesium is another important mineral that is involved in many body processes such as regulating blood glucose, carrying out enzymatic reactions, and metabolizing glucose, protein, fat, and nucleic acids (Awuchi et al., 2020). Pseudocereals contain these minerals in appreciable amounts and can be a good source of food to meet the nutritional requirements of people who follow gluten-free diets. Nevertheless, phosphorus, manganese, and zinc not only help to maintain circulation and blood vessel function in individuals but are also essential for neurotransmitter signaling in the brain, combating conditions like depression, anxiety, and headaches. These minerals are present in significant amounts in buckwheat, and food products made from buckwheat flour can give potential health benefits (Ikeda & Ikeda, 2016). Several studies have indicated that quinoa contains significant quantities of iron, magnesium, and zinc. These components contribute to enhanced postprandial sensitivity, the release of plasma insulin, and anti-hypercholesterolemic effects. In summary, it can be concluded that pseudocereals such as amaranth, quinoa, and buckwheat have an excellent mineral profile and their consumption in the form of different food commodities can be a potential solution to addressing the challenges associated with micronutrient deficiencies and lower their global prevalence.

8.4 Bioavailability of Minerals

An understanding of the bioavailability of the different nutrients present in the food is important to quantify the amount of a particular nutrient that must be supplied through food to meet the recommended daily allowances for different age groups. From the viewpoint of nutrition, bioavailability can be defined as "the fraction of the ingested nutrient capable of being absorbed and available for storage/utilization" (Rousseau et al., 2020). It includes two terms, namely bioaccessibility and bioactivity. When food is consumed, the nutrients get released from the matrix and converted into simpler absorbable forms during the digestion process. This is followed by their absorption into the bloodstream and transportation to the tissues for various physiological functions. Bioaccessibility is the fraction of nutrients which is released from the food matrix and is readily available for absorption, whereas bioactivity is that fraction which is absorbed, transported to the cells, and utilized for different body functions (Alegría-Torán et al., 2015). Bioaccessibility and bioactivity can be assessed with the help of *in-vitro* and *in-vivo* experiments, respectively. There are several factors that can influence the bioavailability of nutrients, such as the form of the food (raw or processed), the presence of antinutrients (e.g., phytates, saponins, polyphenols), and physiological factors (age, gender, pathological conditions). Antinutritional compounds bind with the minerals and form an antinutrient-mineral complex which hampers their bioavailability. Apart from this, the bioaccessibility and bioactivity of the minerals are also affected by the plant cell structure. The cell wall can act as a physical barrier and impede the bioavailability of minerals (Glahn et al., 2016).

8.5 Anti-nutritional Factors Affecting the Bioavailability of Minerals

From the above discussion, it is evident that amaranth, buckwheat, and quinoa have very high mineral contents. However, when these products are consumed in raw or unprocessed form, the bioavailability of these macro- and micronutrients is significantly less. One of the possible reasons for their low bioavailability is the presence of anti-nutritional factors, which occur naturally in these pseudocereals. These anti-nutritional factors are responsible for low nutritional value. Anti-nutritional factors can be defined as naturally occurring substances that, when ingested, affects the utilization and absorption of nutrients such as proteins, minerals, and vitamins by binding to them and, thereby, converting them into an unabsorbable form (Jain et al., 2009).

Although the nutritional profile of pseudocereals is somewhat superior to that of cereal grains, the presence of antinutrients, like phytates, mineral inhibitors, oxalates, saponins, and tannins, impairs the bioavailability of vital minerals. Phytate is an organic compound (especially myo-inositol hexaphosphate) that is naturally found in all plants and is a major antinutrient found in pseudocereals. It functions as the storage form of phosphorus within the plant cells. At physiological pH, phytic acid possesses a negative charge, which gives it a strong affinity to bind with positively charged metal ions, demonstrating strong chelating behavior toward minerals such as calcium, zinc, and iron, to form insoluble and indigestible complexes (Kumar et al., 2010). Hence, the presence of phytate in any food product plays a vital role in determining the absorption of minerals present in that food. The degradation of phytate is crucial to facilitating the bio-assimilation of minerals. Phytase (EC 3.1.3.26) is the enzyme responsible for the hydrolysis of phytate, and, therefore, it can help in the liberation of bound minerals. If the molar ratio of phytate/Fe is higher than five, the bioavailability of zinc is significantly reduced, and the bioavailability of iron can be affected if this ratio is higher than one. The levels of phytates were found to be higher in quinoa than in other pseudocereals or cereals. It has been seen that the bioavailability of minerals remains unaffected if the phytate: iron ratio <1, phyate: zinc ratio is <5, and phytate: calcium ratio <0.17 (Hurrell, 2004; Iglesias-Puig et al., 2015; Umeta et al., 2005). When the ratios exceed the critical values, it indicates a low bioavailability of the mineral present in the grain. Hence, if these molar ratios are higher than the critical values, phytate can have a significant impact on the bio-assimilation of essential minerals in food products.

Saponins are another type of anti-nutritional compound responsible for the bitter flavor of quinoa. They have the ability to form stable complexes with zinc and iron in the intestinal tract and interfere with their absorption into the body (Mir et al., 2018). Oxalates are organic acids found in many common food commodities. They also possess anti-nutritional properties, particularly by binding with the minerals and hindering their absorption into the body and thereby, adversely affecting their bioavailability. One possible reason for this observation could be the binding of dietary fiber, as well as oxalic acid, with minerals leading to the formation of a dietary fiber-mineral-oxalate complex. During the digestion process, this ternary complex is potentially more resistant to breakdown by digestive enzymes compared to the binary complex of oxalate-mineral or the dietary fiber-mineral bond. Oxalates bind with calcium leading to the formation of calcium oxalate, which is a leading cause of stone formation in the kidneys and gall bladder (Jancurová et al., 2009). This also lowers calcium bioavailability.

8.6 Processing Methods for Enhancing the Bioavailability of Minerals

Processing operations, such as dehulling, soaking, size reduction, cooking, steaming, and fermentation, help to increase the palatability of the pseudocereals. Furthermore, these methods have been proven beneficial for reducing the concentration of antinutrients and enhancing the bioavailability of minerals present in different grains (Dhull et al., 2022b, 2023). The effect of different processing techniques on the anti-nutritional and bioavailability of different minerals in amaranth, buckwheat, and quinoa is outlined briefly in Table 8.3. The most common methods used to reduce the antinutrients from the pseudocereals include soaking and dehulling. Some levels of phytates, saponins, and mineral inhibitors are located in the outer layers of these grains. Removal of these outer layers during the dehulling process can help to reduce the amount of antinutrients to some extent. Apart from this, fermentation is another method which has been extensively explored as a potential technology for the degradation of anti-nutritional factors in pseudocereals.

Fermentation is a very ancient processing method which was first used for the purpose of food preservation. Nowadays, it is also applied to improve nutritional (phytochemicals, minerals) and organoleptic (aroma, flavor, color) attributes of different food commodities and enhance their overall palatability. Apart from enhancing these parameters, fermentation is also utilized to decrease the content of antinutrients like phytates and tannins and increase the mineral content available for ready absorption (Dhull et al., 2020a, 2020b). Lactic acid fermentation of pseudocereals is one of the preferred methods to carry out the fermentation of pseudocereals. Throughout the process of fermentation, lactic acid bacteria (LAB) generate organic acids, particularly lactic acid, that lead to a decline in pH (Rollán et al., 2019). This acidic environment promotes the activation of endogenous phytase. During the process of hydrolysis of phytates by phytase enzyme, divalent minerals, such as iron, calcium, zinc, and magnesium, are released in their free forms which are readily absorbed into the bloodstream. Apart from this, lower myo-inositol phosphates are also formed which exhibit potential antioxidant activity and confer benefits to human health.

Germination is the most widely used and effective method employed to enhance the bioavailability of minerals in different pseudocereals. Seed germination is a series of events that marks the commencement of a plant's life cycle. It commences when a dormant, desiccated seed imbibes water and terminates with the emergence of the radicle through the seed coat. During germination, several changes take place inside the grain, like structural breakdown, hydrolysis of macromolecules, activation of enzymes, and mechanical changes. The phytase enzyme is activated during germination which metabolizes phytic acid, reducing the anti-nutritional content and increasing the mineral's bioavailability (Liu et al., 2022). From the viewpoint of anti-nutritional factors, cooking has no significant effect on the degradation of phytic acid.

Table 8.3 Processing Technologies to Improve Mineral Bioavailability in Pseudocereals

Processing Technology	Pseudocereal(s)	Processing Conditions	Effect on Mineral Bioavailability	References
Germination	Quinoa	Soaking in water for 6 h, followed by draining and covering seeds with humid paper. Germination was carried out at 20°C and 100% RH for 7 days.	The content of phytic acid was found to decrease from 32 to 74% while a 45%, 26%, and 15% increase was seen in Zn, Ca, and Fe concentration, respectively, after germination.	Maldonado-Alvarado et al., 2023
	Quinoa	Steeping of grains in water at 25°C for 12 h (seed: water = 1:5). Germination was conducted for 72 h on a wet jute cloth.	Germination decreased concentrations of phytic acid, saponins, and tannins by 50%, 59.6%, and 11.23%, respectively. A 20.25%, 39.43%, and 49.04% increase in Zn, Fe, and Ca concentration, respectively, was also observed in germinated grains.	Darwish et al., 2021
	Quinoa, amaranth, buckwheat	Soaking of grains in water at room temperature for 16 h in the dark. Soaked grains were germinated at 25°C for 24, 48, and 72 h.	The phytic acid and tannin concentrations was reduced by 29.57% and 32.3% in amaranth, 47.57% and 27.08% in quinoa and 17.42% and 59.91% in buckwheat. The concentrations of Cu, Fe, Zn, and Mn increased significantly with germination time.	Thakur et al., 2021
	Amaranth	Soaking in water overnight followed by subsequent soaking in formaldehyde solution and then distilled water. Germination was carried out at 28°C for 3 days.	A significant decrease in antinutrients was observed after germination (34.6%, 25.3%, and 21.9% reduction in tannin, oxalate, and saponin concentration, respectively). Increases in the bioavailability of minerals like Mg (25%), Na (52.7%), Ca (15.7%), and Fe (21.8%) occurred.	Olawoye & Gbadamosi, 2017

(Continued)

Table 8.3 (Continued) Processing Technologies to Improve Mineral Bioavailability in Pseudocereals

Processing Technology	Pseudocereal(s)	Processing Conditions	Effect on Mineral Bioavailability	References
Fermentation	Quinoa, amaranth	Fermentation was carried out using the starter culture, *Lactobacillus plantarum* 299v, for 48 h.	The availability of minerals was enhanced, exhibiting an increase ranging from 1.7 to 4.6 times, while the molar ratios of phytate to minerals decreased by a factor of 1.5 to 4.2 following the degradation of phytate, which was amplified by 1.8 to 4.2 times in fermented flours.	Castro-Alba et al., 2019
	Amaranth	Natural fermentation: mixing of flour with water followed by fermenting the mixture at room temperature for 48 h.	The ash content increased by 14%. Magnesium and iron concentrations increased after fermentation. A decrease in the IP6: Fe molar ratio was seen, indicating enhanced bioavailability of Fe.	Amare et al., 2016
Fermentation + dry roasting	Quinoa	Fermentation was carried out using the starter culture, *Lactobacillus plantarum* 299v, at 30°C for 10 h. The dried and fermented sample was then subjected to dry roasting at 120°C for 3 min.	The % phytate reduction was significantly higher in treated grains (73%) compared with raw grains (30%). The bioavailability of Zn improved from low to moderate as the molar ratio Phy: Zn decreased notably from 25.4 to 7.14.	Castro-Alba et al., 2019
Puffing	Amaranth	Puffing was done in an aluminum pan at 160°C for 30 s.	The puffing process led to an increment in the bioaccessibility of iron and calcium	Burgos et al., 2018
Popping	Amaranth	Grains were popped for 10–15 s in a hot clay pan.	The ash content increased by 10%. A decrease in the IP6: Fe molar ratio was seen, indicating enhanced bioavailability of Fe.	Amare et al., 2016

(Continued)

Table 8.3 (Continued) Processing Technologies to Improve Mineral Bioavailability in Pseudocereals

Processing Technology	Pseudocereal(s)	Processing Conditions	Effect on Mineral Bioavailability	References
Roasting and boiling	Quinoa	Roasting was done at 190°C for 3 min using a hotplate. Boiling was carried out in tap water for 20 min and a seed:water of 1:4)	Both cooking methods resulted in a significant increase in the bioavailability of minerals like calcium, iron, and zinc.	Repo-Carrasco-Valencia et al., 2010
Cooking	Quinoa, amaranth, buckwheat	Grains were cooked in water at 100°C for 15 min.	Higher bioaccessibilities for Cu, Fe, Mg, and Mn were observed in cooked amaranth and quinoa. A 3.5 fold increase in bioaccessibility of Zn and Mn in buckwheat was also seen.	Motta et al., 2022
Extrusion	Amaranth	Single screw extruder at 15% feed moisture. 200 rpm screw speed, ˜50°C temperature, 3 mm die diameter.	The extrusion process increased the bioavailability of calcium and led to more absorption in the bones of rats.	Ferreira & Arêas, 2010

8.7 Future Prospects

Very often, pseudocereals, such as amaranth, buckwheat, and quinoa, are either known as the grains of poor people or people are unaware of the health and nutritional benefits associated with them. In order to popularize these pseudocereals, more efforts are needed to raise awareness among people about their benefits. Considering the prevalence of micronutrient deficiencies in the world, these underutilized crops have the potential to help achieve both food and nutrition security, especially in developing countries. In future, more research works must be carried out to decipher the interrelationship between the processing methods and the bioavailability of minerals. This will aid food scientists in developing novel food products based on these pseudocereals, with improved nutritional profiles and organoleptic characteristics. In order to make pseudocereals a part of our regular diet, it becomes crucial to tailor their properties and gain an in-depth insight into the structural changes required for increasing their palatability and consumer acceptability. There exists limited knowledge on the technological performance of the pseudocereals, which is essential for the commercialization of food products. There have been only limited research studies conducted to investigate the effects of different thermal treatments on the bioavailability of minerals, and hence, a more comprehensive approach is needed to fully understand the potential of pseudocereals in meeting the recommended daily allowances of minerals for different age groups and genders.

Additionally, laboratory studies must be coupled with *in-vivo* studies to fully understand the factors influencing the bioavailability of minerals. Furthermore, with the advent of genetic technology, identification and mapping of the genes that might affect the bioavailability of minerals could help to eradicate the anti-nutritional properties of pseudocereals and thus improve their nutritional profile. Looking forward, it would be essential to investigate the influence of different novel techniques, such as cold plasma, high-pressure processing, electric field, and microwave treatment, on mineral bioaccessibility. This way, it would be possible to optimize the processing conditions to enhance the mineral bioavailability and develop the strategies to use them as ingredients for developing food products for different nutritional requirements.

8.8 Conclusion

The reintroduction of neglected and underutilized pseudocereals into the food basket might be a potential alternative for farmers, consumers, and the food industry, to provide a sustainable source of nutritious, as well as cheap crops. From a nutritional point of view, pseudocereals possess an excellent nutritional profile. Hence, their incorporation into our regular diet might not only help to combat nutrient deficiency but also provide health benefits. The mineral profile of amaranth, buckwheat, and quinoa is somewhat superior to common cereal grains like wheat, rice, and maize. They contain calcium, iron, magnesium, zinc, and many other minerals in appreciable amounts. However, they also contain a diverse array of anti-nutritional compounds like phytates, saponins, and tannins, which hinder the bioavailability of minerals and impair their absorption into the body. Processing technologies such as germination, fermentation, cooking, roasting, puffing, and popping have been employed to reduce the antinutrients and improve the bioavailability of minerals. Future research works need to focus on integrating *in-vitro* research findings with *in-vivo* studies to establish an in-depth understanding of the factors affecting the bioavailability of minerals. This would further help the scientists to focus on the strategies required to enhance the mineral's bioaccessibility of pseudocereals.

References

Alegría-Torán, A., Barberá-Sáez, R., & Cilla-Tatay, A. (2015). Bioavailability of minerals in foods. In *Handbook of Mineral Elements in Food*, 41–67. https://doi.org/10.1002/9781118654316.ch3

Alvarez-Jubete, L., Arendt, E. K., & Gallagher, E. (2010). Nutritive value of pseudocereals and their increasing use as functional gluten-free ingredients. *Trends in Food Science and Technology*, *21*(2), 106–113. https://doi.org/10.1016/j.tifs.2009.10.014

Amare, E., Mouquet-Rivier, C., Rochette, I., Adish, A., & Haki, G. D. (2016). Effect of popping and fermentation on proximate composition, minerals and absorption inhibitors, and mineral bioavailability of *Amaranthus caudatus* grain cultivated in Ethiopia. *Journal of Food Science and Technology*, *53*, 2987–2994. https://doi.org/10.1007%2Fs13197-016-2266-0

Awuchi, C., Godswill, V., Somtochukwu, E., & Kate, C. (2020). Health benefits of micronutrients (vitamins and minerals) and their associated deficiency diseases: A systematic review. *International Journal of Food Sciences*, *3*(1). https://doi.org/10.47604/ijf.1024

Bekkering, C. S., & Tian, L. (2019). Thinking outside of the cereal box: Breeding underutilized (pseudo) cereals for improved human nutrition. *Frontiers in Genetics*, *10*, 1289. https://doi.org/10.3389/fgene.2019.01289

Bonjour, J. P., Guéguen, L., Palacios, C., Shearer, M. J., & Weaver, C. M. (2009). Minerals and vitamins in bone health: The potential value of dietary enhancement. *British Journal of Nutrition*, *101*(11), 1581–1596. https://doi.org/10.1017/S0007114509311721

Burgos, V. E., Binaghi, M. J., de Ferrer, P. A. R., & Armada, M. (2018). Effect of precooking on antinutritional factors and mineral bioaccessibility in kiwicha grains. *Journal of Cereal Science*, *80*, 9–15. https://doi.org/10.1016/j.jcs.2017.12.014

Castro-Alba, V., Lazarte, C. E., Perez-Rea, D., Carlsson, N. G., Almgren, A., Bergenståhl, B., & Granfeldt, Y. (2019). Fermentation of pseudocereals quinoa, canihua, and amaranth to improve mineral accessibility through degradation of phytate. *Journal of the Science of Food and Agriculture*, *99*(11), 5239–5248. https://doi.org/10.1002/jsfa.9793

Castro-Alba, V., Lazarte, C. E., Perez-Rea, D., Sandberg, A. S., Carlsson, N. G., Almgren, A., ... Granfeldt, Y. (2019). Effect of fermentation and dry roasting on the nutritional quality and sensory attributes of quinoa. *Food Science & Nutrition*, *7*(12), 3902–3911. https://doi.org/10.1002/fsn3.1247

Chawla, P., Bhandari, L., Dhull, S. B., Sadh, P. K., Sandhu, S. P., Kaushik, R., & Navnidhi (2017). Biotechnological aspects for enhancement of mineral bioavailability from cereals and legumes. In *Plant Biotechnology: Recent Advancements and Developments*, (Eds. Suresh Kumar Gahlawat, Raj Kumar Salar, Priyanka Siwach, Joginder Singh Duhan, Suresh Kumar, Pawan Kaur), Springer, Singapore, 87–100. https://doi.org/10. 1007/978-981-10-4732-9_5

Darwish, A. M., Al-Jumayi, H. A., & Elhendy, H. A. (2021). Effect of germination on the nutritional profile of quinoa (Cheopodium quinoa Willd.) seeds and its anti-anemic potential in Sprague–Dawley male albino rats. *Cereal Chemistry*, 98(2), 315–327. https://doi.org/10.1002/cche.10366

Dhull, S. B., Kidwai, M. K., Noor, R., Chawla, P., & Rose, P. K. (2022a). A review of nutritional profile and processing of faba bean (*Vicia faba* L.). *Legume Science*, 4(3), e129. https://doi.org/10.1002/leg3.129

Dhull, S. B., Kinabo, J., & Uebersax, M. A. (2023). Nutrient profile and effect of processing methods on the composition and functional properties of lentils (*Lens culinaris* Medik): A review. *Legume Science*, 5(1), e156. https://doi.org/10.1002/leg3.156

Dhull, S. B., Punia, S., Kidwai, M. K., Kaur, M., Chawla, P., Purewal, S. S., Sangwan, M., & Palthania, S. (2020a). *Solid-state fermentation of lentil (Lens culinaris L.) with Aspergillus awamori*: Effect on phenolic compounds, mineral content, and their bioavailability. *Legume Science*, 2(3), e37. https://doi.org/10.1002/leg3.37

Dhull, S. B., Punia, S., Kumar, R., Kumar, M., Nain, K. B., Jangra, K., & Chudamani, C. (2020b). Solid state fermentation of fenugreek (*Trigonella foenum-graecum*): Implications on bioactive compounds, mineral content and in vitro bioavailability. *Journal of Food Science and Technology*, 1–10. https://doi.org/10.1007/s13197-020-04704-y

Dhull, S. B., Kidwai, M. K., Siddiq, M., & Sidhu, J. S. (2022b). Faba (broad) bean production, processing, and nutritional profile. In *Dry Beans and Pulses* (eds M. Siddiq & M.A. Uebersax). https://doi.org/10.1002/9781119776802.ch14

Ferreira, T. A., & Arêas, J. A. G. (2010). Calcium bioavailability of raw and extruded amaranth grains. *Food Science and Technology*, 30, 532–538. https://doi.org/10.1590/S0101-20612010000200037

Glahn, R. P., Tako, E., Cichy, K., & Wiesinger, J. (2016). The cotyledon cell wall and intracellular *matrix are factors that limit iron bioavailability of the common bean (Phaseolus vulgaris)*. *Food and Function*, 7(7), 3193–3200. https://doi.org/10.1039/c6fo00490c

Han, X., Ding, S., Lu, J., & Li, Y. (2022). Global, regional, and national burdens of common micronutrient deficiencies from 1990 to 2019: A secondary trend analysis based on the Global Burden of Disease 2019 study. *EClinicalmedicine*, 44. https://doi.org/10.1016/j.eclinm.2022.101299

Hurrell. (2004). Phytic acid degradation as a means of improving iron absorption. *International Journal for Vitamin and Nutrition Research*, 74(6), 445–452.https://doi.org/10.1024/0300-9831.74.6.445

Iglesias-Puig, E., Monedero, V., & Haros, M. (2015). Bread with whole quinoa flour and bifidobacterial phytases increases dietary mineral intake and bioavailability. *LWT*, 60(1), 71–77. https://doi.org/10.1016/j.lwt.2014.09.045

Ikeda, K., & Ikeda, S. (2016). Factors important for structural properties and quality of buckwheat products. In *Molecular Breeding and Nutritional Aspects of Buckwheat*, (eds. Meiliang Zhou, Ivan Kreft, Sun-Hee Woo, Nikhil Chrungoo, Gunilla Wieslander), Academic Press, US, 193–202. https://doi.org/10.1016/B978-0-12-803692-1.00015-8

Jain, A. K., Kumar, S., & Panwar, J. D. S. (2009). Antinutritional factors and their detoxification in pulses - a review. *Agricultural Reviews, 30*(1), 64–70.

Jancurová, M., Minarovičová, L., & Dandár, A. (2009). Quinoa - a review. *Czech Journal of
Food Sciences, 27*(2), 71–79. https://doi.org/10.17221/32/2008-CJFS

Joshi, D. C., Chaudhari, G. V., Sood, S., Kant, L., Pattanayak, A., Zhang, K., Fan, Y.,
Janovská, D., Meglič, V., & Zhou, M. (2019). Revisiting the versatile buckwheat:
Reinvigorating genetic gains through integrated breeding and genomics approach.
Planta, 250(3), 783–801. https://doi.org/10.2307/48702267

Kumar, V., Sinha, A. K., Makkar, H. P. S., & Becker, K. (2010). Dietary roles of phytate
and phytase in human nutrition: A review. *Food Chemistry, 120*(4), 945–959. https://
doi.org/10.1016/j.foodchem.2009.11.052

Liu, S., Wang, W., Lu, H., Shu, Q., Zhang, Y., & Chen, Q. (2022). New perspectives on
physiological, biochemical and bioactive components during germination of edible
seeds: A review. *Trends in Food Science and Technology, 123*, 187–197. https://doi.org/
10.1016/j.tifs.2022.02.029

Maldonado-Alvarado, P., Pavón-Vargas, D. J., Abarca-Robles, J., Valencia-Chamorro, S., &
Haros, C. M. (2023). Effect of germination on the nutritional properties, phytic acid
content, and phytase activity of quinoa (*Chenopodium quinoa* Willd). *Foods, 12*(2),
389. https://doi.org/10.3390/foods12020389

Malik, M., Sindhu, R., Dhull, S. B., Bou-Mitri, C., Singh, Y., Panwar, S., & Khatkar, B.
S. (2023). Nutritional composition, functionality, and processing technologies for
Amaranth. *Journal of Food Processing and Preservation, 2023*, 1753029. https://doi.org/
10.1155/2023/1753029

Martínez-Villaluenga, C., Peñas, E., & Hernández-Ledesma, B. (2020). Pseudocereal
grains: Nutritional value, health benefits and current applications for the development of gluten-free foods. *Food and Chemical Toxicology, 137*, 111178. https://doi.org/
10.1016/j.fct.2020.111178

Mir, N. A., Riar, C. S., & Singh, S. (2018). Nutritional constituents of pseudo cereals and
their potential use in food systems: A review. *Trends in Food Science and Technology,
75*, 170–180. https://doi.org/10.1016/j.tifs.2018.03.016

Motta, C., Castanheira, I., Matos, A. S., Nascimento, A. C., Assunção, R., Martins, C.,
& Alvito, P. Effect of cooking on the content and bioaccessibility of minerals in
pseudocereals. *Biology and Life Sciences Forum, 17*(1), 17. https://doi.org/10.3390/
blsf2022017017

Nagar, P., Engineer, R., & Rajput, K. (2022). Review on Pseudo-Cereals of India. In V.
Y. Waisundara (Ed.), *Pseudocereals* (Vol. 151). IntechOpen. https://doi.org/10.5772/
intechopen.101834

Olawoye, B. T., & Gbadamosi, S. O. (2017). Effect of different treatments on in vitro protein digestibility, antinutrients, antioxidant properties and mineral composition of
Amaranthus viridis seed. *Cogent Food & Agriculture, 3*(1), 1296402. https://doi.org/10.
1080/23311932.2017.1296402

Pirzadah, T. B., & Malik, B. (2020). Pseudocereals as super foods of 21st century: Recent
technological interventions. *Journal of Agriculture and Food Research, 2*, 100052.
https://doi.org/10.1016/J.JAFR.2020.100052

Punia, S., & Dhull, S. B. (2019). Chia seed (*Salvia hispanica* L.) mucilage (a heteropolysaccharide): Functional, thermal, rheological behaviour and its utilization. *International
Journal of Biological Macromolecules, 140*, 1084–1090. https://doi.org/10.1016/j.
ijbiomac.2019.08.205

Reguera, M., & Haros, C. M. (2016). Structure and composition of kernels. In *Pseudocereals: Chemistry and Technology*, 28–48. https://doi.org/10.1002/9781118938256.ch2

Repo-Carrasco-Valencia, R. A., Encina, C. R., Binaghi, M. J., Greco, C. B., & Ronayne de Ferrer, P. A. (2010). Effects of roasting and boiling of quinoa, kiwicha and kañiwa on composition and availability of minerals in vitro. *Journal of the Science of Food and Agriculture, 90*(12), 2068–2073. https://doi.org/10.1002/jsfa.4053

Rollán, G. C., Gerez, C. L., & Leblanc, J. G. (2019). Lactic fermentation as a strategy to improve the nutritional and functional values of pseudocereals. *Frontiers in Nutrition, 6.* https://doi.org/10.3389/fnut.2019.00098

Rousseau, S., Kyomugasho, C., Celus, M., Hendrickx, M. E. G., & Grauwet, T. (2020). Barriers impairing mineral bioaccessibility and bioavailability in plant-based foods and the perspectives for food processing. *Critical Reviews in Food Science and Nutrition, 60*(5), 826–843. https://doi.org/10.1080/10408398.2018.1552243

Schlemmer, U., Frølich, W., Prieto, R. M., & Grases, F. (2009). Phytate in foods and significance for humans: Food sources, intake, processing, bioavailability, protective role and analysis. *Molecular Nutrition and Food Research, 53*(2), 330–375. https://doi.org/10.1002/mnfr.200900099

Shewry, P. R., & Hey, S. J. (2015). The contribution of wheat to human diet and health. *Food and Energy Security, 4*(3), 178–202. https://doi.org/10.1002/fes3.64

Shiferaw, B., Smale, M., Braun, H. J., Duveiller, E., Reynolds, M., & Muricho, G. (2013). Crops that feed the world 10. Past successes and future challenges to the role played by wheat in global food security. *Food Security, 5*(3), 291–317. https://doi.org/10.1007/s12571-013-0263-y

Tanwar, B., Goyal, A., Irshaan, S., Kumar, V., Sihag, M. K., Patel, A., & Kaur, I. (2019). Quinoa. In *Whole Grains and Their Bioactives: Composition and Health*, 269–305. https://doi.org/10.1002/9781119129486.ch10

Thakur, P., Kumar, K., & Dhaliwal, H. S. (2021). Nutritional facts, bio-active components, and processing aspects of pseudocereals: A comprehensive review. *Food Bioscience, 42*, 101170. https://doi.org/10.1016/j.fbio.2021.101170

Thakur, P., Kumar, K., Ahmed, N., Chauhan, D., Rizvi, Q. U. E. H., Jan, S., ... Dhaliwal, H. S. (2021). Effect of soaking and germination treatments on nutritional, anti-nutritional, and bioactive properties of amaranth (*Amaranthus hypochondriacus L.*), quinoa (*Chenopodium quinoa L.*), and buckwheat (*Fagopyrum esculentum L.*). *Current Research in Food Science, 4*, 917–925. https://doi.org/10.1016/j.crfs.2021.11.019

Umeta, M., West, C. E., & Fufa, H. (2005). Content of zinc, iron, calcium, and their absorption inhibitors in foods commonly consumed in Ethiopia. *Journal of Food Composition and Analysis, 18*(8), 803–817. https://doi.org/10.1016/j.jfca.2004.09.008

Vega-Gálvez, A., Miranda, M., Vergara, J., Uribe, E., Puente, L., & Martínez, E. A. (2010). Nutrition facts and functional potential of quinoa (*Chenopodium quinoa* Willd.), an ancient Andean grain: A review. *Journal of the Science of Food and Agriculture, 90*(15), 2541–2547. https://doi.org/10.1002/jsfa.4158

Venkatesh, U., Sharma, A., Ananthan, V. A., Subbiah, P., & Durga, R. (2021). Micronutrient's deficiency in India: A systematic review and meta-analysis. *Journal of Nutritional Science, 10.* https://doi.org/10.1017/jns.2021.102

Wijngaard, H. H., & Arendt, E. K. (2006). Buckwheat. *Cereal Chemistry, 83*(4), 391–401. https://doi.org/10.1094/CC-83-0391

Zhu, F. (2016). Chemical composition and health effects of Tartary buckwheat. *Food Chemistry, 203*, 231–245. https://doi.org/10.1016/j.foodchem.2016.02.050

Chapter 9

Health Benefits of Pseudocereals

K. Sneha, Ashwani Kumar, Mukul Kumar

9.1 Introduction

The cereals are monocotyledonous plants of the Poaceae family, also known as the grass family, whereas grains or edible seeds that are not from members of the Poaceae but are used as flour for staple foods are pseudocereals and are dicotyledonous species (Martínez-Villaluenga et al., 2020; Mir et al., 2018). Although these seeds resemble cereal grains, amaranth (*Amaranthus* spp. L.; family Amaranthaceae), buckwheat (*Fagopyrum esculentum* Moench.; Polygonaccac), chia (*Salvia hispanica* L.; Lamiaceae), and quinoa (*Chenopodium quinoa* Willd.; Amaranthaceae) do not belong to the Poaceae family and are termed pseudocereals (Bekkering and Tian, 2019). There are also some less well-known pseudocereals, such as chan (*Hyptis suaveolens*; Lamiaceae), jícaro seed (*Crescentia alata*; Bignoniaceae), ojoche (*Brosimum alicastrum*; Moraceae), Andean lupin (*Lupinus mutabilis*; Fabaceae) (Acuña-Gutiérrez et al., 2019), kaniwa (*Chenopodium pallidicaule*; Aamaranthaceae), and kiwicha (*Amaranthus* caudatus; Amaranthaceae) (Repo-Carrasco-Valencia et al., 2010).

Amaranth is an annual plant with broad leaves and grows to a height of 0.5 to 3 m depending on the species. The amaranth seeds are lentil-shaped and are 1.0 mm in diameter. The color of the seeds varies from black, brown, red, golden, and yellow to white due to the presence of betalain pigments (Berghofer and Schoenlechner, 2002; Malik et al., 2023). Two species of buckwheat are closely related and cultivated worldwide, namely common buckwheat (*F. esculentum* Moench) and Tartary buckwheat (*Fagopyrum tataricum* Gaertn.) (Janssen et al., 2017). The hull, spermoderm, endosperm, and embryo make up the buckwheat kernel, which is triangular

DOI: 10.1201/9781003325277-9

and 4–9 mm long (Alvarez-Jubete et al., 2010). Chia is an annual plant, originating from southern Mexico and northern Guatemala, producing a large number of dry, indehiscent fruits that are often referred to as seeds. In pre-Columbian times, these seeds were either white or black but chia seeds sold today have coats that vary in color from black to speckled black to white (de Falco et al., 2017; Punia and Dhull, 2019). Quinoa seeds are tiny, rounded, and flattened, with a diameter of roughly 1.5 mm (Jancurová et al., 2009).

9.2 Nutritional Composition

There has been a need for a healthier replacement for common cereals. Pseudocereals have better nutritional composition. Table 9.1 lists the nutritional composition of a few common pseudocereals.

9.2.1 Protein

Protein quality and quantity are higher in pseudocereals than in cereals. Pseudocereal seeds contain high concentrations of lysine, an essential amino acid that is limiting in cereals. Amaranth and quinoa are significant for children's nutrition because they contain high levels of arginine and histidine, both of which are vital for infants and children (Mir et al., 2018). The pseudocereal amaranth has an average protein composition of 13.1% to 21.0% (Malkar et al., 2009), whereas rice and wheat have 9.1% and 10.5% to 16%, respectively (Repo-Carrasco et al., 2003). Neotropical pseudocereals, such as jícaro seeds and Andean lupin, have higher protein concentrations of 44% and 40%, respectively (Acuña-Gutiérrez et al., 2019).

The primary seed protein components of pseudocereals, such as quinoa, are albumins and globulins (44–77 % of total protein), which is higher in concentration than the prolamins (0.5–7.0 %) (Jancurova et al., 2009). The amino acid profiles of pseudocereals, even though qualitatively similar to common cereals, have higher concentrations of essential amino acids, such as methionine and tryptophan, in the range of 316–670 mg/100 g and 582–610 mg/100 g, respectively, compared with values in rice of 161 and 209 mg/100 g (Pal et al., 2016; Motta et al., 2019, Nitrayova et al., 2014). Glutamic acid is the most abundant amino acid in all pseudocereals, accounting for 13.2% of amaranth and 21.2% of buckwheat (Motta et al., 2019). The amino acid composition of some of the common pseudocereals is given in Table 9.2.

Chia has a protein digestibility of 78.9%, which is lower than amaranth (90%) but equivalent to casein and beans, which have digestibilities of 88.6% and 77.5%, respectively. This is greater than the digestibility values for maize (66.6%), rice (59.4%), and wheat (52.7%) (Grancieri et al., 2019). The protein digestibility, Protein Efficiency Ratio (PER), Net Protein Utilization (NPU), and biological

Table 9.1 Nutritional Composition of Several Common Pseudocereals

Pseudocereal	Moisture (%)	Dietary Fibre (%)	Carbohydrate (%)	Protein (%)	Fat (%)	Ash (%)	Ref
Amaranth	10.3	20.6–27.34	65–75	14.8–16.5	5.6–8.1	2.8	Reguera and Haros, 2017; Malkar et al., 2009
Buckwheat	10–15	29.5	65–78	9.7–15	1.4–4.7	1.3–3.0	Reguera and Haros, 2017; Vršková et al., 2013
Quinoa	4.8	14.2	67–74	8–22	2–10	2.4–3.8	Reguera and Haros, 2017; Jancurova et al., 2009
Chia	3.5–6.7	18–30	26–41	15–25	32.2–36.8	4–5	Prathyusha et al., 2019
Kaniwa	10.1	12.5	63.4	18.8	7.6	4.1	Repo-Carrasco-Valencia et al., 2010
Kiwicha	11.3	5.8	61.8	11.6	7.5	1.7	Repo-Carrasco-Valencia et al., 2010

Table 9.2 Essential Amino Acid Composition of Pseudocereals

Pseudocereal	Histidine mg/100g	Isoleucine mg/100g	Leucine mg/100g	Lysine mg/100g	Methionine mg/100g	Phenylalanine mg/100g	Threonine mg/100g	Tryptophan mg/100g	Valine mg/100g	Ref
Amaranth	398	406	731	559	335	684	452	588	453	Motta et al., 2019
Buckwheat	422	470	902	624	316	813	510	582	595	Motta et al., 2019
Quinoa	431	420	795	546	361	633	446	577	486	Motta et al., 2019
Chia	633	775	1514	1183	467	1021	795	610	940	Ding et al., 2018
Rice	340	566	1030	425	161	510	503	209	895	McKevith et al., 2016

value (BV) of buckwheat seed protein were found to be 68.97%, 2.69%, 59.77%, and 86.33%, respectively (Vršková et al., 2013).

9.2.2 Carbohydrate

Carbohydrate is an important component in cereals and pseudocereals, and contributes 50% to 70% of total energy requirements of the global population. Carbohydrate consists of three groups based on their polymerization, namely sugars (monosaccharides, disaccharides, polyols), oligosaccharides, and polysaccharide (starch and non-starch) (Sindhu et al., 2019). Most pseudocereals have a starch content ranging from 55.1% to 70.4% (Reguera and Haros, 2017). The starch content of amaranth seeds ranges from 65% to 75%, with 4% to 5% dietary fiber. The sugar composition of amaranth was sucrose (0.58 to 0.75 g/100 g), glucose (0.34 to 0.42 g/100 g), fructose (0.12 to 0.17 g/100 g), maltose (0.24 to 0.28 g/100 g), raffinose (0.39 to 0.48 g/100 g), stachyose (0.15 to 0.130 g/100 g), and inositol (0.02 to 0.04 g/100 g) (Venskutonis and Kraujalis, 2013). Buckwheat seeds have a carbohydrate content of 73.3% due to the presence of the pericarp, with 55.8% starch. The starch content in tartary buckwheat was slightly higher than in common buckwheat, at 57.4% (Wijngaard and Arendt, 2006).

Quinoa has a carbohydrate content of 67–74%, with starch comprising 52–60% and other carbohydrates, such as monosaccharides (2%) and disaccharides (2.3%), crude fiber (2.5–3.9%), and pentosans(2.9–3.6%), in small amounts (Jancurová et al., 2009). Chia seeds consist of 26% to 41% carbohydrate and 18% to 30% dietary fiber, of which 5% to 10% is soluble fiber (Prathyusha et al., 2019). Chia seeds absorb water to form a gel structure with carbohydrates (34%) and crude fiber (58%) (Coorey et al., 2014). Kaniwa has a similar carbohydrate content to quinoa. It has a total sugar content, glucose, fructose, maltose, and sucrose content of 6.50%, 1.80%, 0.40%, 1.70%, and 2.60%, respectively (Repo-Carrasco-Valencia and Arana, 2017). The carbohydrate content of jícaro seed is the lowest of all pseudocereals, at 5.5%, and the carbohydrate content of Andean lupin is 32.9%, which is similar to that of chia seeds (Acuña-Gutiérrez et al., 2019).

9.2.3 Lipid

The total lipid content of amaranth seed ranges between 5.7% and 9%, of which 6.4% is glycolipids such as monogalactosyl and galactosyl compounds, while 3.6% is phospholipids, such as phosphatidylcholine, phosphatidylethanolamine, and phosphatidylinositol. Amaranth seed oil has a high level of unsaturation, with 25% oleic acid, 50% linoleic acid, and 1% linolenic acid (Gamel et al., 2007). Buckwheat wholegrain has a lipid content of 45.51–62.64 g/kg. The most abundant fatty acid is linoleic acid in the range of 37.23–47.57%, followed by oleic acid, with a range of 20.96–36.47% in common buckwheat, whereas Tartary buckwheat had 35.54–44.90% of linoleic acid and 26.20–40.76% oleic acid (Sinkovič et al., 2020). The lipid content of chia seed is about 40% (Prathyusha et al., 2019), of which 88% is unsaturated fatty acid, comprising 56.98% α-linolenic acid and 21.51% linoleic acid (Ding et al., 2018).

Linoleic acid, which accounts for 45.8–49.6% of fatty acids in kaniwa, is followed by oleic acid, which accounts for 25.8–27.9%, and linolenic acid, which accounts for 5.7–6.8% (Huamaní et al., 2020). The lipid composition of chan seed ranges between 14.3% and 33.59% of total dry weight (Acuña-Gutiérrez et al., 2019), which contains 76.13% linoleic acid and 10.83% oleic acid (Bachheti et al., 2015). Jicaro seeds contain 38% lipids, including fatty acids like oleic, linoleic, palmitic, stearic, and -linolenic acids, and Andean lupin consists of about 20% oil (Acuña-Gutiérrez et al., 2019).

9.2.4 Minerals and Vitamins

Minerals, such as calcium, magnesium, and iron, are low in gluten-free foods and the gluten-free diet. These and other essential minerals may be found adequately in the pseudocereals amaranth, quinoa, and buckwheat (Alvarez-Jubete et al., 2009). Iron, magnesium, and potassium content of pseudocereals are in the range of 2.88–7.35 mg/100 g, 196–328 mg/100 g, and 510–559 mg/100 g (Mota et al., 2016). The iron, calcium, and zinc contents of kaniwa are 4.9, 2.15, 29.7 mg/100 g, while kiwicha has corresponding contents of 5.0, 1.25, and 27.9 mg/100 g, respectively. The calcium content is considerably lower than that of other pseudocereals (Repo-Carrasco-Valencia et al., 2010). The calcium, iron, magnesium, and potassium contents of the common pseudocereals are given in Table 9.3.

Pseudocereals are rich in B vitamins. The thiamine concentrations of amaranth, buckwheat, quinoa, and chia are 0.1, 0.46, 0.38, and 0.6 mg/100 g, respectively, compared with rice at 0.4 mg/100 g (Bekkering and Tian, 2019; McKevith, 2004). The vitamin A composition of amaranth, quinoa, and chia seeds is 2, 14, and 54 IU, whereas rice and wheat had no significant amount of this vitamin (Bekkering and Tian, 2019). The vitamin contents of the common pseudocereals are given in Table 9.4.

9.2.5 Phenolic Compounds

Polyphenols are organic compounds that have large numbers of phenolic structural units and are found in plants and plant-based foods (Dhull et al., 2019). Most of these compounds have antioxidant properties among other health benefits for humans (Gordillo-Bastidas et al., 2016; Kaur et al., 2018; Dhull et al., 2020a). The total quantity of phenolic acids in amaranth grains ranged from 16.8 to 59.7 mg/100 g, with 7% to 61% being soluble phenolic acids (Venskutonis and Kraujalis, 2013). Buckwheat flour and seed hulls had total phenolic concentrations of 313.0 and 333.0 mg/100 g dry weight, respectively, and a higher flavonoid concentration in the hulls than in the flour, namely 45.6 to 9.8 mg/100 g dry weight, respectively. The flavonoids include rutin, quercetin, and hyperoside with 5.205, 0.608, and

Table 9.3 Mineral Content of Some Pseudocereals

Pseudocereal	Calcium (mg/100 g)	Iron (mg/100 g)	Magnesium (mg/100 g)	Potassium (mg/100 g)	Ref
Amaranth	200	7.3	328	552	Mota et al., 2016
Buckwheat	17.5	2.8	240	510	Mota et al., 2016
Quinoa	77	4.2	196	510	Mota et al., 2016
Chia	631	7.72	335	407	Bekkering and Tian, 2019
Kaniwa	29.7	4.9	-	-	Repo-Carrasco-Valencia et al., 2010
Kiwicha	27.9	5.0	-	-	Repo-Carrasco-Valencia et al., 2010
Wheat	15	1.17	22	107	McKevith, 2004
Rice	28	0.8	25	115	McKevith, 2004

Table 9.4 Vitamin Content of Some Pseudocereals

Pseudocereal	Vitamin A (IU)	Vitamin B₁ (mg/100 g)	Vitamin B₂ (mg/100 g)	Vitamin B₃ (mg/100 g)	Vitamin E (mg/100 g)	Ref
Amaranth	2	0.1	0.23	2.8	2.1	Bekkering and Tian, 2019
Buckwheat	0	0.46	0.14	1.8	5.46	Bekkering and Tian, 2019
Quinoa	14	0.38	0.39	1.06	5.37	Jancurova et al., 2009
Chia	54	0.6	0.2	8.8	8.2	Bekkering and Tian, 2019
Rice	0	0.4	0.02	5.8	0.11	McKevith, 2004

1.616 mg/100 g dry weight in the hull and 2.275, 0.153, and 0.198 mg/100 g dry weight in the flour, respectively (Quettier-Deleu et al., 2000).

White and black chia seeds have total phenolic levels of 3.52 ± 0.08 and 3.42 ± 0.06 mg gallic acid-equivalents/g defatted chia seeds, respectively (Tunçil and Çelik, 2019). In another study, the total phenolic and total flavonoid concentrations of defatted chia seed flour were 14.22 ± 0.22 06 mg gallic acid eq./g and 8.45 ± 0.8 mg catechin eq./g (Rahman et al., 2017). The total phenolic and total flavonoid yield from quinoa seed was 102.86 mg gallic acid eq./100 g dried seeds, and 26.93 mg quercetin eq./100 g dried seeds (Carciochi et al., 2015). Huamani et al. (2020) assessed the phenolic compound of kaniwa to be 1.4–1.9 mg eq. gallic acid/g of total phenolics, 1.5–2.0 mg eq. catechin/g total flavonoids and 2.3–42 mg/100 g of betalains, contributing to its antioxidant properties.

9.3 Health Benefits

Pseudocereals have been found to aid against obesity, diabetes, fatty liver, and child malnutrition. They are also used in the production of gluten-free foods and lactose-free milk. These pseudocereals are rich in bioactive compounds that can be categorized as nutraceuticals (Martínez-Villaluenga et al., 2020). Phytochemicals are bioactive plant components that have health-promoting effects on humans, with the potential of being incorporated into foods or food supplements as nutraceuticals (Dillard and German, 2000; Dhull and Sandhu, 2018). Plant secondary metabolites such as these include a vast range of chemical entities, such as polyphenols, flavonoids, steroidal saponins, organosulfur compounds, and vitamins (Forni et al., 2019; Dhull et al., 2020b, 2021). Preclinical research suggests that phytochemicals present in pseudocereals, such as tocopherols, alkylresorcinols, and flavonoids, are associated with health benefits such as oxidative stress prevention and reduction, as well as anti-inflammatory, antihypertensive, and cardiovascular disease prevention (Pojić and Tiwari, 2021; Henrion et al., 2021).

9.3.1 Lower Blood Cholesterol

The fiber in pseudocereals can help to reduce cholesterol synthesis in the liver by inhibiting cholesterol absorption, while binding to bile acids promotes cholesterol catabolism or fiber fermentation in the colon, promoting short-chain fatty acids that help to reduce cholesterol synthesis in the liver (Priego-Poyato, 2021). In a study, rat feed substituted with quinoa seed flour was found to effectively reduce serum total cholesterol by 26%, LDL by 57%, and triglycerides by 11% (Paśko et al., 2010). Quinoa cereal bars administered to a group of 22 subjects were studied for 30 days. The study indicated that 67.5% of the subjects had a reduction in total cholesterol, 55.9% in triglycerideconcentrations, and 66% showed a reduction in LDL. The average total cholesterol, triglycerides, and LDL (low-density lipoprotein) cholesterol reduced to 158.28 ± 19.03, 89.31 ± 43.26, and 121.31 ± 43.11 mg.dL^{-1}, respectively, in the 30 days after quinoa bar administration, from 175.76 ± 28.41, 101.83 ± 84.07, and 152.65 ± 39.97 mg.dL^{-1}, respectively (Farinazzi-Machado et al., 2012).

The total cholesterol-lowering property of Tartary buckwheat protein was studied on four different hamster groups fed with 24% of Tartary buckwheat protein, 24% rice protein, 24% wheat protein, or a control diet. The control diet resulted in a 37% decrease in total cholesterol, whereas the buckwheat diet had a decrease of 45%, and the rice protein and wheat protein diets had decreases of only 10 and 13%, respectively (Zhang et al., 2017). The effect of buckwheat was studied on hamsters fed a high-fat high-cholesterol diet along with buckwheat or sprouted buckwheat. The study showed that high-fat high-cholesterol diet fed with sprouted buckwheat or buckwheat feed reduced the total serum cholesterol to 164 mg/dl or 179 mg/dl, respectively compared to the control with high-fat high-cholesterol feed having a total serum cholesterol level of 217 mg/dl (Lin et al., 2008). The cholesterol-lowering property of quinoa biscuits in humans was studied for four weeks in adults administered 15 g of biscuits with 60% quinoa. The quinoa-induced decrease in total cholesterol was 5% (5.83 ± 1.24 vs. 5.53 ± 1.07 mmol/L), while LDL cholesterol showed a 7% decrease (3.61 ± 0.96 vs. 3.36 ± 0.87 mmol/L), and a 3% decrease in LDL-to-cholesterol ratio (3.60 ± 1.00 vs. 3.50 ± 0.90) in comparison to their initial levels (Pourshahidi et al., 2020). A clinical study on hypercholesterolemic human subjects administered 59 g of amaranth per day for 28 days resulted in a decrease in total cholesterol level of 45% (Meier et al., 2000).

Moderately hypercholesterolemic patients were subjected to 50 g of defatted amaranth snacks and their plasma lipid levels were studied for two months. There was a significant reduction of HDL cholesterol of 15.2% compared with a reduction in the placebo of 4% (Chávez-Jáuregui et al., 2010).

9.3.2 Antioxidant

Antioxidants help in scavenging free radicals and reduce the risk of degenerative diseases (Kaur et al., 2018; Dhull et al., 2020a). Antioxidants, such as polyphenols, anthocyanins, flavonoids, phytates, avenanthramides, and beta carotenes, are among the phytochemicals with nutraceutical activities that have been found in pseudocereals (Malleshi et al., 2021). Buckwheat is high in flavonoids such as glycosidic forms of quercetin a flavonol, glycosidic forms of apigenin and luteolin (primary flavones), and phenolic acids, making it one of the richest grain sources of polyphenols (Rocchetti et al., 2019). Antioxidants isolated from buckwheat (*F. esculentum*) hull extract were identified as protocatechuic acid (13.4% of dried hulls), 3,4-dihydroxybenzaldehyde (6.1%), hyperin (5%), rutin (4.3%), and quercetin (2.5%) (Watanabe et al., 1997). Canihua (*Chenopodium pallidicaule*) is an annual pseudocereal which is starch-rich, somewhat similar to the related quinoa. Canihua contains antioxidant and polyphenolic compounds such as vanillic acid, kaempferol, catechin gallate, catechin, ferulic acid, quercetin, resorcinol, and 4-methyl resorcinol. Resorcinol was found to be the major antioxidant in the water-soluble extract of canihua (Peñarrieta et al., 2008).

Pseudocereals, such as buckwheat, amaranth, quinoa and their products, were studied for their free phenol contents. The free phenol ranges from 12.4 to 678.1 mg gallic acid-equivalents (GAE)/100 g, of which buckwheat products had the highest levels of 146.8–678.1 mg GAE/100 g, while quinoa and amaranth products had considerably lower levels of up to 226.1 mg GAE/100 g. The total antioxidant capacity (TAC) of buckwheat products, quinoa products, and amaranth products, using the DPPH method, was 167.3–473.9 mg Trolox-equivalents (TE)/100 g, 78.2–100.6 mg TE/100 g and 25.0–69.7 TE/100 g, respectively. Similarly, the TAC of buckwheat products, quinoa products, and amaranth products, using the ABTS method, were 876.9–3524.8 mg TE/100 g, 738.9–984.5 mg TE/100 g and 118.2–431.4 mg TE/100 g, respectively (Škrovánková et al., 2020). The antioxidant activity of chia seeds found in various regions using the 2,2-azino-bis(3-ethylbenz othiazoline-6-sulphonic acid) (ABTS) method showed that the antioxidant capacity of the chia seeds ranged from 1.46 mM Trolox Equivalent Antioxidant Activity (TEAC)/g to 1.05 mM TEAC/g (Kačmárová et al., 2016).

9.3.3 Anti-carcinogenic

Buckwheat's anticancer effects are attributed to its polysaccharides, flavonoids, phenylpropanoids, lectins, proteins, and peptides, that induce cytotoxicity, differentiation, apoptosis, and growth inhibition in a range of cancer cell lines (Bastida et al., 2019). The natural flavonoid quercetin (quercetin-3-rhamnoside) is a glycoside found in buckwheat in quantities between 0.01 and 0.05% in Tartary buckwheat, and between 0.54 and 1.80% in common buckwheat. It has been proven that quercetin can prevent the development of malignant cells in leukemia, breast, liver, ovarian, colorectal, gastric, and endometrial malignancies (Das et al., 2019). The cytotoxic properties of buckwheat hull extract were assessed using the mitochondrial cytotoxic test (MTT) on human HeLa cervical cancer cells. At a dosage of 100 g/ml, the growth inhibitory effect of common buckwheat on HeLa cancer cells was up to 50% (Danihelová et al., 2013). A study on female rates has shown that buckwheat protein extract can regulate serum estradiol levels and reduce mammary tumor development (Kayashita et al., 1999). The impact of consumption of buckwheat protein product on 1,2-dimethylhydrazine (DMH)-induced colon carcinoma in male rats was investigated. The incidence of colonic cancer was reduced by 47% when buckwheat protein was consumed. Consumption of buckwheat protein products decreased the frequency of colon adenocarcinomas and dramatically decreased cell proliferation and c-myc and c-fos protein expression in the colonic epithelium. This implies that a dietary buckwheat protein product reduces cell proliferation and thereby protects from DMH-induced colon carcinogenesis (Liu et al., 2001).

In a study, heat denaturation of amaranth seeds, followed by simulated gastrointestinal digestion, resulted in the release of essential amino acids and peptides with high antioxidant activity. The antitumor efficacy of these hydrolysates was tested in breast cancer cells *in vitro*. For the cytotoxicity effects, different concentrations of amaranth seed protein hydrolysate (ASP-HD) were tested for 24 hours. When cells were treated with ASP-HD at a concentration of 48.3±0.2 g/ml, the results showed that cell proliferation was inhibited in a concentration-dependent pattern. This demonstrates that amaranth seed protein hydrolysate has anticancer properties (Taniya et al., 2020). The anticancer effects of alcalase, trypsin, and pepsin from amaranth protein hydrolysates were investigated in an *in vitro* study. On MCF-7, A549, and HEK 293 cancer cell lines, the MTT cytotoxicity test revealed that the trypsin hydrolysate had a significant anticancer effect (Ramkisson et al., 2020).

A study used cancer cell growth inhibition to determine the anticancer efficacy of chia seed. A strong dosage-dependent cytotoxic impact was seen in human breast epithelial MCF-7 cells when exposed to chia seed at various concentrations of exposure over 24 hours. The cell viability at concentrations of 100, 300, and 1000 μl/mg were 71.7, 49.4 and 39.6%, respectively, which demonstrates that chia seed extract has anti-cancer properties (Mutar and Alsadooni, 2019). Human cancer cells were used to test chia seed oil's anticancer properties. The anticancer activity of chia seed oil in an MTT test indicated cytotoxic effectiveness of up to 90%. It reduced chronic myelogenous cells (CM leukemia) growth by up to 67% in a trypan blue assay. The trypan blue assay on MCF-7 (human breast tumor) cells demonstrated a reduction of cell proliferation of up to 47% (Gazem et al., 2017).

9.3.4 Antidiabetic

Diabetes is a metabolic illness in which the body experiences hyperglycemic and hypoglycemia states as a result of changes in glucose, lipid, and protein metabolism. This condition is treatable or can be managed with a switch to coarse grains such as pseudocereals in the regular diet since it includes complex carbohydrates and natural inhibitors of the enzymes that cause diabetes (Malleshi et al., 2021). Pseudocereals, such as buckwheat, have a low glycemic index, which help to improve insulin resistance and glycemic control. A study was conducted by substituting white wheat bread products for breakfast with pseudocereal buckwheat-based food to evaluate blood glucose levels. The peak blood glucose levels found in subjects were 174 mg/dl after 120 minutes on buckwheat-based food, compared on white wheat bread with 263 mg/dl after 90 minutes of consumption. The glycemic index of buckwheat is 26.8 whereas the white wheat bread has a glycemic index of 100 (Gabrial et al., 2016).

The effects of a buckwheat concentrate containing D-chiro-inositol (D-CI) , an insulin mediator component, on hyperglycemia and glucose tolerance in streptozotocin-induced diabetic (STZ) rats were studied. The antihyperglycemic effects of chemically synthesized D-CI were investigated. Buckwheat contains a significant amount of D-CI, which can help lower blood glucose levels in diabetic individuals. The rats were fed different buckwheat concentrates containing 10 or 20 mg of D-CI/kg of body weight, and both were effective in lowering serum glucose concentrations by 12–19% after 90 and 120 minutes of administration (Kawa et al., 2003).

The effect of amaranth seed extract was studied against non-insulin-dependent diabetes in rats. It was observed that a single oral dose of 2000 mg/kg body weight improved glucose tolerance in rats. It was also observed that oral administration of 1000 mg/kg body weight for 21 days improved glucose tolerance and reduced glycated hemoglobin levels by 19.83% in type 2 diabetic Goto-Kakizaki rats (Zambrana et al., 2018). Peptides extracted from quinoa flour were found to inhibit enzymes involved in carbohydrate metabolism such as dipeptidyl peptidase IV (DPP-IV) and α-glucosidase, which reduces glucose absorption in the blood (Vilcacundo et al., 2017). The impact of quinoa seed extract on blood glucose was studied for five weeks. The study revealed that the blood glucose levels in quinoa-fed rats were lowered by 10% (Paśko et al., 2010).

9.3.5 Other Health Benefits

Pseudocereals have a high protein content and a good amino acid balance, so that they may be used as a replacement for those who can only eat a small number of foods rich in proteins and amino acids, such as meat, eggs, and milk, owing to intolerance or other factors (Zandona et al., 2020). Pseudocereals are nutritionally acceptable substitutes for starch replacement in products including gluten-free breads, spaghetti, breakfast cereals, and cookies (Henrion et al., 2021). Amaranth seeds have a protein content of 16.5%, primarily albumins and globulins, which is higher than in most cereals. Amaranth seeds have a high concentration of essential amino acids and a more balanced amino acid composition than cereal grains. It is also rich in riboflavin, vitamin E, calcium, magnesium, and iron. Prolamins, the proteins toxic to celiac patients, make up just 0.7 and 1.3% of total proteins in whole and defatted amaranth flours, respectively, whereas they make up about 30–35% of total proteins in other gluten-containing cereals (Ballabio et al., 2011).

Dehulled quinoa flour contains 15.6% protein, 6.3% of which is lysine, values which are comparable to soybean protein. Quinoa protein can be matched to the milk protein casein due to the similarity in protein efficiency ratio, protein digestibility, and nitrogen balance (Ranhotra et al., 1993). Quinoa has a high biological value (83%) due to its high protein content, which contains all of the necessary amino acids. Minerals, such as calcium, iron, zinc, magnesium, and manganese, are also abundant in quinoa (Amador et al., 2014).

The predominant dietary fiber in pseudocereals is pectin, unlike non-cellulose dietary fiber polysaccharides, cellulose, and β-glucans found in cereals (Ciudad-Mulero et al., 2019). Quinoa has a total dietary fiber content of 10% whereas amaranth has total dietary fibre value of 11%. About 78 % of the dietary fiber in both these pseudocereals is insoluble. Galacturonic acid, arabinose, galactose, xylose, and glucose are the major components of insoluble fiber from quinoa and amaranth. Soluble dietary fiber makes up 22% of total dietary fiber in both pseudocereals, which are higher than in wheat and maize (15%) (Lamothe et al., 2015).

Pseudocereals are rich in essential micronutrients, unlike cereals such as corn, rice, and wheat. Pseudocereals have nearly double the mineral concentrations of ordinary cereals (Pirzadah and Malik, 2020).

Postmenopausal women who consumed 25 g of milled chia seeds per day for seven weeks had substantial increases in plasma α-linolenic acid (ALA) and eicosapentaenoic acid (EPA) of 135% and 30%, respectively. This could be a source of fatty acids alternative to the conventional routes of fish and fish products (Jin et al., 2012).

9.4 Conclusion

Pseudocereals are good sources of essential nutrients such as protein, fiber, vitamins, minerals, and phytochemicals. Pseudocereals can be used as a supplement or replacement for the common cereals as their nutritional composition makes them a superior source. Their uses in the prevention and management of diseases like obesity, diabetes, hypertension, hypercholesterolemia, and cancer have been studied. It has also been found that pseudocereals are gluten-free and have potential antioxidant capacity. There are several other minor pseudocereals, such as chan, jícaro seeds, ojoche, Andean lupine, kaniwa, and kiwicha among others, that need to be studied further.

References

Acuña-Gutiérrez, C., Campos-Boza, S., Hernández-Pridybailo, A., and Jiménez, V. M. (2019). Nutritional and industrial relevance of particular Neotropical pseudo-cereals. In Piatti, C., Graeff-Hönninger, S. and Khajehei, F. (Eds.), *Food Tech Transitions: Reconnecting Agri-Food, Technology and Society* (pp. 65–80). Springer. https://doi.org/10.1007/978-3-030-21059-5_4

Alvarez-Jubete, L., Arendt, E. K., and Gallagher, E. (2009). Nutritive value and chemical composition of pseudocereals as gluten-free ingredients. *International Journal of Food Sciences and Nutrition*, 60(sup4), 240–257. https://doi.org/10.1080/09637480902950597

Alvarez-Jubete, L., Arendt, E. K., and Gallagher, E. (2010). Nutritive value of pseudocereals and their increasing use as functional gluten-free ingredients. *Trends in Food Science and Technology*, 21(2), 106–113. https://doi.org/10.1016/j.tifs.2009.10.014

Amador, M. M. D. L., Montilla, C. I. M., and Martín, S. C. (2014). Alternative grains as potential raw material for gluten–free food development in the diet of celiac and gluten–sensitive patients. *Austin Journal of Food and Nutrition*, 2(3), 1–9. http://hdl.handle.net/11441/67174

Bachheti, R. K., Rai, I., Joshi, A., and Satyan, R. S. (2015). Chemical composition and antimicrobial activity of *Hyptis suaveolens* Poit. seed oil from Uttarakhand State, India. *Oriental Pharmacy and Experimental Medicine*, 15(2), 141–146. https://doi.org/10.1007/s13596-015-0184-8

Ballabio, C., Uberti, F., Di Lorenzo, C., Brandolini, A., Penas, E., and Restani, P. (2011). Biochemical and immunochemical characterization of different varieties of amaranth (*Amaranthus* L. ssp.) as a safe ingredient for gluten-free products. *Journal of Agricultural and Food Chemistry*, 59(24), 12969–12974. https://doi.org/10.1021/jf2041824

Bastida, J. A. G., Llopis, J. M. L., and Zielinski, H. (2019). Buckwheat. In Johnson, J. and Wallace, T. C. (Eds.), *Whole Grains and Their Bioactives: Composition and Health* (pp. 251–268). Wiley. https://doi.org/10.1002/9781119129486.ch9

Bekkering, C. S., and Tian, L. (2019). Thinking outside of the cereal box: Breeding underutilized (pseudo) cereals for improved human nutrition. *Frontiers in Genetics*, 10, 1289. https://doi.org/10.3389/fgene.2019.01289

Berghofer, E., and Schoenlechner, R. (2002). Grain amaranth. In *Pseudocereals and Less Common Cereals* (pp. 219–260). Springer. https://doi.org/10.1007/978-3-662-09544-7_7

Carciochi, R. A., Manrique, G. D., and Dimitrov, K. (2015). Optimization of antioxidant phenolic compounds extraction from quinoa (*Chenopodium quinoa*) seeds. *Journal of Food Science and Technology*, 52(7), 4396–4404. https://doi.org/10.1007/s13197-014-1514-4

Chávez-Jáuregui, R. N., Santos, R. D., Macedo, A., Chacra, A. P. M., Martinez, T. L., and Arêas, J. A. G. (2010). Effects of defatted amaranth (*Amaranthus caudatus* L.) snacks on lipid metabolism of patients with moderate hypercholesterolemia. *Food Science and Technology*, 30(4), 1007–1010. https://doi.org/10.1590/S0101-20612010000400026

Ciudad-Mulero, M., Fernández-Ruiz, V., Matallana-González, M. C., and Morales, P. (2019). Dietary fiber sources and human benefits: The case study of cereal and pseudocereals. In *Advances in Food and Nutrition Research* (Vol. 90, pp. 83–134). Academic Press. https://doi.org/10.1016/bs.afnr.2019.02.002

Coorey, R., Tjoe, A., and Jayasena, V. (2014). Gelling properties of chia seed and flour. *Journal of Food Science*, 79(5), E859–E866. https://doi.org/10.1111/1750-3841.12444

Danihelová, M. (2013). Cytotoxic and antioxidant activity of buckwheat hull extracts. *Journal of Microbiology, Biotechnology and Food Sciences*, 2(2), 1314–1323.

Das, S. K., Avasthe, R. K., Ghosh, G. K., and Dutta, S. K. (2019). Pseudocereal buckwheat with potential anticancer activity. *Bulletin of Pure and Applied Sciences Section B-Botany*, 38(2), 94–95. https://www.researchgate.net/profile/Shaon_Das4/publication/337949152_Pseudocereal_Buckwheat_With_Potential_Anticancer_Activity/links/5df7a5ad4585159aa482bc7a/Pseudocereal-Buckwheat-With-Potential-Anticancer-Activity.pdf

de Falco, B., Amato, M., and Lanzotti, V. (2017). Chia seeds products: An overview. *Phytochemistry Reviews*, 16(4), 745–760. https://doi.org/10.1007/s11101-017-9511-7

Dhull, S. B., and Sandhu, K. S. (2018). Wheat-fenugreek composite flour noodles: Effect on functional, pasting, cooking and sensory properties. *Current Research in Nutrition and Food Science Journal*, 6(1), 174–182.

Dhull, S. B., Kaur, M., and Sandhu, K. S. (2020a). Antioxidant characterization and in vitro DNA damage protection potential of some Indian fenugreek (*Trigonella foenum-graecum*) cultivars: Effect of solvents. *Journal of Food Science and Technology*, 57, 3457–3466.

Dhull, S. B., Punia, S., Kidwai, M. K., Kaur, M., Chawla, P., Purewal, S. S., Sangwan, M., and Palthania, S. (2020b). Solid-state fermentation of lentil (*Lens culinaris* L.) with *Aspergillus awamori*: Effect on phenolic compounds, mineral content, and their bioavailability. *Legume Science*, 2(3), 1–12.

Dhull, S. B., Punia, S., Kumar, R., Kumar, M., Nain, K. B., Jangra, K., and Chudamani, C. (2021). Solid state fermentation of fenugreek (*Trigonella foenum-graecum*): Implications on bioactive compounds, mineral content and in vitro bioavailability. *Journal of Food Science and Technology*, 58, 1927–1936.

Dhull, S. B., Punia, S., Sandhu, K. S., Chawla, P., Kaur, R., and Singh, A. (2019). Effect of debittered fenugreek (*Trigonella foenum-graecum* L.) flour addition on physical, nutritional, antioxidant, and sensory properties of wheat flour rusk. *Legume Science*, 2(1), 1–9.

Dillard, C. J., and German, J. B. (2000). Phytochemicals: Nutraceuticals and human health. *Journal of the Science of Food and Agriculture*, 80(12), 1744–1756. https://doi.org/10.1002/1097-0010(20000915)80:12%3C1744::AID-JSFA725%3E3.0.CO;2-W

Ding, Y., Lin, H. W., Lin, Y. L., Yang, D. J., Yu, Y. S., Chen, J. W., Wang, S. Y., and Chen, Y. C. (2018). Nutritional composition in the chia seed and its processing properties on restructured ham-like products. *Journal of Food and Drug Analysis*, 26(1), 124–134. https://doi.org/10.1016/j.jfda.2016.12.012

Farinazzi-Machado, F. M. V., Barbalho, S. M., Oshiiwa, M., Goulart, R., and Pessan Junior, O. (2012). Use of cereal bars with quinoa (*Chenopodium quinoa* Willd.) to reduce risk factors related to cardiovascular diseases. *Food Science and Technology*, 32(2), 239–244. https://doi.org/10.1590/S0101-20612012005000040

Forni, C., Facchiano, F., Bartoli, M., Pieretti, S., Facchiano, A., D'Arcangelo, D., Norelli, S., Valle, G., Nisini, R., Beninati, S., Tabolacci, C., and Jadeja, R. N. (2019). Beneficial role of phytochemicals on oxidative stress and age-related diseases. *BioMed Research International*, 2019, 1–16. https://doi.org/10.1155/2019/8748253

Gabrial, S. G., Shakib, M. C. R., and Gabrial, G. N. (2016). Effect of pseudocereal-based breakfast meals on the first and second meal glucose tolerance in healthy and diabetic subjects. *Open Access Macedonian Journal of Medical Sciences*, 4(4), 565. https://doi.org/10.3889%2Foamjms.2016.115

Gamel, T. H., Mesallam, A. S., Damir, A. A., Shekib, L. A., and Linssen, J. P. (2007). Characterization of amaranth seed oils. *Journal of Food Lipids*, 14(3), 323–334. https://doi.org/10.1111/j.1745-4522.2007.00089.x

Gazem, R. A. A., Puneeth, H. R., Shivmadhu, C., and Madhu, A. C. S. (2017). In vitro anticancer and anti-lipoxygenase activities of chia seed oil and its blends with selected vegetable oils. *Asian Journal of Pharmaceutical and Clinical Research*, 10(10). http://doi.org/10.22159/ajpcr.2017.v10i10.19450

Gordillo-Bastidas, E., Díaz-Rizzolo, D. A., Roura, E., Massanés, T., and Gomis, R. (2016). Quinoa (*Chenopodium quinoa* Willd), from nutritional value to potential health benefits: An integrative review. *Journal of Nutrition and Food Sciences*, 6(497). http://doi.org/10.4172/2155-9600.1000497

Grancieri, M., Martino, H. S. D., and Gonzalez de Mejia, E. (2019). Chia seed (Salvia hispanica L.) as a source of proteins and bioactive peptides with health benefits: A review. *Comprehensive Reviews in Food Science and Food Safety*, 18(2), 480–499. https://doi.org/10.1111/1541-4337.12423

Henrion, M., Labat, E., and Lamothe, L. (2021). Pseudocereals as healthy grains. In Pojić, M. and Tiwari, U. (Eds.), *Innovative Processing Technologies for Healthy Grains* (pp. 37–59). Wiley Blackwell Pub. https://doi.org/10.1002/9781119470182.ch3

Huamaní, F., Tapia, M., Portales, R., Doroteo, V., Ruiz, C., and Rojas, R. (2020). Proximate analysis, phenolics, betalains, and antioxidant activities of three ecotypes of kañiwa (Chenopodium pallidicaule aellen) from Perú. *Pharmacologyonline*, 1, 229–236.

Jancurová, M., Minarovičová, L., and Dandar, A. (2009). Quinoa–a rewiev. *Czech Journal of Food Sciences*, 27(2), 71–79. https://www.agriculturejournals.cz/publicFiles/32_2008-CJFS.pdf

Janssen, F., Pauly, A., Rombouts, I., Jansens, K. J., Deleu, L. J., and Delcour, J. A. (2017). Proteins of amaranth (*Amaranthus* spp.), buckwheat (*Fagopyrum* spp.), and quinoa (*Chenopodium* spp.): A food science and technology perspective. *Comprehensive Reviews in Food Science and Food Safety*, 16(1), 39–58. https://doi.org/10.1111/1541-4337.12240

Jin, F., Nieman, D. C., Sha, W., Xie, G., Qiu, Y., and Jia, W. (2012). Supplementation of milled chia seeds increases plasma ALA and EPA in postmenopausal women. *Plant Foods for Human Nutrition*, 67(2), 105–110. https://doi.org/10.1007/s11130-012-0286-0

Kawa, J. M., Taylor, C. G., and Przybylski, R. (2003). Buckwheat concentrate reduces serum glucose in streptozotocin-diabetic rats. *Journal of Agricultural and Food Chemistry*, 51(25), 7287–7291. https://doi.org/10.1021/jf0302153

Kayashita, J., Shimaoka, I., Nakajoh, M., Kishida, N., and Kato, N. (1999). Consumption of a buckwheat protein extract retards 7, 12-dimethylbenz [α] anthracene-induced mammary carcinogenesis in rats. *Bioscience, Biotechnology, and Biochemistry*, 63(10), 1837–1839. https://doi.org/10.1271/bbb.63.1837

Lamothe, L. M., Srichuwong, S., Reuhs, B. L., and Hamaker, B. R. (2015). Quinoa (*Chenopodium* quinoa W.) and amaranth (*Amaranthus caudatus* L.) provide dietary fibres high in pectic substances and xyloglucans. *Food Chemistry*, 167, 490–496. https://doi.org/10.1016/j.foodchem.2014.07.022

Lin, L. Y., Peng, C. C., Yang, Y. L., and Peng, R. Y. (2008). Optimization of bioactive compounds in buckwheat sprouts and their effect on blood cholesterol in hamsters. *Journal of Agricultural and Food Chemistry*, 56(4), 1216–1223. https://doi.org/10.1021/jf072886x

Liu, Z., Ishikawa, W., Huang, X., Tomotake, H., Kayashita, J., Watanabe, H., and Kato, N. (2001). A buckwheat protein product suppresses 1, 2-dimethylhydrazine-induced colon carcinogenesis in rats by reducing cell proliferation. *The Journal of Nutrition*, 131(6), 1850–1853. https://doi.org/10.1093/jn/131.6.1850

Kačmárová, K., Lavová, B., Socha, P., and Urminská, D. (2016). Characterization of protein fractions and antioxidant activity of Chia seeds (*Salvia hispanica* L.). *Potravinarstvo Slovak Journal of Food Sciences*, 10(1), 78–82. https://doi.org/10.5219/563

Kaur, P., Dhull, S. B., Sandhu, K. S., Salar, R. K., and Purewal, S. S. (2018). Tulsi (*Ocimum tenuiflorum*) seeds: In vitro DNA damage protection, bioactive compounds and antioxidant potential. *Journal of Food Measurement and Characterization*, 12(3), 1530–1538.

Maier, S. M., Turner, N. D., and Lupton, J. R. (2000). Serum lipids in hypercholesterolemic men and women consuming oat bran and amaranth products. *Cereal Chemistry*, 77(3), 297–302. https://doi.org/10.1094/CCHEM.2000.77.3.297

Malik, M., Sindhu, R., Dhull, S. B., Bou-Mitri, C., Singh, Y., Panwar, S., and Khatkar, B. S. (2023). Nutritional composition, functionality, and processing technologies for amaranth. *Journal of Food Processing and Preservation*, 2023, 1753029. https://doi.org/10.1155/2023/1753029

Martínez-Villalueng, C., Peñas, E., and Hernández-Ledesma, B. (2020). Pseudocereal grains: Nutritional value, health benefits and currentapplications for the development of Gluten-Free Foods. *Food and Chemical Toxicology*, 137, 111178. https://doi.org/10.1016/j.fct.2020.111178

Malleshi, N. G., Agarwal, A., Tiwari, A., and Sood, S. (2021). Nutritional quality and health benefits. In *Millets and Pseudo Cereals* (pp. 159–168). Woodhead Publishing. https://doi.org/10.1016/B978-0-12-820089-6.00009-4

McKevith, B. (2004). Nutritional aspects of cereals. *Nutrition Bulletin*, 29(2), 111–142. https://doi.org/10.1111/j.1467-3010.2004.00418.x

Mir, N. A., Riar, C. S., and Singh, S. (2018). Nutritional constituents of pseudo cereals and their potential use in food systems: A review. *Trends in Food Science and Technology*, 75, 170–180. https://doi.org/10.1016/j.tifs.2018.03.016

Mlakar, S. G., Turinek, M., Jakop, M., Bavec, M., and Bavec, F. (2009). Nutrition value and use of grain amaranth: Potential future application in bread making. *Agricultura*, 6(4), 43–53.

Mota, C., Nascimento, A. C., Santos, M., Delgado, I., Coelho, I., Rego, A., Matos, A. S., Torres, D., and Castanheira, I. (2016). The effect of cooking methods on the mineral content of quinoa (Chenopodium quinoa), amaranth (Amaranthus sp.) and buckwheat (Fagopyrum esculentum). *Journal of Food Composition and Analysis*, 49, 57–64. https://doi.org/10.1016/j.jfca.2016.02.006

Motta, C., Castanheira, I., Gonzales, G. B., Delgado, I., Torres, D., Santos, M., and Matos, A. S. (2019). Impact of cooking methods and malting on amino acids content in amaranth, buckwheat and quinoa. *Journal of Food Composition and Analysis*, 76, 58–65. https://doi.org/10.1016/j.jfca.2018.10.001

Mutar, H. A., and Alsadooni, J. F. K. (2019). Antioxidant and anti-cancer activity of chia seed extract in breast cancer cell line. *Annals of Tropical Medicine and Health*, 22(8), 173–181. http://doi.org/10.36295/ASRO.2019.220818

Nitrayová, S., Brestenský, M., Heger, J., Patráš, P., Rafay, J., and Sirotkin, A. (2014). Amino acids and fatty acids profile of chia (Salvia hispanica L.) and flax (Linum usitatissimum L.) seed. *Slovak Journal of Food Sciences*, 8(1), 72–76.

Pal, P., Kaur, P., Singh, N., Kaur, A., Misra, N. N., Tiwari, B. K., Cullen, P. J., and Virdi, A. S. (2016). Effect of nonthermal plasma on physico-chemical, amino acid composition, pasting and protein characteristics of short and long grain rice flour. *Food Research International*, 81, 50–57. http://dx.doi.org/10.1016/j.foodres.2015.12.019

Paśko, P., Zagrodzki, P., Bartoń, H., Chłopicka, J., and Gorinstein, S. (2010). Effect of quinoa seeds (Chenopodium quinoa) in diet on some biochemical parameters and essential elements in blood of high fructose-fed rats. *Plant Foods for Human Nutrition*, 65(4), 333–338. https://doi.org/10.1007/s11130-010-0197-x

Peñarrieta, J. M., Alvarado, J. A., Åkesson, B., and Bergenståhl, B. (2008). Total antioxidant capacity and content of flavonoids and other phenolic compounds in canihua (Chenopodium pallidicaule): An Andean pseudocereal. *Molecular Nutrition and Food Research*, 52(6), 708–717. https://doi.org/10.1002/mnfr.200700189

Pirzadah, T. B., and Malik, B. (2020). Pseudocereals as super foods of 21st century: Recent technological interventions. *Journal of Agriculture and Food Research. Energy (Kcal)*, 355(345), 346. https://doi.org/10.1016/j.jafr.2020.100052

Pojić, M., and Tiwari, U. (2021). Processing technologies for healthy grains: Introduction. In Pojić, M. and Tiwari, U. (Eds.), *Innovative Processing Technologies for Healthy Grains* (pp. 1–8). Wiley Blackwell Pub. https://doi.org/10.1002/9781119470182.ch1

Pourshahidi, L. K., Caballero, E., Osses, A., Hyland, B. W., Ternan, N. G., and Gill, C. I. (2020). Modest improvement in CVD risk markers in older adults following quinoa (Chenopodium quinoa Willd.) consumption: A randomized-controlled crossover study with a novel food product. *European Journal of Nutrition*, 1–11. https://doi.org/10.1007/s00394-019-02169-0

Prathyusha, P., Kumari, B. A., Suneetha, W. J., and Srujana, M. N. S. (2019). Chia seeds for nutritional security. *Journal of Pharmacognosy and Phytochemistry*, 8(3), 2702–2707.

Priego-Poyato, S., Rodrigo-Garcia, M., Escudero-Feliu, J., Garcia-Costela, M., Lima-Cabello, E., Carazo-Gallego, A., Morales-Santana, S., Josefa Leon, J., and Jimenez-Lopez, J. C. (2021). Current advances research in nutraceutical compounds of legumes, pseudocereals and cereals. https://www.intechopen.com/online-first/current-advances-research-in-nutraceutical-compounds-of-legumes-pseudocereals-and-cereals

Punia, S., and Dhull, S. B. (2019). Chia seed (*Salvia hispanica* L.) mucilage (a heteropolysaccharide): Functional, thermal, rheological behaviour and its utilization. *International Journal of Biological Macromolecules*, 140, 1084–1090.

Quettier-Deleu, C., Gressier, B., Vasseur, J., Dine, T., Brunet, C., Luyckx, M., Cazin, M., Cazin, J. C., Bailleul, F., and Trotin, F. (2000). Phenolic compounds and antioxidant activities of buckwheat (Fagopyrum esculentum Moench) hulls and flour. *Journal of Ethnopharmacology*, 72(1–2), 35–42. https://doi.org/10.1016/S0378-8741(00)00196-3

Rahman, M. J., de Camargo, A. C., and Shahidi, F. (2017). Phenolic and polyphenolic profiles of chia seeds and their *in-vitro* biological activities. *Journal of Functional Foods*, 35, 622–634. https://doi.org/10.1016/j.jff.2017.06.044

Ramkisson, S., Dwarka, D., Venter, S., and Mellem, J. J. (2020). *In vitro* anticancer and antioxidant potential of *Amaranthus cruentus* protein and its hydrolysates. *Food Science and Technology*, 40, 634–639. https://doi.org/10.1590/fst.36219

Ranhotra, G. S., Gelroth, J. A., Glaser, B. K., Lorenz, K. J., and Johnson, D. L. (1993). Composition and protein nutritional quality of quinoa. *Cereal Chemistry*, 70, 303–303. https://www.cerealsgrains.org/publications/cc/backissues/1993/Documents/70_303.pdf

Reguera, M., and Haros, C. M. (2017). Structure and composition of kernels. In Haros C. M. and Schoenlechner, R. (Eds.), *Pseudocereals: Chemistry and Technology* (pp. 28–48). Wiley Blackwell.

Repo-Carrasco-Valencia, R. A., Encina, C. R., Binaghi, M. J., Greco, C. B., and Ronayne de Ferrer, P. A. (2010). Effects of roasting and boiling of quinoa, kiwicha and kañiwa on composition and availability of minerals in vitro. *Journal of the Science of Food and Agriculture*, 90(12), 2068–2073. https://doi.org/10.1002/jsfa.4053

Repo-Carrasco, R., and Arana, J. V. (2017). Carbohydrates of kernels. In Haros C. M. and Schoenlechner, R. (Eds.), *Pseudocereals: Chemistry and Technology* (pp. 49–64). Wiley Blackwell.

Repo-Carrasco, R., Espinoza, C., and Jacobsen, S. E. (2003). Nutritional value and use of the Andean crops quinoa (Chenopodium quinoa) and kañiwa (Chenopodium pallidicaule). *Food Reviews International*, 19(1–2), 179–189. http://dx.doi.org/10.1081/FRI-120018884

Rocchetti, G., Miragoli, F., Zacconi, C., Lucini, L., and Rebecchi, A. (2019). Impact of cooking and fermentation by lactic acid bacteria on phenolic profile of quinoa and buckwheat seeds. *Food Research International*, 119, 886–894. https://doi.org/10.1016/j.foodres.2018.10.073

Sindhu, R. I. T. U., and Khatkar, B. S. (2019). Pseudocereals nutritional composition functional properties and food applications. *Food Bioactives: Functionality and Applications in Human Health*, 1, 410. https://books.google.co.in/books?id=Tna7DwAAQBAJ &lpg=PA129&ots=Rafw-HG7uu&dq=pseudocereal%20carbohydrate%20composition&lr&pg=PA134#v=onepage&q=pseudocereal%20carbohydrate%20composition&f=false

Sinkovič, L., Kokalj, D., Vidrih, R., and Meglič, V. (2020). Milling fractions fatty acid composition of common (*Fagopyrum esculentum* Moench) and Tartary (*Fagopyrum tataricum* (L.) Gaertn.) buckwheat. *Journal of Stored Products Research*, 85, 101551. https://doi.org/10.1016/j.jspr.2019.101551

Škrovánková, S., Válková, D., and Mlček, J. (2020). Polyphenols and antioxidant activity in pseudocereals and their products. *Potravinarstvo Slovak Journal of Food Sciences*. https://doi.org/10.5219/1341

Taniya, M. S., Reshma, M. V., Shanimol, P. S., Krishnan, G., and Priya, S. (2020). Bioactive peptides from amaranth seed protein hydrolysates induced apoptosis and antimigratory effects in breast cancer cells. *Food Bioscience*, 35, 100588. https://doi.org/10.1016/j.fbio.2020.100588

Tunçil, Y. E., and Çelik, Ö. F. (2019). Total phenolic contents, antioxidant and antibacterial activities of chia seeds (Salvia hispanica L.) having different coat color. AkademikZiraatDergisi, 8(1), 113–120. https://doi.org/10.29278/azd.593853

Venskutonis, P. R., and Kraujalis, P. (2013). Nutritional components of amaranth seeds and vegetables: A review on composition, properties, and uses. *Comprehensive Reviews in Food Science and Food Safety*, 12(4), 381–412. https://doi.org/10.1111/1541-4337.12021

Vilcacundo, R., Martínez-Villaluenga, C., and Hernández-Ledesma, B. (2017). Release of dipeptidyl peptidase IV, α-amylase and α-glucosidase inhibitory peptides from quinoa (*Chenopodium quinoa* Willd.) during in vitro simulated gastrointestinal digestion. *Journal of Functional Foods*, 35, 531–539. https://doi.org/10.1016/j.jff.2017.06.024

Vršková, M., Bencová, E., Foltys, V., Havrlentová, M., and Cicová, I. (2013). Protein quality evaluation of naked oat (*Avena nuda* l.) and buckwheat (*Fagopyrum esculentum* Moench.) by biological methods and PDCAAS method. *The Journal of Microbiology, Biotechnology and Food Sciences*, 2(1), 2079–2086.

Watanabe, M., Ohshita, Y., and Tsushida, T. (1997). Antioxidant compounds from buckwheat (*Fagopyrum esculentum* Moench) hulls. *Journal of Agricultural and Food Chemistry*, 45(4), 1039–1044. https://doi.org/10.1021/jf9605557

Wijngaard, H., and Arendt, E. K. (2006). Buckwheat. *Cereal Chemistry*, 83(4), 391–401. https://doi.org/10.1094/CC-83-0391

Zambrana, S., Lundqvist, L. C., Veliz, V., Catrina, S. B., Gonzales, E., and Östenson, C. G. (2018). Amaranthus caudatus stimulates insulin secretion in Goto-Kakizaki rats, a model of diabetes mellitus type 2. *Nutrients*, 10(1), 94. https://doi.org/10.3390/nu10010094

Zandona, L., Lima, C., and Lannes, S. (2020). Plant-based milk substitutes: Factors to lead to its use and benefits to human health. In *Milk Substitutes*. IntechOpen. https://www.intechopen.com/books/milk-substitutes-selected-aspects/plant-based-milk-substitutes-factors-to-lead-to-its-use-and-benefits-to-human-health

Zhang, C., Zhang, R., Li, Y. M., Liang, N., Zhao, Y., Zhu, H., He, Z., Liu, J., Hao, W., Jiao, R., Ma, K. Y., and Chen, Z. Y. (2017). Cholesterol-lowering activity of Tartary buckwheat protein. *Journal of Agricultural and Food Chemistry*, 65(9), 1900–1906. https://doi.org/10.1021/acs.jafc.7b00066

Chapter 10

Innovative Processing Technologies for Pseudocereals

Ayan Sarkar, Debapam Saha, Abhishek Pradhan, Punyadarshini Punam Tripathy

10.1 Introduction

Pseudocereals are a group of non-grass plants that are often used as substitutes for traditional cereals, such as wheat, rice, and maize. Although they are not true cereals, they are similar in nutritional composition and are often used in the same way. Some examples of pseudocereals include quinoa, amaranth, and buckwheat. These plants are often gluten-free and are rich in protein, fiber, and many other important components (Henrion et al., 2020; Malik et al., 2023; Punia & Dhull, 2019). To retain these nutrients in pseudocereal raw materials during processing for the finished products, nowadays some innovative processing technologies are employed for pseudocereals to enhance their functionality and increase their food product utilization (see Figure 10.1). For example, some processing methods have been developed to remove the bitter taste of quinoa, making it more palatable for consumers. Other technologies are being used to improve the textural and sensory qualities of pseudocereal-based products, such as gluten-free bread or pasta (Cabrera-Chávez et al., 2012; Chaple et al., 2023; De Arcangelis et al., 2020; Makdoud & Rosentrater, 2017). Thus, innovative processing technologies are important for pseudocereals because they can help to increase their market potential, provide consumers with healthier food options, and contribute to the sustainability of the food industry.

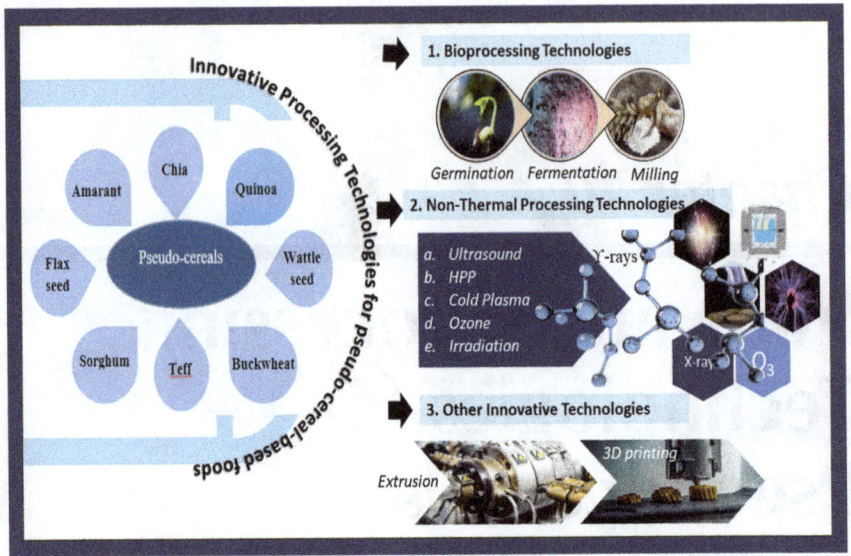

Figure 10.1   Different innovative processing technologies for pseudocereals.

There are various examples of innovative processing technologies including extrusion, high-pressure processing (HPP), cold plasma, ultrasound, ozone, 3D printing, milling, etc., which are used in different industries, including food processing, pharmaceuticals, and chemical manufacturing. The future of innovative processing of pseudocereals is promising, as there is high demand for health-promoting and sustainable food products (Sasthri et al., 2020). These innovative processing technologies are going to help in the precision processing, development of functional food ingredients, scaling up of sustainable processing, valorisation of by-products, etc. The future of innovative processing of pseudocereals is exciting, as it offers new opportunities to develop healthy, sustainable, and functional food products while contributing to the development of a more resilient and equitable food system (Agregán et al., 2022).

10.2 Different Innovative Techniques Used in the Processing of Pseudocereals

With the increasing demand for alternative grains to substitute for traditional cereals, there is a growing interest in developing innovative processing techniques for pseudocereals to improve their quality and sensory characteristics while retaining their nutritional profile intact. In this section, some of the latest techniques used in pseudocereal processing, including bioprocessing, size reduction, different non-thermal processing, extrusion cooking, and 3D printing, will be explored, and are shown in Table 10.1. These techniques not only enhance the sensory properties of pseudocereals, but also augment their nutritional quality and functional properties, making them an attractive option to serve in several food industries.

10.2.1 Bioprocessing

Bioprocessing is a set of techniques that use living organisms or their derivatives to modify or transform raw materials into high-value products. In recent years, bioprocessing of pseudocereals has attracted considerable attention due to its potential to generate value-added products with enriched nutritional quality and functionality (Sasthri et al., 2020). Pseudocereals, like quinoa, buckwheat, and amaranth, are rich in bioactive compounds (saponins, phytosterols, and flavonoids), which have potential health benefits (Ruales & Nair, 1994) and can be used as a source of functional food ingredients for nutraceutical industries. However, these compounds can also affect the sensory profile and shelf life stability of pseudocereal products (Zhang et al., 2017). Bioprocessing techniques, such as microbial fermentation (Lin et al., 2020; Dhull et al., 2020a, 2020b), enzymatic hydrolysis (Xu et al., 2021), and sprouting and germination (Tuan et al., 2018), can modify the composition, molecular interactions and morphology of different grains and increase the availability of bioactives with decrease in concentrations of antinutrients. In the next section, various bioprocessing techniques used for pseudo-cereals and their potential applications in the food industries are discussed.

10.2.1.1 Soaking and Germination

Soaking and germination are the processes by which a seed begins to germinate and transform into a new plant. In the soaking process, the seed absorbs water and swells, and the embryo, which is inside the seed, grows into a new plant. Germination involves the emergence of the root and shoot from the seed radicle and plumule, respectively, that eventually grows into a new plant. Basically, sprouting and germination of grains involve the activation of enzymes in the seed which leads to an increase in the bioavailability of essential nutrients like proteins, vitamins, and minerals, while reducing the levels of antinutrients, such as phytates and tannins. (Zhou et al. (2019) and Dhull et al. (2022a) reported that germination of buckwheat increased the protein content and antioxidant activity. Additionally, Giménez-Bastida et al. (2015) found that germination of pseudocereals improved the nutritional profile and functional properties of the grains, making them suitable for use in healthier food products. Table 10.1 demonstrates the effect of different soaking and germination times and temperature treatments on the overall nutrient landscape and organoleptic properties of different pseudocereals.

Soaking and germination of grains have been used as a therapeutic approach for several ailments due to increased bioavailability of nutrients and the breakdown of

Table 10.1 Different Innovative Processing Techniques for Pseudocereals

Pseudocereals/ products	Techniques Used	Effect on Physicochemical, Functional, and Sensory Properties	References
Black quinoa	UAE performed at 20°C for 10 min and 1.20 g/ ml sample-to-solvent ratio.	Increases in gallic acid, protocatechuic acid, vanillic acid, catechin, flavonoids, and other bioactive compounds were observed.	Melini & Melini, 2021
Buckwheat, chia, quinoa and teff	LFUS treatment given at 40% amplitude for 30–40 min using ultrasonic bath.	The hydration rate and time lag phase are improved by LFUS. It also increases the extraction of phenolic compounds, alters the structure of starch, which affects its pasting abilities, and partially denatures proteins, which promotes their interfacial characteristics and peptide availability.	Estivi et al., 2022
Buckwheat	Single and double HPP treatment (600 MPa, 30 min) in 135 L vessel; combined with a soaking (at 40°C, 4 h) pre-treatment.	Microstructural disruptions on the starch granules were observed with the ability to produce a stable foam and an emulsion layer. Decreases in gelatinization enthalpy and peak viscosity and increased AOX and higher mineral retention were noted.	Gutiérrez et al., 2022

(Continued)

Table 10.1 (Continued) Different Innovative Processing Techniques for Pseudocereals

Pseudocereals/ products	Techniques Used	Effect on Physicochemical, Functional, and Sensory Properties	References
Amaranth and quinoa	HPP treatment was given, varying the holding time (10–30 min), temperature (40–60°C) and pressure (100–600 MPa).	The starch granules swelled after the HPP treatment; RS content increased in quinoa but dropped down in amaranth compared with the native flour.	Linsberger-Martin et al., 2012
Buckwheat	CP treatment of 50–52 Pa and 1500 W; 15, 30, 45, 60, 90, and 120 s; low-pressure radiofrequency system using oxygen as the feed gas.	CP treatment above 60 s negatively affected the germination of seeds while it also significantly reduced the fungal frequency and diversity.	Mravlje et al., 2021
Quinoa	ACP treatment was given at 37.2 kHz and 20°C in different combinations: 5 min at 50 kV, 10 min at 50 kV, 5 min at 60 kV, and 10 min at 60 kV.	ACP treatment significantly affected rheological, thermal, hydration and morphological characteristics of the flour and a voltage dependent decrease in enthalpy was observed that otherwise increase with exposure time at constant voltage.	Zare et al., 2022
Amaranth	Soaking in water for 6–24 h.	Increases protein digestibility and reduces anti-nutritional factors.	Ajay & Pradyuman, 2018

(Continued)

Table 10.1 (Continued) Different Innovative Processing Techniques for Pseudocereals

Pseudocereals/ products	Techniques Used	Effect on Physicochemical, Functional, and Sensory Properties	References
Amaranth	Soaking in 0.5% NaHCO$_3$ for 8 hours.	Increases protein digestibility and decreases viscosity.	Kunyanga et al., 2012
Amaranth	Combination of soaking and sprouting.	Synergistic effect on nutritional improvement, further increasing protein digestibility, TPC and AOX.	Thakur et al., 2021
Buckwheat and quinoa	Soaking in water for 12 h.	Increases TPC, reduces phytic acid content, increases AOX, and improves *in-vitro* protein digestibility.	Lintschinger et al., 1997
Quinoa	Germination under different conditions (light, temperature, and humidity).	Affects nutrient composition, AOX, and sensory properties of quinoa sprouts.	Strenske et al., 2017
Teff	Germination for 24–72 h.	Increased TPC, protein, and fiber content, AOX, and flavonoid content.	Woldetsadik et al., 2020
Teff	Germination at 25°C for 48 h.	Increased protein content and significantly reduced phytic acid content.	Omary et al., 2012

Note: UAE: ultrasound-assisted extraction; LFUS: low-frequency ultrasound treatment; HPP: high-pressure processing; CP: cold pl asma; ACP: atmospheric cold plasma; AOX: anti-oxidant; RS: resistant starch; ACP: autologous conditioned plasma; TPC: total phenolic content.

antinutrients (Dhull et al., 2022b). Soaking treatment for around 12 h could result in increased water absorption capacity (WAC), improved water solubility index, and increased emulsifying properties, along with reduced viscosity. However, no significant alterations were observed in microstructures although increased molecular weight of proteins was reported by Lahuta et al. (1995). On increasing the sprouting time to a maximum of three days, remarkable rises in WAC, pH, antioxidant capacity, and α-glucosidase inhibitory activities were observed. Thereafter, microstructural analysis revealed smaller particle size and increased porosity with strong protein–protein interactions (Li et al., 2017). Furthermore, on increasing the germination time to five days, pseudocereals like buckwheat and quinoa exhibited higher titratable acidity, improved oil absorption capacity (OAC) and solubility, and increased protein-protein interactions (Le et al., 2021; Tai et al., 2021). Liu et al. (2022) reported that sprouting and germination of cereal and legume grains not only enhanced the nutritional properties but also contributed to the development of novel food products. Moreover, the germination could also alter the flavor, for example, with increased metabolic activity resulting in the production of reducing sugars (Lasekan & Lasekan, 2012). Overall, soaking, sprouting, and germination are natural and cost-effective methods that can improve the nutritional quality, functionality, and sensory characteristics of pseudocereals (Dhull et al., 2023).

10.2.1.2 Fermentation

Fermentation of pseudocereals is a bioprocess that involves the use of microorganisms to biologically transform the carbohydrates and proteins present in them into simpler compounds through anerobic metabolism (Dhull et al., 2020a). This process is characterized by the creation of organic acids, gases, flavor compounds, nutritional components (such as γ aminobutyric acid and bioactive peptides), and the decrease in concentration of antinutritional components (phytates or saponins) (Castro-Alba et al., 2019; Rollán et al., 2019, Dhull et al., 2020b).

Fermentation of pseudocereals can be achieved through the utilization of various microorganisms, like lactic acid bacteria (LAB), filamentous fungi, and yeasts. During the fermentation process, the microorganisms consume the available carbohydrates and produce lactic acid and other organic acids, which lower the pH of the product and also resist the growth of spoilage microorganisms which helps to enhance the shelf life of the product. The LAB used during the process also initiate the esterase activity that aims to improve the extraction of polyphenols and subsequently increase the antioxidant capacity (AOX) (Cantatore et al., 2019).

In addition, fermentation of pseudocereals can improve the nutritional benefit of the product by enhancing the bioavailability of certain nutrients, such as minerals and free amino acids (Chawla et al., 2017). These released amino acids lead to the formation of volatile compounds during baking and other post-processing methods and markedly trigger the aroma and sensory profile of the product (Corsetti & Settanni, 2007). For instance, an appreciable physical, sensory, and textural profile was achieved at 25% incorporation of fermented millets and pseudocereals in composite bread (Pradhan & Tripathy, 2022). Furthermore, the reduction of anti-nutritional factors (ANFs) in pseudocereals goes through a complex process and, for that, it is very necessary to understand the mechanism of the metabolism of ANFs. Interestingly, it has been found that fermentation was more efficient in achieving phytate reduction when applied to flour material rather than whole grains (Castro-Alba et al., 2019); a phytate loss of up to 64–93% occurred in pseudocereal flours; whereas, in the case of whole grains, losses of only 12–51% were observed. This was possibly because of endogenous phytase activity in several pseudocereals rather than the phytase formed by the added microorganisms and also owing to the higher surface area-to-volume ratio associated with flours, which would increase rates of metabolism.

Fermentation can improve the protein digestibility of pseudocereals like amaranth and also lead to changes in the grains' structure and texture, making it more palatable. The effect of fermentation is more apparent in the microstructural surface morphology of the pseudocereal flours. The proteolytic degradation occurring during the fermentation process loosens the protein-starch matrix that would otherwise present compactly inside the protein coat within the cell. As a result, starch molecules are released from the globular protein blocks, and consequently the temperature-induced swelling of starch is promoted that would otherwise be impaired in the fermented flour dough (Elkhalifa et al., 2006).

Incorporation of fermented pseudocereals in commercial bakery and extruded products has been an ongoing trend to develop functional products with tailored ingredients. The textural properties of pasta, such as hardness as well as fracturability, were found to be lower in the case of fermented quinoa flour than with native quinoa flour, but resilience and cohesiveness seemed to be improved. These changes can be achieved by proteolysis occurring during fermentation (Rizzello et al., 2017). This further can be explained by the increments of β-sheets protein configuration, compared with α-helix structures, thereby boosting the surface hydrophobicity of the fermented flour, with enhanced OAC and emulsifying properties. Several studies have been carried out on the fermentation of quinoa and other pseudocereals, like amaranth and buckwheat, which are good probiotic carriers. Huang et al. (2020) claimed that fermentation of buckwheat not only increased the levels of essential amino acids, minerals, vitamins, and phenols, but also improved its flavour and texture dramatically. The fermented buckwheat products tasted more acidic and sour, which was much preferred by Korean consumers. Interestingly, the texture of fermented buckwheat was found to be softer and less chewy than the non-fermented ones.

10.2.2 Innovations in Primary Pseudocereal Processing

The primary processing of pseudocereals involves the conversion of raw, harvested crops into edible products. This includes a range of activities, such as cleaning, dehulling, milling, and sorting. The primary processing of pseudocereals also has implications for their nutritional value, techno-functional properties, and organoleptic characteristics. Therefore, understanding primary processing is vital for developing high-quality pseudocereal-based end products that meet the growing demand for healthy and sustainable food choices.

10.2.2.1 Dry Milling

Dry milling is a type of primary processing of pseudocereals that involves size reduction by grinding the whole grain into flour or meal without the addition of any liquid. The process of dry milling typically involves passing the raw grains through a series of rollers or mills that crush and grind the grains to the desired particle size. The resulting flour or meal can be used as a base ingredient in different food products, such as bread, pasta, or snacks. In fact, dry milling of pseudocereals is often preferred over wet milling due to its lower energy requirement, reduced water usage, and enhanced shelf life of the product. Additionally, dry milling can preserve the intrinsic nutritional value and flavor of the grain since there is no exposure to water or heat. However, dry milling can also result in a lower yield and fine dust production that can pose a health hazard for workers in the milling industries (Leewatchararongjaroen & Anuntagool, 2016). Several studies have investigated the effect of dry milling on the quality and nutritional parameters of pseudocereals. Ballester-Sánchez et al. (2020) compared the physicochemical and sensory properties of quinoa flour obtained by dry and wet milling. Dry milling resulted in higher protein and fat content, while wet milling produced flour with higher ash content, in addition, dry milled quinoa flour had better sensory attributes like color, flavor, and texture (Ballester-Sánchez et al., 2020). Wang et al. (2023) described the effect of different milling techniques on the nutritional profile of quinoa flour. They found that dry-milled amaranth flour had better pasting properties and more uniform particle size distribution, along with higher protein content and AOX. Overall, dry milling plays a crucial role in the manufacture of high-quality, gluten-free products. However, careful consideration of the milling techniques and its effects on the evaluation of final product quality is necessary.

10.2.2.2 Other Milling Methods

In addition to dry milling, there are several other milling methods that can be used to process pseudocereals. These methods include wet milling, stone milling, and air classifier milling. Wet milling involves soaking the grains in water before milling, that consequently increases the yield of the milling process and results in a finer particle size. This method can also remove some of the bitter or astringent compounds present in some pseudocereals, such as saponins in quinoa. However, wet milling can result in a mild decrease in nutritional value of the final product on account of the loss of some water-soluble vitamins, minerals, carotenoids and anthocyanins. Stone milling is a traditional method of milling that involves size reduction through crushing the grains between two stone discs or plates. This method is often used for small-scale or artisanal production of flour, and is claimed to preserve the flavor and texture of the grains. But it can result in a variable particle size distribution accompanied by lower yield than is achieved by other milling methods. Air classifier milling, on the other hand, is associated with separating the different components of the grains based on their particle size and density, using air flow. This method can result in a more uniform particle size distribution and higher yield compared with other methods. Additionally, air classifier milling can separate the bran and germ from the endosperm of the grains, allowing for the production of enriched flour with a superior nutrient landscape.

Several studies have focused on the effect of various milling methods on the overall composition of pseudocereal grains. (Wang et al., 2023), comparing the physicochemical and nutritional properties of quinoa flour produced by three different methods: dry milling, stone milling, and air classifier milling. The study concluded that air classifier milling produced flour with higher protein and fiber content along with a more uniform particle size distribution, compared with the other methods. Wang et al. (2023) evaluated the effects of wet and dry milling on the biochemical and antioxidant properties of amaranth flour. Interestingly, it was reported that dry milling was associated with the retention of more antioxidant capacity, whereas wet-milled flour retained better protein and ash content. Thus, the different types of milling methods came up with their own sets of pros and cons. It is important that selection of the appropriate one should be based on the desired properties of the final product as well as the scale of production.

10.2.3 Different Non-thermal Technologies Used in the Pseudocereal Processing

Thermal processing is the most popular way of processing pseudocereal crops, because it has the potential to inactivate specific enzymes and bacteria. However, it has been discovered that, under extreme conditions, this technology can cause the loss of organoleptic qualities and nutrient bioavailability. The innovative strategy to preserve the quality of the pseudocereal products is to employ non-thermal technologies to reduce the loss of these attributes. Technologies for non-thermal preservation include ultrasound (US), cold plasma (CP), high-pressure processing (HPP), and ozone.

10.2.3.1 Ultrasonication (US)

Ultrasonication is a mechanical wave which has a frequency higher than 20 kHz, or above human audible ranges, and is commonly used for purposes such as cleaning, emulsification, dispersion, extraction, and sonochemistry. It can be divided into three types: low-frequency ultrasonication (20–100 kHz), sonochemistry range (100 kHZ–2 MHz), and medical sonication (5–10 MHz) (Mason & Peters, 2002). It is an emergent non-thermal technology that is applied to increase yield of the product, and to shorten treatment times (Bhargava et al., 2021; Mohammadi Ziarani et al., 2020).

The cavitation phenomenon is primarily responsible for the effects of ultrasound on liquid systems. Tensile forces exerted on the medium during the negative pressure half-wave (rarefaction) cause an increase in the spacing between molecules. The liquid deteriorates and empty bubbles emerge once the cavitation threshold is achieved. A cavity rapidly bursts and releases energy when it can no longer withstand the surrounding liquid pressure and, due to this, localized heat generation can happen (Estivi et al., 2022). A liquid begins to cavitate at a lower sound pressure as the temperature increases because of changes in the surface tension and viscosity. The pressure differential will be reduced and bubble implosions will be cushioned, but the water vapor pressure will maximize and generate vapor-filled bubbles.

In the processing of cereals, the effects of ultrasound are primarily focused on the extraction of bioactive compounds, starch and protein modifications, and fermentation enhancement, enzyme activation, etc. (Janve et al., 2015; Kaur & Gill, 2019; Liang et al., 2017; Zhang et al., 2015).

Ultrasound is used because it is energy-saving, with relatively low costs, helping in the use of different processes, including analysis, extraction, modification of chemical and technological properties, and emulsification (Awad et al., 2012), as well as to achieve microbial decontamination (Schmidt et al., 2019).

Ultrasound accelerates hydration and also the time span of the lag phase while performing as a pre-treatment for extracting, fermenting, and germination of pseudocereals. Furthermore, it enhances and accelerates sprouting by promoting the hydration rate, which helps to release germination promoters and remove germination inhibitors. Therefore, using ultrasound during the germination of cereals and pseudocereals under stressful conditions, that encourages the production of antioxidants, could be a simple way of accelerating the synthesis of bioactive compounds, such as phenolics.

10.2.3.2 High-pressure Processing (HPP)

HPP is an emerging non-thermal technology that is basically used for the preservation of fruit juices (Bender & Schönlechner, 2020). During high-pressure processing, food is typically treated at high pressures (typically more than 200 MPa), which, in addition to microbial inactivation, cause structural and textural alterations. Starch gelatinization and protein polymerization are key factors that influence these alterations (Vallons et al., 2011). HPP can also enhance the functional properties of the gluten-free products (Balasubramaniam et al., 2016). With increasing pressure levels, barley and wheat starches have also shown a loss of bi-refringence and an increase in granule damage (Estrada-Girón et al., 2005). Crystalline areas would only partially disintegrate as a result of HPP-induced gelatinization, unless extreme pressure conditions are reached (Balakrishna et al., 2020).

The protein-starch matrix of whole buckwheat grains showed some structural changes as a result of the HPP treatment (600 MPa for 30 min). After a four-hour pre-treatment with soaking at 40°C, the degree of modifications became more noticeable. The resulting flours had diminished emulsion activity and foaming ability, but improved water absorption capacity. Furthermore Kieffer et al. (2007) found that HPP can increase the formation of protein networks in the pseudocereals. So, HPP could be a very efficient non-thermal technology which can be used for the processing of pseudocereals.

10.2.3.3 Cold Plasma (CP)

Cold plasma, an emerging non-thermal technology, has so far been used at various stages along cereal processing for a variety of purposes, like improved germination, toxin degradation, microbial decontamination, and structural alterations in biopolymers for improved functionality. Plasma is an ionized gas composed of electrons, radicals, atoms, ions and other molecules that coexist with UV photons as well as visible light. Although plasma has no net charge, these elements give it special properties that make it conduct electricity (Marti et al., 2017; Marti et al., 2018). Protonotariou et al. (2015) indicated that CP might be used as an innovative and alternative method for seed priming and additional processing in the food sector.

However, the majority of studies examining plasma-induced modifications to grain, flour, and dough structure are driven by refining wheat flour's bread-making potential, wheat's biopolymer modifications, or the safety considerations of applying plasma to decontaminate other grains. Few studies have been conducted using plasma to improve the bread-making properties of wheat grain or other cereals. Several preliminary studies have demonstrated that CP treatment is only effective in raising the capacity of weak flours to produce bread. The impact of CP treatment on the changes of functional characteristics in the cereal processing is mostly caused by changes in time and voltage (Chaple et al., 2023).

The structures of proteins and starches can be dramatically influenced by plasma active species produced by cold plasma therapy. Comparing all treated samples to the control sample, it was discovered that the proteins of all treated samples formed more cross-links, a phenomenon which was also accelerated by prolonging the treatment period. The production of hydrogen cross-links is also influenced by cold plasma treatment, depending on the voltage used. Cold plasma treatment of quinoa flour also creates intermolecular connections, such as starch-starch, starch-protein, or protein-protein. According to Sarkar et al. (2023), CP treatment is an emerging method that can be used to improve industrial uses while lowering the anti-nutritional qualities of millets and other cereal crops. Additionally, combining various technologies, such as plasma-activated water and heat-moisture treatments, can open up new opportunities for the use of alternative grains in breadmaking (Shi et al., 2022).

10.2.3.4 Ozonation

Ozone treatment is a very important novel technology and economical method that can be used in the cereal industry for (i) fumigating stored grains, (ii) microbial disinfection, (iii) treating cereals that are contaminated with mycotoxins, and (iv) achieving changes in the physicochemical properties of the main cereal components (such as starch and protein). However, it may affect the rheological and textural characteristics of the end products, and also the color, storage, and germination capacity of the grains (Tiwari et al., 2010; Zhu, 2018). Granella et al. (2018) reported finding drying and ozone technology to be very useful to reduce microbial decontamination. Ozone treatment could reduce the fungal count by 92.86%, representing a decrease from 1.87 to 0.13 colony-forming units (cfu)/g, and also retain the physiological qualities of the wheat grains following ozone exposure for 45 minutes at 50°C drying temperature (Granella et al., 2018). The inactivation of *Fusarium graminearum* and the reduction of deoxynivalenol contamination were shown when the grains were exposed to 60 mmol/mol of ozone for 120 minutes (Savi et al., 2014) and the decrease in zearalenone and ochratoxin A also found at the same ozone exposure of contaminated maize (Qi et al., 2016). This is because mycotoxin production is frequently linked to fungal contamination of grains.

10.2.3.5 Irradiation

Fundamentally, irradiations of food is localized to be of one of three kinds, viz. X rays, gamma rays (γ-rays), and electron beams (EB). These ionizing radiations are dosed into food as per specific applications based on their characteristics and the sources they refer to (Table 10.2). Apart from these, some non-ionizing radiation treatments, like UV irradiation, of food are among the most acceptable food processing treatments these days, due to their *modus operandi*, effective results, and economical management (Pillai, 2016).

Pseudocereals, like quinoa, buckwheat, etc., have the properties of grains, face similar problems of pests and diseases as rice, during their storage and processing. This storage loss accounts for 10% of the total loss of pseudocereals each year (Luo et al., 2019). Electron beam irradiation (EBI), a new sterilisation technique with advantages of energy savings, higher efficiency, low temperature increase, safety, and dependability, is a viable alternative to conventional storage techniques, such as low-temperature storage, air conditioning, and chemical storage. Thus, EBI has been extensively used for the storage, sterilization, and delay of maturity in the industries of cereals and pseudocereals. In fact, the lipase activity in grains can be effectively inhibited with EBI, along with reduction in total microbial count and improvement in quality and shelf life during storage (Luo et al., 2019). The dosage

Table 10.2 Classification of Irradiation Treatment

Sl No.	Types of Radiation	Sources of Radiation	Characteristics
1.	Gamma rays	Gamma rays are released by radioactive versions of the elements cesium (cesium-137) and cobalt (cobalt-60).	• High penetrating power • Higher efficiency • Lower throughput
2.	X-rays	A high-energy electron stream that is directed at a target substance—typically one of the heavy metals—reflects the electrons back into the food, producing X-rays.	• Higher penetrating power • Lower efficiency • Higher throughput
3.	Electron beams	A stream of extremely fast electrons that are launched into food by an electron accelerator from an electron beam (also known as an e-beam), which is comparable to X-rays.	• Low penetrating power. • High efficiency. • High throughput.

and rate of irradiation with which grains and pseudocereals were treated largely depends on the amount of moisture present in them. Luo et al. (2021) reported a significant inhibition of bacterial count, lipase activity, and free fatty acid (FFA) content during storage following irradiation. Irradiation treatment at 8 kGy dosage resulted in a decrease in phytic acid concentration by 19%, while an increase in the total phenolic content (TPC) of 33% was noted.

A reduction in microbial load in buckwheat grains on exposure to low-energy electrons (soft electrons) was observed by Hayashi et al. (1998). Sometimes, a combination treatment of aqueous chlorine dioxide, fumaric acid, CO_2-enriched atmospheric packaging, and ultraviolet radiation (UV-C) is also given to pseudocereal sprouts for the reduction of total aerobic bacteria, yeast, and mold, without significantly affecting the organoleptic properties. However, an increase in the concentration of rutin was observed after treatment with very high dosages, which in turn increased the bitterness of the food (Chun & Song, 2014). In a comparative investigation (Orsák et al., 2001), the impact of UV, microwave, and irradiation on three buckwheat samples was observed. Various effects on polyphenol and rutin content were observed, depending on the irradiation technique and dose used. Additionally, sprouts exposed to light-emitting diodes (LEDs) had higher levels of rutin and flavone-C-glycosides (Hossen, 2007).

As a result, irradiation may extend the shelf life of food while maintaining its sensory quality, improving its microbiological quality, and augmenting its nutritional value due to the rise in bioactive substances in products generated from pseudocereal plants. Despite the fact that the general public still has little understanding of irradiation, interest in buying safety-enhanced irradiated food is rising, especially as people become more aware of its advantages and hazards).

10.2.4 Other Innovative Technologies

Other technologies, such as extrusion and 3D printing, are more useful for the processing of pseudocereal crops than the non-thermal technologies discussed above.

10.2.4.1 Extrusion

The English word "extrusion" originates from the Latin term "extrude", signifying "to thrust out" or "force out". Extrusion refers to a technique whereby materials like metal, plastic, or a blend of food components are compelled through a die or an aperture possessing the required shape, all the while undergoing elevated temperatures and pressures. In the food processing industry, this procedure is known as food extrusion (Riaz, 2019). The extrusion technique is used in many areas of food preparation. This is due to the result of various factors, such as automated control, uninterrupted operation, enhanced productivity and capacity, versatility, adaptability, superior product quality, and a diverse selection of predefined or customizable product forms and attributes. The manufacturing of cereal-based snacks has become more reliant on the extrusion process, which has been shown to be both an effective and a time-saving method. Snacks extruded using ingredients that are rich in starch, such as refined flour or corn grits, are detrimental to one's health because of their elevated fat content, caloric density, and sugar content. As a result of this reasoning, gluten-free grains such as sorghum, corn, rice, and millet have gained widespread usage in the manufacture of extruded snacks (Woomer & Adedeji, 2021). The application of extrusion cooking to pseudocereals induces various structural modifications. These include the starch gelatinization, crystalline structure disruption, and starch molecule fragmentation. Additionally, denaturation of protein and the formation of inclusion complexes between lipids and amylose have been reported. The process of extrusion induces physicochemical modifications in the extrudates, thereby expanding the potential uses of the native flour, such as the underutilized pseudocereals (Jafari et al., 2017).

Several researchers, as referenced in Table 10.3, have investigated the development of pseudocereal-based extrudates. A study employed a blend of rice flour and amaranth flour at a ratio of 75:25 to manufacture pasta enriched with minerals and fiber (Cabrera-Chávez et al., 2012). The pasta that was manufactured using gluten-free-based ingredients exhibited enhanced firmness, which can be attributed to the extrusion cooking technique. Another study involved the creation of a

Table 10.3   Major Findings in Pseudocereal-based Extruded Products

Products	Composition	Findings	References
Extruded snack	Amaranth, quinoa, and corn.	The extrudates that consisted of amaranth and quinoa exhibited the most elevated sectional expansion index.	Sajid Mushtaq et al., 2021
Pasta	Rice flour and amaranth flour in varying proportions.	Improved protein digestibility and textural characteristics were observed in the developed pasta, which was fortified with fiber and minerals.	Cabrera-Chávez et al., 2012
Pasta	Buckwheat and rice flour.	Decreased cooking loss and less sticking.	De Arcangelis et al., 2020
Spaghetti	Maize, broad bean, quinoa flour (50:30:20).	Quinoa increased protein utilization and nutrition.	Giménez et al., 2016
Extruded snack	Amaranth, buckwheat, and millet.	Rise in bulk density and nutritional enhancement	Brennan et al., 2012
Pasta	Quinoa, amaranth, and rice.	The pasta that was developed exhibited superior acceptability, although its texture was characterised by stickiness.	Makdoud & Rosentrater, 2017
Extruded snacks	Corn, quinoa, kaniwa, and amaranth.	Enhanced nutritional and structural characteristics.	Ayala-Soto et al., 2015

high-quality pasta-like product made from pseudocereals (Giménez et al., 2016). This was achieved through the addition of maize, broad beans, and quinoa in a ratio of 50:30:20. Gluten-free spaghetti, containing quinoa, increased nutrition, net protein utilization, and fiber content.

Food extrusion technology has several advantages over alternative food processing techniques. These advantages include the following. (1) Through the utilization of a variety of ingredients and raw materials, the production of a vast array of food products becomes possible through the modification of extruder feed materials or the adjustment of extruder operating conditions. The product exhibited a notable degree of adaptability, rendering it suitable for the development of innovative and original food items. This quality effectively addresses the needs of consumers who seek novel food products. (2) It enables the production of diverse items of varying appearance, color, shape, and texture that are otherwise challenging to attain through alternative techniques. (3) It has high energy efficiency, low processing cost, and a smaller footprint compared with alternative systems. Furthermore, it possesses the capability to convert by-products and waste from food processing into desirable goods.

One of the main disadvantages associated with extrusion processing is the relatively high initial installation costs. The successful execution of the process requires meticulous selection of die assembly and process parameters to prevent failure and the generation of undesired by-products. Another prerequisite is the possession of technical expertise. Although there are certain drawbacks associated with food extrusion, it is apparent that its benefits outweigh them.

10.2.4.2 3D Printing

3D food printing (3DFP) is a new technique that could tailor food to individual preferences, primarily in terms of texture and form. This is a robotic building process involving the input of food elements and their extrusion along a predetermined way which results in the deposition of successive layers (Sun et al., 2018). The steps involved in the generative procedure for 3D printing is shown in Figure 10.2. The fused deposition technique is utilized in the extrusion 3D printing of food products, where a syringe-based extruder is employed to inject a food slurry, which is highly viscous (Derossi et al., 2019). Extrusion-based printing relies on food material qualities such as moisture, rheology, crosslinking processes, and thermal properties.

The categorization of extrusion 3D printing encompasses three main types: soft-element extrusion, melted-element extrusion, and gel-forming extrusion. Soft-element extrusion has been employed in additive manufacturing to create three-dimensional structures by depositing and fusing self-supporting layers of various materials, such as meat paste, dough, and processed cheese. The viscosity of the material plays an important role in enabling extrusion through a fine nozzle, while also providing adequate support to the structure post-deposition.

Three forms of materials that are appropriate for melted-element extrusion include paste-like, powder (or solid pieces), and filament (which is infrequently

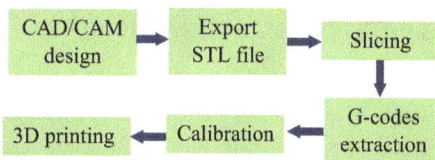

CAD/CAM design → Export STL file → Slicing → G-codes extraction → Calibration → 3D printing

Figure 10.2 Critical variables and steps in 3D food printing.

used in food-related contexts). Maintaining appropriate temperature control is a crucial factor in achieving printability during melting extrusion-based 3D printing of pastes that are predominantly composed of amorphous fat or sugar.

In gel-forming extrusion, the successful extrusion of materials that form hydrogels is reliant on the rheological characteristics of the polymer and the process by which the formation of gel occurs. Initially, the polymer solution needs to exhibit viscoelastic properties, and subsequently transitioning into self-supporting gels before the successive layers are laid out.

10.2.4.3 Application of 3D Printing Technology on Pseudocereal-based Bakery Products

Pseudocereals, namely quinoa, amaranth, and buckwheat, have garnered attention in contemporary times owing to their nutritional advantages and gluten-free nature. Three-dimensional printing technology has been employed to augment the nutrition of food products through the integration of pseudocereals.

A study involved the development of 3D printed pasta, utilizing quinoa flour as a means of enhancing the nutritional value and increasing the protein content. The pasta exhibited favourable mechanical characteristics and satisfactory sensory qualities (Torres Vargas et al., 2021). A novel approach was devised for the extrusion-based 3D printing of pizza dough, utilizing a blend of quinoa flour, brown rice flour, and plantain flour. Further comparative analysis was conducted on the physical characteristics of the subject matter in relation to the pizza dough and crust that are currently available on the market (Dey et al., 2023). Another study presents an analysis of the influence of the 3D printing on the microstructure, moisture content, and sensory perception of gluten-free snacks. These snack-bites were prepared using lupin and lupin/chickpea flour combined with pea protein isolates. Furthermore, comparative testing using lupin formulations demonstrated that the 3D-printed snack products exhibited increased stiffness and a more brittle texture compared with conventionally produced snacks (Agarwal et al., 2022). The primary objective of digitalized food printing is to enhance the nutritional profile of current food products by minimising or eliminating undesirable components and substituting them with healthier alternatives. In this context, certain vegetable oils that are abundant in omega-3 fatty acids, such as chia seed oil and flax seed oil, may be utilized as a substitute for a portion of the fats (Sun et al., 2018).

10.2.4.4 Benefits, Drawbacks, and Difficulties of 3D Printing of Foods

The 3D printing utilization technology in the food industry presents numerous potential benefits. The 3D food printing process is considerably intricate beyond its visual appearance. Optimization of various conditions is imperative in the 3D printing of foods, encompassing the appropriate mechanical force application, digital recipe design, and judicious selection of feeding elements (Yang et al., 2017). Occasionally, ambient temperature can influence the passage rate of food product through the extruder nozzle (Lipson & Kurman, 2013). Other challenges of 3D food printing include quicker printing at a higher resolution, the creation of accurate food texture, and the acceptability of the 3D printed products' new value chain.

10.3 Conclusion

Innovative processing technologies have the potential to transform the pseudocereal market and pave the way for new and exciting products. Pseudocereals like quinoa, amaranth, and buckwheat have been lauded for their exceptional nutritional value, but they have yet to reach their full potential due to a lack of up-to-date processing methods. Micronutrient bioavailability, protein content, and functional qualities like emulsification, gelling, and foaming may all be improved with the use of various innovative technologies. Cold plasma, ultrasound, irradiation, extrusion, and 3D printing are among the cutting-edge processing techniques that have been successfully applied to pseudocereals. These techniques offer numerous advantages, including increased efficiency, improved product consistency, and decreased production costs. Furthermore, these technologies enable the development of novel and enticing foods that satisfy the requirements of consumers pursuing healthier and more sustainable options. The pseudocereal industry can achieve sustained growth and meet consumer demands for healthier and more sustainable food options by fully utilizing these technological advancements.

References

Agarwal, D., Wallace, A., Kim, E. H.-J., Wadamori, Y., Feng, L., Hedderley, D., & Morgenstern, M. P. (2022). Rheological, structural and textural characteristics of 3D-printed and conventionally-produced gluten-free snack made with chickpea and lupin flour. *Future Foods, 5*, 100134.

Agregán, R., Guzel, N., Guzel, M., Bangar, S. P., Zengin, G., Kumar, M., & Lorenzo, J. M. (2022). The effects of processing technologies on nutritional and anti-nutritional properties of pseudocereals and minor cereal. *Food and Bioprocess Technology, 16*(5), 961–986.

Ajay, S., & Pradyuman, K. (2018). Optimization of gluten free biscuit from foxtail, copra meal and amaranth. *Food Science and Technology, 39*(1), 43–49.

Awad, T., Moharram, H., Shaltout, O., Asker, D., & Youssef, M. (2012). Applications of ultrasound in analysis, processing and quality control of food: A review. *Food Research International, 48*(2), 410–427.

Ayala-Soto, F. E., Serna-Saldívar, S. O., Welti-Chanes, J., & Gutierrez-Uribe, J. A. (2015). Phenolic compounds, antioxidant capacity and gelling properties of glucoarabinoxylans from three types of sorghum brans. *Journal of Cereal Science, 65*, 277–284.

Balakrishna, A. K., Wazed, M. A., & Farid, M. (2020). A review on the effect of high pressure processing (HPP) on gelatinization and infusion of nutrients. *Molecules, 25*(10), 2369.

Balasubramaniam, V., Barbosa-Cánovas, G. V., & Lelieveld, H. (2016). *High Pressure Processing of Food: Principles, Technology and Applications.* New York: Springer.

Ballester-Sánchez, J., Fernández-Espinar, M. T., & Haros, C. (2020). Isolation of red quinoa fibre by wet and dry milling and application as a potential functional bakery ingredient. *Food Hydrocolloids, 101*, 105513.

Bender, D., & Schönlechner, R. (2020). Innovative approaches towards improved gluten-free bread properties. *Journal of Cereal Science, 91*, 102904.

Bhargava, N., Mor, R. S., Kumar, K., & Sharanagat, V. S. (2021). Advances in application of ultrasound in food processing: A review. *Ultrasonics Sonochemistry, 70*, 105293.

Brennan, M. A., Menard, C., Roudaut, G., & Brennan, C. S. (2012). Amaranth, millet and buckwheat flours affect the physical properties of extruded breakfast cereals and modulates their potential glycaemic impact. *Starch-Stärke, 64*(5), 392–398.

Cabrera-Chávez, F., de la Barca, A. M. C., Islas-Rubio, A. R., Marti, A., Marengo, M., Pagani, M. A., Bonomi, F., & Iametti, S. (2012). Molecular rearrangements in extrusion processes for the production of amaranth-enriched, gluten-free rice pasta. *LWT, 47*(2), 421–426.

Cantatore, V., Filannino, P., Gambacorta, G., De Pasquale, I., Pan, S., Gobbetti, M., & Di Cagno, R. (2019). Lactic acid fermentation to re-cycle apple by-products for wheat bread fortification. *Frontiers in Microbiology, 10*, 2574.

Castro-Alba, V., Lazarte, C. E., Perez-Rea, D., Carlsson, N. G., Almgren, A., Bergenståhl, B., & Granfeldt, Y. (2019). Fermentation of pseudocereals quinoa, canihua, and amaranth to improve mineral accessibility through degradation of phytate. *Journal of the Science of Food and Agriculture, 99*(11), 5239–5248.

Chaple, S., Sarangapani, C., Dickson, S., & Bourke, P. (2023). Product development and X-ray microtomography of a traditional white pan bread from plasma functionalized flour. *LWT, 174*, 114326.

Chawla, P., Bhandari, L., Dhull, S. B., Sadh, P. K., Sandhu, S. P., Kaushik, R., & Navnidhi (2017). Biotechnological aspects for enhancement of mineral bioavailability from cereals and legumes. In Suresh Kumar Gahlawat, Raj Kumar Salar, Priyanka Siwach, Joginder Singh Duhan, Suresh Kumar & Pawan Kaur (Eds.), *Plant Biotechnology: Recent Advancements and Developments*, 87–100. Singapore: Springer.

Chun, H. H., & Song, K. B. (2014). Optimisation of the combined treatments of aqueous chlorine dioxide, fumaric acid and ultraviolet-C for improving the microbial quality and maintaining sensory quality of common buckwheat sprout. *International Journal of Food Science and Technology, 49*(1), 121–127.

Corsetti, A., & Settanni, L. (2007). Lactobacilli in sourdough fermentation. *Food Research International, 40*(5), 539–558.

De Arcangelis, E., Cuomo, F., Trivisonno, M. C., Marconi, E., & Messia, M. C. (2020). Gelatinization and pasta making conditions for buckwheat gluten-free pasta. *Journal of Cereal Science, 95*, 103073.

Derossi, A., Caporizzi, R., Ricci, I., & Severini, C. (2019). Chapter 3. critical variables in 3D food printing. In F. C. Godoi, B. R. Bhandari & M. Zhang (Eds.), *Fundamentals of 3D Printing and Applications*, vol. 3, 9780128145647. London: Elsevier Inc .

Dey, S., Maurya, C., Hettiarachchy, N., Seo, H.-S., & Zhou, W. (2023). Textural characteristics and color analyses of 3D printed gluten-free pizza dough and crust. *Journal of Food Science and Technology, 60*(2), 453–463.

Dhull, S. B., Kidwai, M. K., Noor, R., Chawla, P., & Rose, P. K. (2022a). A review of nutritional profile and processing of faba bean (*Vicia faba* L.). *Legume Science*, *4*(3), e129.

Dhull, S. B., Kinabo, J., & Uebersax, M. A. (2023). Nutrient profile and effect of processing methods on the composition and functional properties of lentils (*Lens culinaris* Medik): A review. *Legume Science*, *5*(1), e156.

Dhull, S. B., Punia, S., Kidwai, M. K., Kaur, M., Chawla, P., Purewal, S. S., Sangwan, M., & Palthania, S. (2020b). Solid-state fermentation of lentil (*Lens culinaris* L.) with *Aspergillus awamori*: Effect on phenolic compounds, mineral content, and their bioavailability. *Legume Science*, *2*(3), e37.

Dhull, S. B., Punia, S., Kumar, R., Kumar, M., Nain, K. B., Jangra, K., & Chudamani, C. (2020c). Solid state fermentation of fenugreek (*Trigonella foenum-graecum*): Implications on bioactive compounds, mineral content and in vitro bioavailability. *Journal of Food Science and Technology*, *58*, 1927–1936.

Dhull, S. B., Kidwai, M. K., Siddiq, M., & Sidhu, J. S. (2022b). Faba (broad) bean production, processing, and nutritional profile. In M. Siddiq & M. A. Uebersax (Eds.), *Dry Beans and Pulses*. https://doi.org/10.1002/9781119776802.ch14.

Elkhalifa, A. E. O., Bernhardt, R., Bonomi, F., Iametti, S., Pagani, M. A., & Zardi, M. (2006). Fermentation modifies protein/protein and protein/starch interactions in sorghum dough. *European Food Research and Technology*, *222*(5–6), 559–564.

Estivi, L., Brandolini, A., Condezo-Hoyos, L., & Hidalgo, A. (2022). Impact of low-frequency ultrasound technology on physical, chemical and technological properties of cereals and pseudocereals. *Ultrasonics Sonochemistry*, *86*, 106044. https://doi.org/10.1016/j.ultsonch.2022.106044.

Estrada-Girón, Y., Swanson, B., & Barbosa-Cánovas, G. (2005). Advances in the use of high hydrostatic pressure for processing cereal grains and legumes. *Trends in Food Science and Technology*, *16*(5), 194–203.

Giménez, M. A., Drago, S. R., Bassett, M. N., Lobo, M. O., & Samman, N. C. (2016). Nutritional improvement of corn pasta-like product with broad bean (*Vicia faba*) and quinoa (*Chenopodium quinoa*). *Food Chemistry*, *199*, 150–156.

Giménez-Bastida, J. A., Piskuła, M., & Zieliński, H. (2015). Recent advances in development of gluten-free buckwheat products. *Trends in Food Science and Technology*, *44*(1), 58–65.

Granella, S. J., Christ, D., Werncke, I., Bechlin, T. R., & Coelho, S. R. M. (2018). Effect of drying and ozonation process on naturally contaminated wheat seeds. *Journal of Cereal Science*, *80*, 205–211.

Gutiérrez, Á. L., Rico, D., Ronda, F., Martín-Diana, A. B., & Caballero, P. A. (2022). Development of a gluten-free whole grain flour by combining soaking and high hydrostatic pressure treatments for enhancing functional, nutritional and bioactive properties. *Journal of Cereal Science*, *105*, 103458. https://doi.org/10.1016/j.jcs.2022.103458.

Hayashi, T., Takahashi, Y., & Todoriki, S. (1998). Sterilization of foods with low-energy electrons ("soft-electrons"). *Radiation Physics and Chemistry*, *52*(1–6), 73–76.

Henrion, M., Labat, E., & Lamothe, L. (2020). Pseudocereals as healthy grains: An overview. *Innovative Processing Technologies for Healthy Grains*, 37–59.

Hossen, M. Z. (2007). Light emitting diodes increase phenolics of buckwheat (*Fagopyrum esculentum*) sprouts. *Journal of Plant Interactions*, *2*(1), 71–78.

Huang, Z.-R., Chen, M., Guo, W.-L., Li, T.-T., Liu, B., Bai, W.-D., Ai, L.-Z., Rao, P.-F., Ni, L., & Lv, X.-C. (2020). *Monascus purpureus*-fermented common buckwheat protects against dyslipidemia and non-alcoholic fatty liver disease through the regulation of liver metabolome and intestinal microbiome. *Food Research International,* *136,* 109511.

Jafari, M., Koocheki, A., & Milani, E. (2017). Effect of extrusion cooking of sorghum flour on rheology, morphology and heating rate of sorghum–wheat composite dough. *Journal of Cereal Science, 77,* 49–57.

Janve, B., Yang, W., & Sims, C. (2015). Sensory and quality evaluation of traditional compared with power ultrasound processed corn (*Zea mays*) tortilla chips. *Journal of Food Science, 80*(6), S1368–S1376.

Kaur, H., & Gill, B. S. (2019). Effect of high-intensity ultrasound treatment on nutritional, rheological and structural properties of starches obtained from different cereals. *International Journal of Biological Macromolecules, 126,* 367–375.

Kieffer, R., Schurer, F., Köhler, P., & Wieser, H. (2007). Effect of hydrostatic pressure and temperature on the chemical and functional properties of wheat gluten: Studies on gluten, gliadin and glutenin. *Journal of Cereal Science, 45*(3), 285–292.

Kunyanga, C. N., Imungi, J. K., Okoth, M. W., Biesalski, H. K., & Vadivel, V. (2012). Total phenolic content, antioxidant and antidiabetic properties of methanolic extract of raw and traditionally processed Kenyan indigenous food ingredients. *LWT - Food Science and Technology, 45*(2), 269–276. https://doi.org/10.1016/j.lwt.2011.08.006.

Lahuta, L., Sojka, E., Login, A., & Socha, A. (1995). Changes in amylolytic active enzymes in germinating legume seeds (Vicia faba L. ssp. minor, Pisum sativum L.) and cereal grains (*Secale* cereale L., *Triticale*). *Biological Bulletin of Poznań, 32.*

Lasekan, O., & Lasekan, A. (2012). Effect of processing and flavour fine-tuning techniques on the volatile flavour constituents of pseudocereals and some minor cereals. *Journal of Food, Agriculture and Environment, 10*(2 Part 1), 73–79.

Le, L., Gong, X., An, Q., Xiang, D., Zou, L., Peng, L., Wu, X., Tan, M., Nie, Z., & Wu, Q. (2021). Quinoa sprouts as potential vegetable source: Nutrient composition and functional contents of different quinoa sprout varieties. *Food Chemistry, 357,* 129752.

Leewatchararongjaroen, J., & Anuntagool, J. (2016). Effects of dry-milling and wet-milling on chemical, physical and gelatinization properties of rice flour. *Rice Science, 23*(5), 274–281.

Li, C., Oh, S.-G., Lee, D.-H., Baik, H.-W., & Chung, H.-J. (2017). Effect of germination on the structures and physicochemical properties of starches from brown rice, oat, sorghum, and millet. *International Journal of Biological Macromolecules, 105*(1), 931–939.

Liang, Q., Ren, X., Ma, H., Li, S., Xu, K., & Oladejo, A. O. (2017). Effect of low-frequency ultrasonic-assisted enzymolysis on the physicochemical and antioxidant properties of corn protein hydrolysates. *Journal of Food Quality.*

Lin, D., Long, X., Huang, Y., Yang, Y., Wu, Z., Chen, H., Zhang, Q., Wu, D., Qin, W., & Tu, Z. (2020). Effects of microbial fermentation and microwave treatment on the composition, structural characteristics, and functional properties of modified okara dietary fiber. *LWT, 123,* 109059.

Linsberger-Martin, G., Lukasch, B., & Berghofer, E. (2012). Effects of high hydrostatic pressure on the RS content of amaranth, quinoa and wheat starch. *Starch-Stärke, 64*(2), 157–165.

Lintschinger, J., Fuchs, N., Moser, H., Jäger, R., Hlebeina, T., Markolin, G., & Gössler, W. (1997). Uptake of various trace elements during germination of wheat, buckwheat and quinoa. *Plant Foods for Human Nutrition*, *50*(3), 223–237. https://doi.org/10. 1007/BF02436059.

Lipson, H., & Kurman, M. (2013). *Fabricated: The New World of 3D Printing*. John Wiley & Sons.

Liu, S., Wang, W., Lu, H., Shu, Q., Zhang, Y., & Chen, Q. (2022). New perspectives on physiological, biochemical and bioactive components during germination of edible seeds: A review. *Trends in Food Science and Technology*, *123*, 187–197.

Luo, X., Du, Z., Yang, K., Wang, J., Zhou, J., Liu, J., & Chen, Z. (2021). Effect of electron beam irradiation on phytochemical composition, lipase activity and fatty acid of quinoa. *Journal of Cereal Science*, *98*, 103161.

Luo, X., Li, Y., Yang, D., Xing, J., Li, K., Yang, M., Wang, R., Wang, L., Zhang, Y., & Chen, Z. (2019). Effects of electron beam irradiation on storability of brown and milled rice. *Journal of Stored Products Research*, *81*, 22–30.

Makdoud, S., & Rosentrater, K. (2017). Development and testing of gluten-free pasta based on rice, quinoa and amaranth flours. *Agricultural and Biosystems Engineering, Food SCIENCE and Human Nutrition Journal of Food Research*, *6*(4), 0887.

Malik, M., Sindhu, R., Dhull, S. B., Bou-Mitri, C., Singh, Y., Panwar, S., & Khatkar, B. S. (2023). Nutritional composition, functionality, and processing technologies for amaranth. *Journal of Food Processing and Preservation*, *2023*, 1753029. https://doi.org /10.1155/2023/1753029.

Marti, A., Cardone, G., Nicolodi, A., Quaglia, L., & Pagani, M. A. (2017). Sprouted wheat as an alternative to conventional flour improvers in bread-making. *LWT*, *80*, 230–236.

Marti, A., Cardone, G., Pagani, M. A., & Casiraghi, M. C. (2018). Flour from sprouted wheat as a new ingredient in bread-making. *LWT*, *89*, 237–243.

Mason, T. J., & Peters, D. (2002). *Practical Sonochemistry: Power Ultrasound Uses and Applications*. Cambridge: Woodhead Publishing.

Melini, V., & Melini, F. (2021). Modelling and optimization of ultrasound-assisted extraction of phenolic compounds from black quinoa by response surface methodology. *Molecules*, *26*(12), 3616. https://www.mdpi.com/1420-3049/26/12/3616.

Mohammadi Ziarani, G., Kheilkordi, Z., & Gholamzadeh, P. (2020). Ultrasound-assisted synthesis of heterocyclic compounds. *Molecular Diversity*, *24*(3), 771–820.

Mravlje, J., Regvar, M., Starič, P., Mozetič, M., & Vogel-Mikuš, K. (2021). Cold plasma affects germination and fungal community structure of buckwheat seeds. *Plants*, *10*(5), 851.

Omary, M. B., Fong, C., Rothschild, J., & Finney, P. (2012). Effects of germination on the nutritional profile of gluten-free cereals and pseudocereals: A review. *Cereal Chemistry Journal*, *89*(1), 1–14.

Orsák, M., Lachman, J., Pivec, V., Hamouz, K., & Vejvodová, M. (2001). Changes of selected secondary metabolites in potatoes and buckwheat caused by UV, gamma- and microwave irradiation. *Rostlinna Vyroba*, *47*(11), 493–500.

Pillai, S. D. (2016). Introduction to electron-beam food irradiation. *Chemical Engineering Progress*, *112*(11), 36–44.

Pradhan, A., & Tripathy, P. P. (2022). Effect of little millet (*Panicum miliare*) on physical, rheological, nutritional and microstructural properties of bread. *Journal of Food Processing and Preservation*, *46*(8), e16782.

Protonotariou, S., Mandala, I., & Rosell, C. M. (2015). Jet milling effect on functionality, quality and in vitro digestibility of whole wheat flour and bread. *Food and Bioprocess Technology*, *8*(6), 1319–1329.

Punia, S., & Dhull, S. B. (2019). Chia seed (*Salvia hispanica* L.) mucilage (a heteropolysaccharide): Functional, thermal, rheological behaviour and its utilization. *International Journal of Biological Macromolecules, 140*, 1084–1090.

Qi, L., Li, Y., Luo, X., Wang, R., Zheng, R., Wang, L., Li, Y., Yang, D., Fang, W., & Chen, Z. (2016). Detoxification of zearalenone and ochratoxin A by ozone and quality evaluation of ozonised corn. *Food Additives and Contaminants: Part A, 33*(11), 1700–1710.

Riaz, M. N. (2019). Food extruders. In M. Kutz (Eds.), *Handbook of Farm, Dairy and Food Machinery Engineering* (pp. 483–497). London: Elsevier .

Rizzello, C. G., Lorusso, A., Russo, V., Pinto, D., Marzani, B., & Gobbetti, M. (2017). Improving the antioxidant properties of quinoa flour through fermentation with selected autochthonous lactic acid bacteria. *International Journal of Food Microbiology, 241*, 252–261.

Rollán, G. C., Gerez, C. L., & LeBlanc, J. G. (2019). Lactic fermentation as a strategy to improve the nutritional and functional values of pseudocereals. *Frontiers in Nutrition, 6*, 98.

Ruales, J., & Nair, B. (1994). Properties of starch and dietary fibre in raw and processed quinoa (*Chenopodium quinoa* Willd) seeds. *Plant Foods for Human Nutrition, 45*(3), 223–246.

Sajid Mushtaq, B., Zhang, W., Al-Ansi, W., Ul Haq, F., Rehman, A., Omer, R., Mahmood Khan, I., Niazi, S., Ahmad, A., & Ali Mahdi, A. (2021). A critical review on the development, physicochemical variations and technical concerns of gluten free extrudates in food systems. *Food Reviews International, 39*(5), 2806–2834.

Sarkar, A., Niranjan, T., Patel, G., Kheto, A., Tiwari, B. K., & Dwivedi, M. (2023). Impact of cold plasma treatment on nutritional, antinutritional, functional, thermal, rheological, and structural properties of pearl millet flour. *Journal of Food Process Engineering, 46*(5), e14317.

Sasthri, V. M., Krishnakumar, N., & Prabhasankar, P. (2020). Advances in conventional cereal and pseudocereal processing. *Innovative Processing Technologies for Healthy Grains*, 61–81.

Savi, G. D., Piacentini, K. C., Bittencourt, K. O., & Scussel, V. M. (2014). Ozone treatment efficiency on *Fusarium graminearum* and deoxynivalenol degradation and its effects on whole wheat grains (*Triticum aestivum* L.) quality and germination. *Journal of Stored Products Research, 59*, 245–253.

Schmidt, M., Zannini, E., & Arendt, E. K. (2019). Screening of post-harvest decontamination methods for cereal grains and their impact on grain quality and technological performance. *European Food Research and Technology, 245*(5), 1061–1074.

Shi, M., Wang, F., Ji, X., Yan, Y., & Liu, Y. (2022). Effects of plasma-activated water and heat moisture treatment on the properties of wheat flour and dough. *International Journal of Food Science and Technology, 57*(4), 1988–1994.

Strenske, A., & Vasconcelos, E. S. d., Egewarth, V. A., Herzog, N. F. M., & Malavasi, M. d. M. (2017). Responses of quinoa (*Chenopodium quinoa* Willd.) seeds stored under different germination temperatures. *Acta Scientiarum. Agronomy, 39, V*(A), 83–88.

Sun, J., Zhou, W., Yan, L., Huang, D., & Lin, L.-y. (2018). Extrusion-based food printing for digitalized food design and nutrition control. *Journal of Food Engineering, 220*, 1–11.

Tai, L., Wang, H.-J., Xu, X.-J., Sun, W.-H., Ju, L., Liu, W.-T., Li, W.-Q., Sun, J., & Chen, K.-M. (2021). Pre-harvest sprouting in cereals: Genetic and biochemical mechanisms. *Journal of Experimental Botany, 72*(8), 2857–2876.

Thakur, P., Kumar, K., Ahmed, N., Chauhan, D., Eain Hyder Rizvi, Q. U., Jan, S., Singh, T. P., & Dhaliwal, H. S. (2021). Effect of soaking and germination treatments on nutritional, anti-nutritional, and bioactive properties of amaranth (*Amaranthus hypochondriacus* L.), quinoa (*Chenopodium quinoa* L.), and buckwheat (*Fagopyrum esculentum* L.). *Current Research in Food Science, 4*, 917–925. https://doi.org/10.1016/j.crfs.2021.11.019.

Tiwari, B., Brennan, C. S., Curran, T., Gallagher, E., Cullen, P., & O'Donnell, C. (2010). Application of ozone in grain processing. *Journal of Cereal Science, 51*(3), 248–255.

Torres Vargas, O. L., Lema Gonzalez, M., & Galeano Loaiza, Y. V. (2021). Optimization study of pasta extruded with quinoa flour (*Chenopodium quinoa* Willd.). *CyTA: Journal of Food, 19*(1), 220–227.

Tuan, P. A., Kumar, R., Rehal, P. K., Toora, P. K., & Ayele, B. T. (2018). Molecular mechanisms underlying abscisic acid/gibberellin balance in the control of seed dormancy and germination in cereals. *Frontiers in Plant Science, 9*, 668.

Vallons, K. J., Ryan, L. A., & Arendt, E. K. (2011). Promoting structure formation by high pressure in gluten-free flours. *LWT – Food Science and Technology, 44*(7), 1672–1680.

Wang, C., Cao, H., Wang, P., Dai, Z., Guan, X., Huang, K., Zhang, Y., & Song, H. (2023). Changes of components and organizational structure induced by different milling degrees on the physicochemical properties and cooking characteristics of quinoa. *Food Structure, 36*, 100316.

Woldetsadik, D., Llorent-Martínez, E. J., Ortega-Barrales, P., Haile, A., Hailu, H., Madani, N., Warner, N. S., & Fleming, D. E. B. (2020). Contents of metal(loid)s in a traditional Ethiopian flat bread (injera), dietary intake, and health risk assessment in Addis Ababa, Ethiopia. *Biological Trace Element Research, 198*(2), 732–743. https://doi.org/10.1007/s12011-020-02099-7.

Woomer, J. S., & Adedeji, A. A. (2021). Current applications of gluten-free grains–A review. *Critical Reviews in Food Science and Nutrition, 61*(1), 14–24. https://doi.org/10.1080/10408398.2020.1713724.

Xu, R., Du, H., Wang, H., Zhang, M., Wu, M., Liu, C., Yu, G., Zhang, X., Si, C., Choi, S.-E. (2021). Valorization of enzymatic hydrolysis residues from corncob into lignin-containing cellulose nanofibrils and lignin nanoparticles. *Frontiers in Bioengineering and Biotechnology, 9*, 677963.

Yang, F., Zhang, M., & Bhandari, B. (2017). Recent development in 3D food printing. *Critical Reviews in Food Science and Nutrition, 57*(14), 3145–3153.

Zare, L., Mollakhalili-Meybodi, N., Fallahzadeh, H., & Arab, M. (2022). Effect of atmospheric pressure cold plasma (ACP) treatment on the technological characteristics of quinoa flour. *LWT, 155*, 112898.

Zhang, L., Li, X., Ma, B., Gao, Q., Du, H., Han, Y., Li, Y., Cao, Y., Qi, M., Zhu, Y. (2017). The Tartary buckwheat genome provides insights into rutin biosynthesis and abiotic stress tolerance. *Molecular Plant, 10*(9), 1224–1237.

Zhang, Y., Ma, H., Wang, B., Qu, W., Li, Y., He, R., & Wali, A. (2015). Effects of ultrasound pretreatment on the enzymolysis and structural characterization of wheat gluten. *Food Biophysics, 10*(4), 385–395.

Zhou, X.-L., Chen, Z.-D., Zhou, Y.-M., Shi, R.-H., & Li, Z.-J. (2019). The effect of Tartary buckwheat flavonoids in inhibiting the proliferation of MGC80-3 cells during seed germination. *Molecules, 24*(17), 3092.

Zhu, F. (2018). Effect of ozone treatment on the quality of grain products. *Food Chemistry, 264*, 358–366.

Chapter 11

Machines and Equipment for Pseudocereal Processing

Nikhil Dnyaneshwar Patil, Prince Chawla

11.1 Introduction

Pseudocereals, such as quinoa, amaranth, and buckwheat, have gained significant popularity in recent years due to their nutritional benefits, gluten-free nature, and versatility in various food applications (Khairuddin and Lasekan 2021; Malik et al., 2023; Punia and Dhull 2019). As the demand for pseudocereal-based products continues to rise, the role of machinery in their processing becomes increasingly crucial. Machinery enables the efficient transformation of raw pseudocereal grains into value-added products with consistent quality, texture, and functionality (Prabha et al. 2021). The importance of machinery in pseudocereal processing lies in its ability to streamline and optimize the production chain. Each stage of the processing process requires specific machinery and equipment designed to perform particular tasks (Yazdani-Asrami et al. 2023). From cleaning and grading the raw grains to packaging the final products, machinery ensures the efficiency, accuracy, and safety of the entire process (Liu et al. 2022).

To understand the significance of machinery in pseudocereal processing, it is essential to have an overview of the various processing steps involved. Each step requires specific machinery and equipment tailored to its unique requirements (Sendek et al. 2022). Let's delve deeper into each of these steps. The first step in pseudocereal processing is cleaning and grading (Hernández-Martínez et al. 2022). Raw pseudocereal grains may contain impurities such as stones, dust, husks, or

DOI: 10.1201/9781003325277-11

other foreign materials. Machinery such as destoners, vibrating screens, and air aspirators are used to remove these impurities, ensuring the purity and cleanliness of the grains (Arendt and Zannini 2013).

Destoners utilize density separation techniques to separate heavy impurities from the grains, while vibrating screens and air aspirators employ size and density differences to achieve effective cleaning and preparation of the grains (Kaleemullah et al. 2023). Milling and grinding are essential steps in grain processing, where the grains are transformed into flour, meal, or other suitable forms for further processing or consumption (Dhull et al. 2022a, 2022b, 2023). Hammer mills, pin mills, and ball mills are commonly used machinery in this stage (Hernández-Martínez et al. 2022). Hammer mills use high-speed rotating hammers to impact and pulverize the grains, achieving the desired particle size distribution. Pin mills employ high-speed rotating pins or discs to achieve finer grinding, while ball mills use rotating chambers filled with grinding media to refine the pseudocereal flours or meals to the desired fineness (Xue and Gao 2022). Separation and classification play a vital role in ensuring the uniformity and consistency of pseudocereal products. Machinery such as air classifiers, gravity separators, and magnetic separators are employed in this stage (Pagani et al. 2007). Air classifiers utilize airflow and centrifugal forces to separate pseudocereal grains, based on their size and shape, ensuring the desired particle size distribution (Peariso 2008). Gravity separators separate grains based on their specific gravity, employing fluidization and stratification techniques. Magnetic separators, on the other hand, remove ferrous impurities from the grains using powerful magnets (Srinivas 2022). Mixing and blending machinery is used to achieve uniform distribution of ingredients and ensure consistency in pseudo ereal products. Ribbon blenders, paddle mixers, and fluidized bed mixers are commonly employed. Ribbon blenders utilize a ribbon-like agitator to thoroughly blend the pseudocereal flours, meals, and other ingredients, ensuring homogeneity (Arendt and Zannini 2013). Paddle mixers are suitable for mixing pseudocereal doughs, batters, and pastes, ensuring proper hydration and consistent mixing. Fluidized bed mixers create a fluid-like state within a bed of pseudocereal particles, allowing for gentle mixing and uniform coating of ingredients (Han et al. 2021). Drying is a crucial step in pseudocereal processing, as it removes moisture from the products to enhance shelf stability and prevent microbial growth (Kear and McCandlish 1993). Belt dryers, fluidized bed dryers, and tray dryers are machinery commonly used for drying pseudocereal products. Belt dryers utilize a continuous conveyor belt system and controlled airflow to remove moisture effectively, ensuring the preservation of product quality and shelf stability (Boukid 2021). Fluidized bed dryers employ hot air or gas flow to create fluidization, promoting uniform drying and minimizing drying time. Tray dryers, on the other hand, are used for batch drying of pseudocereal products, with multiple trays stacked in a controlled environment with hot air circulation to achieve the desired drying effect (Majumder et al. 2022). Extrusion and expansion machinery are employed to create various textured products, snacks, or breakfast cereals from pseudocereal doughs or mixtures (Choton et al. 2020). Single-screw extruders, twin-screw extruders, and expansion chambers are commonly used at this stage. Single-screw extruders utilize a single rotating screw to mix, cook, and shape the pseudocereal dough, resulting in a range of desirable textures and forms (Pandey et al. 2021). Twin-screw extruders offer enhanced process control and flexibility, employing two intermeshing screws to convey, mix, and process the pseudocereal materials, achieving precise control over product attributes and texture (Choton et al. 2020). Expansion chambers, also known as puffing machines, utilize heat, pressure, and moisture to puff the pseudocereal grains, resulting in light and crunchy finished products. Packaging is the final step in pseudocereal processing, where the products are properly contained and protected. Bagging machines, filling machines, and sealing machines are utilized in this stage (Subramani et al. 2020). Bagging machines automate the packing process, efficiently filling bags or pouches of various sizes with pseudocereal products. Filling machines accurately dispense pseudocereal products, such as powders, granules, or flakes, into containers or packages (Ronnenberg 2016). Sealing machines create a hermetic seal on the packages, ensuring product freshness, protection, and extended shelf life. Quality control and testing are crucial aspects of pseudocereal processing to ensure product consistency and adherence to quality standards (Coles et al. 2003). Machinery such as moisture analyzers, particle size analyzers, and texture analyzers are used in this stage. Moisture analyzers employ various methods, such as halogen heating or infrared technology, to provide quick and accurate moisture measurements, ensuring product quality and stability (Hadde and Chen 2021). Particle size analyzers utilize laser diffraction or sieving methods to determine the size and distribution of pseudocereal particles, ensuring consistency and uniformity (Farkas et al. 2021). Texture analysers assess the textural attributes of pseudocereal products, providing quantitative data on hardness, chewiness, or crispness, enabling quality control and product development (Olakanmi et al. 2022). Automation and process control systems play a crucial role in optimizing pseudocereal processing operations (Li et al. 2022). Programmable Logic Controllers (PLCs) are utilized to automate and control various processes, ensuring efficient and reliable operations (Yang et al. 2022). Supervisory Control and Data Acquisition (SCADA) systems provide centralized monitoring and control of the pseudocereal processing operations, enabling real-time data analysis and decision-making. Industrial robots are employed for tasks such as material handling, sorting, and packaging, enhancing productivity and reducing human labor requirements (Tariq et al. 2023). Safety and hygiene are paramount in pseudocereal processing facilities (Tariq et al. 2023). Dust collection systems are used to control and remove airborne dust particles generated during processing, ensuring a clean and safe working environment and preventing product contamination (Liu et al. 2014). Sanitation equipment, including cleaning stations, sanitizing sprayers, and foamers, are employed to maintain hygiene and cleanliness in production areas. Personal Protective Equipment (PPE), such as gloves, goggles, masks, and protective clothing, is essential for ensuring the safety of workers in the processing plants (Marriott et al. 2018). Maintenance and repair tools are necessary to ensure the smooth operation and longevity of pseudocereal processing machinery. Hand tools, such as wrenches, screwdrivers, pliers, and hammers, are used for routine maintenance and minor repairs (Meng et al. 2021). Power tools, such as drills, grinders, and saws, provide efficient means for handling more complex repairs or modifications (Black and Temple 2017). Lubrication systems, including grease guns, oilers, and automated lubricators, are employed to ensure proper lubrication of moving parts, reducing friction, wear, and potential breakdowns (Black and Temple 2017).

Throughout this chapter, we will explore each of these processing steps in depth, discussing the machinery and equipment used and their specific functions. Case studies will be presented to provide practical examples of machinery application in pseudocereal processing. By gaining a comprehensive understanding of the role of machinery in pseudocereal processing, we can effectively optimize production, improve product quality, and meet the growing demand for these nutritious and versatile grains.

11.2 Cleaning and Grading Equipment

11.2.1 Destoners

Destoners are machines specifically designed to remove stones and heavy impurities from pseudocereal grains. These impurities can negatively impact the quality of the final product and pose a risk to subsequent processing machinery (Ribeiro et al. 2022). Destoners employ principles of density separation to efficiently separate the pseudocereal grains from the unwanted materials. Destoners operate based on the principle of differential density between the pseudocereal grains and the stones or heavy impurities (Thomas et al. 2017). The machine consists of an inclined vibrating table or deck with adjustable airflow. As the pseudocereal grains are fed onto the vibrating table, they move across the surface due to the shaking motion (Van Heurck 2015). The heavy stones and impurities, being denser than the grains, settle at the bottom of the table, while the lighter grains move to the top. The destoner utilizes a combination of vibration and airflow to assist in the separation process (Srinivas 2022). The vibration helps to spread the pseudocereal grains evenly across the table, while the adjustable airflow creates an upward current that carries the lighter grains to the top, allowing the heavier impurities to sink to the bottom (Hart 2019). This differential movement based on density enables effective separation of stones and heavy impurities from the pseudocereal grains. Destoners are widely used in the processing of pseudocereals to ensure the purity and cleanliness of the grains (Arendt and Zannini 2013). They are employed at the beginning of the processing chain, typically after the initial cleaning step, to remove stones, rocks, and other heavy impurities that may be present. By removing these unwanted materials, destoners help to improve the quality and safety of the pseudocereal grains, reducing the risk of damage to subsequent processing machinery and ensuring that the final products meet quality standards (Shahidi 2005). Destoners have certain advantages like efficient removal of stones and heavy impurities, improved product quality, and protection of downstream machinery. On the other hand, there are also some disadvantages like the limited removal of lighter impurities and potential grain damage (Sarkar and Fu 2022)

11.2.2 Vibrating Screens

Vibrating screens are widely used in pseudocereal processing to separate grains based on their size and shape. These screens consist of multiple decks with varying mesh sizes, allowing for precise classification and removal of undersized or oversized grains (Oates 2008). Vibrating screens operate by utilizing the principle of vibration and sieving. The machine consists of a vibrating screen deck or multiple decks stacked on top of each other (Chen et al. 2020). Each deck is equipped with a mesh with a specific opening size, which determines the size of grains that can pass through it. As pseudocereal grains are fed onto the vibrating screen, they move across the surface due to the vibrating motion imparted by the machine (Gawenda 2021). The vibration causes the grains to stratify, with the smaller grains passing through the openings in the mesh, while the larger grains are retained on the screen surface. The multiple decks with varying mesh sizes allow for the classification of grains into different size fractions (Da Costa et al. 2014). Vibrating screens are commonly used in pseudocereal processing for several purposes, like size classification. Vibrating screens help to separate pseudocereal grains into different-sized fractions, ensuring uniformity and consistency in the final product (Aryafar et al. 2018). The screens are effective in removing undersized or oversized grains, which may affect the texture and quality of the final product (Schuler et al. 1995). Vibrating screens can also be used for scalping, where larger impurities or foreign materials are removed from the pseudocereal grains before further processing (Peariso 2008). Vibrating screens have some advantages like precise size classification, customizable mesh sizes, and high throughput. This method also has several disadvantages, like limited removal of shape-based impurities (Zhao and Guo 2020)

11.2.3 Air Aspirator

Air aspirators are commonly used in pseudocereal processing to remove lightweight impurities, such as dust, chaff, and husk from the grains. These machines utilize air currents to achieve effective cleaning and preparation of the pseudocereal grains (Kaushal and Kumar 2022). Air aspirators operate on the principles of pneumatic conveying and gravity separation. They consist of an enclosed chamber or duct through which the grains and impurities are passed (Wimberly 1983). The controlled airflow creates an upward current, carrying the lighter impurities away from the grains. Dust, chaff, and husk are lifted by the airflow, while the denser grains fall downward due to gravity. This differential movement enables the separation of light impurities from the grains (Peariso 2008). Air aspirators have various uses in pseudocereal processing. They effectively remove dust particles generated during handling and processing, ensuring the cleanliness and safety of the grains (Kaushal and Kumar 2022). Additionally, they separate chaff and husk, which are lighter materials, from the pseudocereal grains, leading to improved quality and purity of the final product. Moreover, air aspirators prepare the grains for further processing by removing lightweight impurities, ensuring that the grains are clean and ready for subsequent steps (Grubben and Soetjipto Partohardjono 1997). Air aspirators offer several advantages. They efficiently remove light impurities, such as dust, chaff, and husk, from the grains, ensuring their cleanliness and purity (Kaushal and Kumar 2022). By eliminating these impurities, air aspirators enhance the overall quality and appearance of the pseudocereal grains and the final products. Furthermore, the removal of airborne dust and impurities reduces the risk of product contamination, contributing to the safety and hygiene of the processing environment (Schoeman and Manley 2019). However, air aspirators also have certain limitations. They are primarily designed for the removal of light impurities and may not effectively remove heavier impurities like stones or metals (Sinha et al. 2023); additional equipment or manual sorting may be required to address these impurities. Additionally, the operation of air aspirators can generate airborne dust particles and produce noise. It is important to ensure proper ventilation and implement noise control measures in the processing facility to minimize any negative effects (Humphries and Vincent 1976).

In conclusion, destoners, vibrating screens, and air aspirators are essential machinery in pseudocereal processing for cleaning, classification, and separation purposes. Each machine employs specific mechanisms to achieve efficient removal of impurities based on density, size, and air currents. Understanding the mechanisms, uses, advantages, and disadvantages of these machines is crucial for optimizing the processing operations and ensuring the quality and purity of pseudocereal grains.

11.3 Milling and Grinding Equipment

11.3.1 Hammer Mills

Hammer mills are widely used in pseudocereal processing for size reduction and grinding of grains into flour or meal. These machines utilize high-speed rotating hammers that impact and pulverize the grains, resulting in the desired particle size distribution (Sasthri et al. 2020). The mechanism involves feeding the grains into the grinding chamber, where they are subjected to repeated hammer impacts. The high rotational speed of the hammers generates forceful impacts that break the grains into smaller particles. The crushed material is then discharged through a screen, controlling the final particle size (Volkhonov et al. 2020). Hammer mills are versatile and can handle a wide range of pseudocereal grains. They are commonly used in flour milling operations, where a consistent particle size is crucial for product quality (Sasthri et al. 2020). The advantages of hammer mills include their high processing capacity, ability to handle a variety of grains, and relatively simple operation. However, they can generate a significant amount of dust during operation, and excessive heat generation can affect the nutritional properties of the grains (Neal and Wright 2013).

11.3.2 Pin Mills

Pin mills are suitable for finer grinding of pseudocereal grains while maintaining the integrity of the grains. These mills employ high-speed rotating pins or discs that impact the grains, resulting in particle size reduction (Arendt and Zannini 2013). The mechanism involves feeding the grains into the grinding chamber, where the pins or discs create intense impact forces. The grains are repeatedly struck, leading to size reduction without excessive heat generation (Kushwaha et al. 2021). Pin mills are particularly useful for producing fine flours or meals for various applications, such as baking or food processing. They offer advantages such as precise control over particle size, minimal heat generation, and the ability to process fragile grains (Scanlon et al. 2018). However, they may have limitations when processing large volumes of grains due to their lower processing capacity compared with hammer mills (Svihus et al. 2004).

11.3.3 Ball Mills

Ball mills are used for refining pseudocereal flours and meals, especially for applications that require fine particle sizes. These mills consist of rotating chambers filled with grinding media, such as steel balls (Andreotti et al. 2013). The mechanism involves the rotation of the chambers, causing the grinding media to cascade and impact the grains, resulting in particle size reduction (Schlem et al. 2021). The refined flour or meal is obtained by controlling the duration of the milling process. Ball mills are commonly used in the production of fine flours and meals for high-end culinary and industrial applications (Amir and McDonagh 2012). They offer advantages such as the ability to achieve precise particle size control, high processing efficiency, and versatility in processing different pseudocereal grains. However, they may require more advanced equipment, and have higher operational and maintenance costs compared to other grinding methods (Arendt and Zannini 2013).

In summary, hammer mills, pin mills, and ball mills are essential machines in pseudocereal processing for size reduction and grinding. Hammer mills use high-speed rotating hammers to impact and pulverize grains, while pin mills employ rotating pins or discs for finer grinding. Ball mills utilize rotating chambers and grinding media for refining flours and meals to achieve fine particle sizes. Understanding the mechanisms, uses, advantages, and disadvantages of these machines is crucial for selecting the appropriate grinding method in pseudocereal processing, ensuring optimal product quality and performance.

11.4 Separation and Classification Equipment

11.4.1 Air Classifiers

Air classifiers are essential equipment used in pseudocereal processing to separate grains based on their size and shape. These classifiers employ airflow and centrifugal forces to efficiently classify the grains into different size fractions, ensuring uniformity and consistency (Valentiuk and Stankevych 2020). The mechanism of air classifiers involves the introduction of the grain mixture into a chamber or a series of chambers through which an air stream is flowing. The airflow carries the grains upward, and centrifugal forces cause the larger and heavier grains to move towards the outer wall, while the smaller and lighter grains remain closer to the center (Jankovic 2015). The classified fractions are then collected through different outlets or chambers. Air classifiers offer several advantages in pseudocereal processing. They provide precise and accurate size classification, ensuring consistent particle sizes in the final product (Shapiro and Galperin 2005). This is crucial for achieving uniformity and quality in applications such as flour milling or food processing. Air classifiers are highly efficient, capable of handling large volumes of grains in a relatively short time (Alam and Saeed 2013). They also have a compact design, making them suitable for integration into processing lines. However, air classifiers may require careful adjustment and optimization to achieve the desired classification efficiency (Lippmann 1990). Additionally, they may generate some dust during operation, necessitating proper dust control measures in the processing facility (Amyotte and Eckhoff 2010).

11.4.2 Gravity Separators

Gravity separators are employed to separate pseudocereal grains based on their specific gravity. These separators utilize principles of fluidization and stratification to achieve efficient separation, enabling the removal of lighter or heavier grains from the desired product stream (Kaushal and Kumar 2022). The mechanism involves the introduction of the grain mixture onto an inclined deck or a vibrating table. As the grains flow along the deck, an upward airflow is applied, causing the grains to stratify based on their specific gravity (Sharma and Kumar, n.d.). Lighter grains tend to rise to the top, while heavier grains settle at the bottom. Adjusting the inclination and airflow rate allows for precise separation of different grain fractions (Yadav et al. 2022). Gravity separators offer several advantages in pseudocereal processing. They provide effective removal of impurities and foreign materials that have different specific gravities from the grains (Liu et al. 2013). This helps in improving the quality and purity of the final product. Gravity separators are versatile and can handle a wide range of pseudocereal grains with varying specific gravities (Offiah et al. 2019). They are relatively simple to operate and require minimal maintenance. However, they may have limitations in separating grains with very similar specific gravities or in handling large volumes of grains efficiently. Additional equipment or multiple passes may be required in such cases (Roughan et al. 2004).

11.4.3 Magnetic Separators

Magnetic separators offer significant advantages in pseudocereal processing. They effectively remove ferrous impurities, preventing contamination, and ensuring the quality and safety of the grains (Peariso 2008). Magnetic separators are highly efficient and can operate continuously without the need for manual intervention. They are relatively easy to install and require minimal maintenance (Daehn et al. 2019). However, it is important to note that magnetic separators are only effective in removing ferrous impurities and may not be suitable for removing non-magnetic impurities (Song et al. 2002). Additionally, they may have limitations in handling high volumes of grains or processing grains with high moisture content (Shun-Ichiro and Paterson 1986).

In summary, air classifiers, gravity separators, and magnetic separators are important machines used in pseudocereal processing for separation purposes. Air classifiers classify grains based on size and shape using airflow and centrifugal forces. Gravity separators separate grains based on specific gravity through fluidization and stratification. Magnetic separators remove ferrous impurities using powerful magnets. Understanding the mechanisms, uses, advantages, and limitations of these separators is crucial for optimizing pseudocereal processing operations and ensuring the purity and quality of the grains.

11.5 Mixing and Blending Equipment

11.5.1 Ribbon Blenders

Ribbon blenders are highly efficient and versatile machines used in the processing of pseudocereal flours, meals, and ingredients. These blenders are specifically designed to achieve homogeneous mixing and blending of the components, ensuring uniform distribution of particles and consistent product quality (Singh et al. 2020). The mechanism of ribbon blenders involves the use of a ribbon-like agitator, which is typically composed of two or more helical ribbons. These ribbons are mounted on a central shaft that rotates within a trough-shaped vessel (Spinks 2020). As the shaft rotates, the ribbons move the materials in a double helical pattern, creating a continuous mixing action. The combination of radial and axial movement of the ribbons results in effective blending of the ingredients (Robinson and Cleary 2012). Ribbon blenders offer several advantages in pseudocereal processing. One of the key advantages is their ability to handle a wide range of viscosities and densities, making them suitable for various pseudocereal applications (Razykov et al. 2011). They can accommodate dry powders and granules, as well as wet or sticky ingredients. The blending action of ribbon blenders ensures that all particles come into contact with one another, promoting consistent mixing and distribution of the ingredients (Muzzio et al. 2004). This is particularly important in applications where uniformity and homogeneity are critical, such as in the production of ready-to-eat pseudocereal products, snack foods, or baked goods (Moes et al. 2008). Furthermore, ribbon blenders are known for their gentle mixing action, which minimizes product degradation or damage. The ribbon agitator design prevents excessive heat generation during blending, helping to preserve the nutritional properties and sensory characteristics of the pseudocereal products (Fu 2008). Additionally, ribbon blenders are relatively easy to operate and maintain, with simple controls and easy access for cleaning.However, there are certain considerations and limitations to be aware of when using ribbon blenders (Bhowmik et al. 2021). For instance, the size and capacity of the blender should be chosen based on the desired batch size and production requirements. Overfilling the blender can negatively impact its blending efficiency and result in uneven mixing (Byrn et al. 2015). It is also important to consider the compatibility of the materials being blended and ensure that the blender is properly cleaned and sanitized between batches to prevent cross-contamination (Brown and Arrowsmith 2015).

In summary, ribbon blenders are versatile and efficient machines used for homogeneous mixing and blending of pseudocereal flours, meals, and ingredients. Their unique ribbon agitator design ensures thorough blending and uniform distribution of particles, resulting in consistent product quality. Understanding the mechanism and advantages of ribbon blenders is essential for optimizing pseudocereal processing operations and achieving desired product characteristics.

11.5.2 Paddle Mixers

Paddle mixers are commonly used in pseudocereal processing for mixing doughs, batters, and pastes. These mixers utilize rotating paddles to combine the ingredients and achieve proper hydration, resulting in consistent and well-mixed products (Siegmann et al. 2021). The mechanism of paddle mixers involves the rotation of paddles within a mixing chamber or bowl. The paddles are positioned at strategic angles and are designed to move the ingredients in a specific pattern (Kaye 1997). As the paddles rotate, they lift and fold the ingredients, creating a kneading and mixing action. This action promotes the dispersion of dry ingredients, hydration of flours, and incorporation of air, resulting in a homogeneous and uniform mixture (Premi and Sharma 2022). Paddle mixers offer several advantages in pseudocereal processing. They are particularly effective in blending doughs and batters, ensuring proper hydration and uniform distribution of ingredients (Elgeti et al. 2015). The paddles provide sufficient mixing and kneading action in the development of pseudocereal doughs, which is essential for the structure and texture of baked products (Singh et al. 2020). Paddle mixers are versatile and can accommodate different batch sizes and viscosities, making them suitable for a wide range of pseudocereal products, such as bread, pasta, and pastries (Boukid 2021). Additionally, paddle mixers are designed to minimize ingredient waste and maintain consistent mixing quality. The rotating paddles scrape the sides and bottom of the mixing chamber, ensuring that all ingredients are thoroughly incorporated and preventing the development of pockets of unmixed material (Stark 2012). Paddle mixers are also relatively easy to operate and clean, with accessible mixing chambers and removable paddles. However, it is important to consider the limitations of paddle mixers in pseudocereal processing (Varzakas et al. 2014). These mixers may have limitations in handling extremely high-viscosity or high-fat mixtures. In such cases, specialized equipment may be required. Additionally, the mixing time and paddle speed should be carefully controlled to avoid overmixing or excessive heat generation, which can affect the quality of the final product (Boukid 2021).

11.5.3 Fluidized Bed Mixers

Fluidized bed mixers are another type of mixing equipment used in pseudocereal processing. These mixers employ the principle of fluidization to mix and blend pseudocereal grains, flakes, or granules (KuShaari et al. 2006). They create a fluid-like state by passing air through the bed of particles, allowing for gentle mixing and uniform coating of ingredients.The mechanism of fluidized bed mixers involves introducing the pseudocereal grains or particles into a chamber or vessel equipped with an air distribution system (Parikh 2017). The air is forced through a permeable membrane or distributor plate at the bottom of the chamber, creating upward airflow. The upward airflow causes the particles to become suspended in the air, creating a fluidized state (Denton and Gilpin-Brown 1961). As a result, the particles move and mix freely within the chamber, facilitating blending and coating of ingredients. Fluidized bed mixers offer several advantages for pseudocereal processing (Muzzio et al. 2004). They provide gentle mixing, minimizing product breakage or damage. This is particularly important when handling fragile or delicate pseudocereal flakes or granules (Arendt and Zannini 2013). The fluidization process allows for efficient and uniform mixing of different particle sizes, ensuring homogeneity in the final product (Venables and Wells 2001). Fluidized bed mixers are also capable of coating particles with liquids or powders, enabling the addition of flavors, colors, or functional ingredients to pseudocereal products (Rani et al. 2023). Furthermore, fluidized bed mixers are known for their high mixing efficiency and short processing time. The continuous movement and interaction of particles within the fluidized bed chamber promote rapid and thorough blending (Zhang et al. 2022). These mixers are also easy to clean and maintain, with simple disassembly and cleaning procedures. However, it is important to consider the limitations of fluidized bed mixers in pseudocereal processing (Aarnisalo et al. 2006). The capacity and airflow rate of the mixer should be carefully controlled to ensure efficient mixing without causing excessive particle entrainment or loss (Altun et al. 2009). The selection of appropriate particle sizes and bed depths is also crucial for achieving the desired mixing results. Additionally, specialized equipment may be required (Shekunov et al. 2007).

11.6 Drying Equipment

11.6.1 Belt Dryers

Belt dryers are widely employed in pseudocereal processing for the drying of grains, flakes, or powders. These dryers utilize a continuous conveyor belt system to transport the pseudocereal products through a drying chamber, where controlled airflow is applied to remove moisture (Teunou and Poncelet 2002). The belt is typically made of a porous material that allows for efficient airflow and moisture evaporation. The mechanism of belt dryers involves the continuous movement of the pseudocereal products on the conveyor belt (Courtois et al. 2001). The belt passes through a series of drying zones, where heated air is circulated over and underneath the belt. The airflow, combined with the prolonged exposure to heat, facilitates the evaporation of moisture from the pseudocereal products. Belt dryers offer several advantages in pseudocereal processing (Kaushal and Kumar 2022). Their continuous operation allows for high production capacity and consistent drying results. The controlled airflow and temperature ensure uniform drying across the entire product surface, minimizing the risk of over-drying or under-drying (Sharma et al. 2009). This promotes the preservation of product quality, including color, texture, and nutritional properties. Belt dryers are also energy efficient, as the heat and airflow can be optimized for efficient moisture removal (Hany et al. 2022).

Furthermore, belt dryers are flexible and versatile, accommodating different pseudocereal products and processing conditions. The conveyor belt speed can be adjusted to control the residence time of the products in the drying chamber, enabling precise drying control (Gullichsen et al. 2016). The design of the drying chamber and airflow distribution system can be customized to suit specific product requirements. This versatility makes belt dryers suitable for a wide range of pseudocereal products, including grains, flakes, and powders (Verica et al. 2015). However, there are certain considerations when using belt dryers in pseudocereal processing. The selection of appropriate drying parameters, such as temperature, airflow rate, and residence time, is crucial to achieve the desired moisture content and product quality (Harle et al. 2007). It is important to avoid excessive drying, which can lead to product degradation or loss of nutritional value. Proper monitoring and control of the drying process are essential to ensure consistent and optimal drying results (Gunasekaran 1999).

11.6.2 Fluidized Bed Dryers

Fluidized bed dryers are specifically designed for the rapid and efficient drying of pseudocereal particles. These dryers utilize hot air or gas flow to create fluidization of the particles, promoting uniform drying and minimizing drying time (Saini et al. 2022). The mechanism of fluidized bed dryers involves the introduction of pseudocereal particles into a drying chamber or bed (Teunou and Poncelet 2002). Hot air or gas is forced through the bed, creating a fluidized state where the particles become suspended and move freely. The fluidization promotes efficient heat and mass transfer, allowing for rapid moisture evaporation from the pseudocereal particles (Sieniutycz 2015). Fluidized bed dryers offer several advantages in pseudocereal processing. One of the key advantages is their ability to achieve high drying rates, resulting in shorter drying times compared with other drying methods (Sozzi et al. 2021). The fluidization process ensures uniform heat distribution and contact between the particles and the drying medium, leading to consistent and thorough drying. This promotes the preservation of product quality, including color, flavor, and nutritional attributes (Hovmand 2020).

Additionally, fluidized bed dryers offer excellent control over the drying process. The airflow rate, temperature, and residence time can be precisely adjusted to meet the specific drying requirements of different pseudocereal products (Aghbashlo et al. 2014). This allows for customization and optimization of the drying parameters, ensuring desired moisture content and product characteristics are achieved. The continuous operation of fluidized bed dryers also contributes to high production capacity and efficiency (Steven et al. 2005). However, there are considerations when using fluidized bed dryers in pseudocereal processing. The selection of appropriate drying temperatures and airflow rates should be carefully determined to prevent excessive drying or thermal degradation of the products (Devahastin and Mujumdar 2003). The particle size and shape of the pseudocereal particles can also affect the fluidization behavior and drying performance. Proper monitoring and control of the drying parameters, as well as regular cleaning and maintenance of the drying chamber, are necessary to ensure optimal drying results and equipment efficiency (El-Emam et al. 2021).

11.6.3 Tray Dryers

Tray dryers are commonly used for batch drying of pseudocereal products. These dryers consist of multiple trays or shelves where the pseudocereal items are placed for drying. The trays are stacked in a controlled environment with hot air circulation to achieve the desired drying effect (Amankwah 2019).The mechanism of tray dryers involves the arrangement of the pseudocereal products on the trays in a single layer. The trays are then loaded into the drying chamber, which is equipped with heating elements and fans (Onyinge 2019). The hot air is circulated over and around the trays, facilitating the evaporation of moisture from the pseudocereal products. The moist air is typically exhausted from the drying chamber to maintain proper airflow and drying conditions (Orrego et al. 2023). Tray dryers offer several advantages in pseudocereal processing. They provide flexibility in terms of batch size and product variety. The trays can be easily loaded and unloaded, allowing for efficient handling and processing of different pseudocereal items (Mujumdar 2004). Tray dryers are also cost-effective and require relatively low initial investment compared with other drying equipment. They are suitable for small- to medium-scale operations or when specific drying requirements need to be met (Iranshahi et al. 2023).

Furthermore, tray dryers offer good control over the drying process. The temperature, airflow, and drying time can be adjusted based on the moisture content and desired final characteristics of the pseudocereal products (Akpinar et al. 2003). This allows for customization and optimization of the drying conditions to achieve the desired moisture content and quality attributes. However, there are certain considerations when using tray dryers in pseudocereal processing. The batch drying process may result in longer drying times compared with continuous drying methods (Santos 2021). Proper monitoring and control of the drying parameters are essential to prevent over-drying or under-drying of the pseudocereal products. Additionally, the uniformity of drying may vary across the trays, requiring careful placement and rotation of the trays during the drying process. Regular cleaning and maintenance of the drying chamber and trays are necessary to ensure hygienic and efficient drying operations (Iranshahi ct al. 2023).

11.7 Extrusion and Expansion Equipment

11.7.1 Single-screw Extruders

Single-screw extruders are commonly used in the food industry for the production of various pseudocereal-based snacks, breakfast cereals, and textured products. These extruders consist of a single rotating screw housed in a barrel (Offiah et al. 2019). The screw is responsible for conveying, mixing, cooking, and shaping the pseudocereal dough as it moves along the barrel. The process begins with the feeding of the pseudocereal dough into the extruder's hopper (Sudhakar et al. 2023). The rotating screw pulls the dough into the barrel and gradually moves it forward. As the dough progresses, it encounters increasing levels of heat and pressure. The combination of heat and pressure causes the dough to cook and undergo physical and chemical transformations (Wilkinson and Ryan 1998). During the extrusion process, the pseudocereal dough is subjected to mechanical shear forces generated by the screw. This shearing action helps in mixing the ingredients uniformly and breaking down any lumps or agglomerates (Singh et al. 2020). It also aids in cooking the dough by distributing the heat evenly and facilitating the gelatinization of starches. The shape and texture of the extruded product can be controlled by using various dies or shaping elements at the end of the extruder (Filli et al. 2014). Single-screw extruders offer advantages such as simplicity, ease of operation, and lower capital costs compared with twin-screw extruders. However, they may have limitations in terms of precise control over the process and product attributes (Martin 2016).

11.7.2 Twin-screw Extruders

Twin-screw extruders provide enhanced process control and flexibility in the processing of pseudocereal-based products. These extruders feature two intermeshing screws that work together to convey, mix, and process the pseudocereal materials (Sudhakar et al. 2023). The twin-screw extrusion process begins with the feeding of the pseudocereal ingredients into the extruder. The intermeshing screws rotate in opposite directions, creating a shearing and kneading effect on the material (Khanpit et al. 2022). This action ensures thorough mixing, dispersion of ingredients, and efficient heat transfer. One key advantage of twin-screw extruders is their ability to offer precise control over product attributes and texture (Hemlata and Tiwari 2016). The design and configuration of the screws can be customized to achieve specific processing objectives. For example, different screw elements, such as kneading blocks, mixing elements, and conveying elements, can be incorporated to achieve desired mixing, cooking, and shaping effects. The twin-screw extrusion process also allows for the incorporation of additional ingredients, such as flavorings, colorants, and nutritional additives, at specific stages of the extruder (Rogers and Bottaci 1997). This flexibility enables the production of a wide range of pseudocereal-based products with varying textures, shapes, and flavors (Shah et al. 2021).

11.7.3 Expansion Chambers

Expansion chambers, also known as puffing machines, are used in the production of puffed pseudocereal products. These machines use a combination of heat, pressure, and moisture to puff the pseudocereal grains, resulting in light and crunchy finished products (Subramani et al. 2020). The expansion chamber consists of a closed vessel or chamber with a controlled atmosphere. The pseudocereal grains, typically in the form of whole or processed grains, are introduced into the chamber. The chamber is then sealed, and heat is applied to increase the temperature inside (Kim et al. 1999). The pressure inside the chamber is also increased by either introducing steam or by trapping the steam released from the pseudocereal grains themselves. The combination of heat and pressure causes the moisture inside the grains to turn into steam, creating internal pressure (Sanna et al. 2011). This pressure builds up until the grain structure cannot contain it, leading to a sudden release of pressure. As the pressure is released, the steam expands rapidly, causing the pseudocereal grains to puff and increase in volume. The sudden expansion creates a light and porous structure in the grains, resulting in a characteristic crispy texture (Mujica et al. 2003). The puffed grains are then cooled and separated from any residual moisture before further processing or packaging. Expansion chambers offer a simple and efficient method for producing puffed pseudocereal products. The parameters, such as temperature, pressure, and moisture content, can be adjusted to achieve the desired puffing characteristics and final product attributes (Mariotti et al. 2006).

11.8 Packaging Equipment

11.8.1 Bagging Machines

Bagging machines are employed to automatically pack pseudocereal products into bags or pouches of various sizes. They ensure accurate filling, sealing, and labelling of the packages, facilitating efficient packaging operations (Singh and Singh 2005).

11.8.2 Filling Machines

Filling machines are used to accurately dispense pseudocereal products, such as powders, granules, or flakes, into containers or packages. They enable precise control over product weight or volume, ensuring consistent and reliable packaging (Sanjeev and Ramesh 2006).

11.8.3 Sealing Machines

Sealing machines are used to hermetically seal packages or containers containing pseudocereal products. They employ heat, pressure, or adhesive methods to create a secure seal, ensuring product freshness, protection, and extended shelf life (Lacroix 2009).

11.9 Quality Control and Testing Equipment

11.9.1 Moisture Analyzers

Moisture analyzers are instrumental in determining the moisture content of pseudocereal products. Accurate moisture measurement is crucial as it directly affects the product's quality, stability, and shelf life (Pirsa et al. 2022). These analyzers utilize different techniques, such as halogen heating or infrared technology, to measure the moisture content quickly and accurately. Halogen heating moisture analyzers work by heating the sample with a halogen lamp, which causes the moisture in the product to evaporate (Razvi et al. 2021). The weight loss during the evaporation process is measured and used to calculate the moisture content. Infrared moisture analyzers, on the other hand, use the principle of infrared absorption to determine the moisture content (Qi et al. 2020). They emit infrared radiation onto the sample, and the amount of radiation absorbed is measured to estimate the moisture content. Moisture analyzers offer several advantages, including fast analysis times, non-destructive testing, and high precision (Shen et al. 2020). By accurately measuring the moisture content, manufacturers can ensure that their pseudocereal products meet the desired quality standards and have the appropriate moisture level for optimal shelf stability (Baiano 2021).

11.9.2 *Particle Size Analyzers*

Particle size analyzers are utilized to measure the particle size distribution of pseudocereal flours, meals, or other processed products. Particle size plays a crucial role in the texture, appearance, and functionality of the products. These analyzers employ laser diffraction or sieving methods to determine the size and distribution of particles (Coțovanu et al. 2022). Laser diffraction analyzers work by passing a laser beam through a dispersed sample of pseudocereal particles. The laser light scatters as it encounters the particles, and the scattering pattern is analyzed to determine the particle size distribution (Sun 2020). Sieving analyzers, on the other hand, separate particles into different-sized fractions, using a series of stacked sieves with progressively smaller openings. The weight of particles retained on each sieve is measured, and the particle size distribution is calculated based on these measurements(Madarász et al. 2023). Accurate particle size analysis is vital for ensuring consistency and uniformity in pseudocereal products. It impacts various characteristics such as texture, flow ability, and cooking properties. By measuring and controlling particle size, manufacturers can optimize product performance and meet specific requirements for different applications (Cappelli et al. 2020).

11.9.3 *Texture Analyzers*

Texture analyzers are used to assess the textural attributes of pseudocereal products, such as hardness, chewiness, or crispness. Texture is a critical sensory attribute that greatly influences consumer perception and acceptance of food products (Xu et al. 2020). These analyzers employ various methods, such as compression or penetration testing, to provide quantitative data on product texture. Compression testing involves applying a controlled force to a sample of the pseudocereal product and measuring the force required to achieve a specific deformation or breakage point. Penetration testing, on the other hand, involves inserting a probe into the product and measuring the resistance encountered during penetration (Khosravi et al. 2020). Texture analyzers provide valuable information for quality control, product development, and process optimization. By accurately measuring textural attributes, manufacturers can ensure consistent product quality, meet consumer preferences, and adjust formulations or processing parameters to achieve the desired texture profiles (Fiorentini et al. 2020).

Overall, moisture analyzers, particle size analyzers, and texture analyzers are essential tools for quality control and product development in the pseudocereal industry. These analytical instruments enable manufacturers to monitor and adjust key parameters to ensure the desired product's attributes, consistency, and consumer satisfaction.

11.10 Automation and Process Control Systems

11.10.1 Programmable Logic Controllers (PLCs)

Programmable Logic Controllers (PLCs) are widely used in pseudocereal processing plants to automate and control various processes. These devices are programmable electronic systems that can monitor inputs, make decisions based on pre-set logic, and control outputs to achieve specific tasks or functions (Starý et al. 2020). PLCs offer the ability to integrate and synchronize different machinery and equipment, ensuring efficient and reliable operation. PLCs are programmed using specialized software, allowing operators to define the desired control logic and set up specific operational parameters (Jayachandran et al. 2021). They can handle complex control sequences, manage inputs from sensors, and activate outputs to control actuators, motors, valves, and other devices. PLCs are capable of handling multiple inputs and outputs simultaneously, enabling precise control and coordination of different stages in pseudocereal processing (Tieu 2021). The use of PLCs in pseudocereal processing plants brings numerous benefits, including improved process efficiency, reduced manual intervention, enhanced product consistency, and increased safety (Benettayeb et al. 2022). They offer flexibility in adapting to different production requirements and can be easily reprogrammed or reconfigured when process changes are needed (Enrique et al. 2022).

11.10.2 Supervisory Control and Data Acquisition (SCADA) Systems

Supervisory Control and Data Acquisition (SCADA) systems are used in pseudocereal processing plants to provide centralized monitoring and control of operations. These systems collect real-time data from various equipment and processes throughout the plant and present it to operators in a graphical interface (Bai et al. 2023). SCADA systems enable operators to monitor the status of equipment, track production metrics, and make informed decisions to optimize process control and efficiency. SCADA systems offer a range of functionalities, including data acquisition, data logging, alarm management, and remote control (Shalko 2022). They provide real-time visualization of process variables, trends, and alarms, allowing operators to identify and address issues promptly. SCADA systems can also generate reports, analyze historical data, and provide insights for process improvement and optimization. By implementing SCADA systems, pseudocereal processing plants can achieve improved operational efficiency, reduced downtime, and enhanced decision-making capabilities (Stojanovic et al. 2019). The centralized monitoring and control enable better coordination between different process units and facilitate proactive maintenance and troubleshooting (Provan et al. 2020).

11.10.3 *Industrial Robots*

Industrial robots have found applications in pseudocereal processing for various tasks, including material handling, sorting, and packaging. These robots offer increased speed, precision, and efficiency compared with manual labor, enhancing productivity and reducing human labor requirements (Dzedzickis et al. 2021). Industrial robots can be programmed to perform repetitive tasks with high accuracy and consistency. In pseudocereal processing, they can handle the movement of raw materials, transfer products between different stages of processing, and assist in packaging and palletizing operations (Liu et al. 2023). By automating these tasks, robots can improve production throughput, minimize product damage, and ensure consistent product quality. Furthermore, industrial robots can be equipped with sensors and vision systems to enable advanced functionalities. For example, they can detect and sort pseudocereal products based on color, size, or quality criteria (Tambare et al. 2021). This automation reduces the reliance on human operators for tedious and physically demanding tasks and allows them to focus on more complex and value-added activities. Overall, the integration of PLCs, SCADA systems, and industrial robots in pseudocereal processing plants leads to enhanced automation, control, and efficiency. These technologies enable streamlined operations, better data management, and increased productivity, contributing to improved product quality and overall plant performance(Mourtzis et al. 2022).

11.10 Case Studies: Machinery in Pseudocereal Processing

11.10.1 *Quinoa Processing Line*

This case study explores the machinery and equipment used in a quinoa processing line, covering cleaning, milling, separation, drying, and packaging operations specific to quinoa processing (Angeli et al. 2020).

11.10.2 Amaranth Flour Mill

This case study focuses on the machinery and equipment employed in an amaranth flour mill, including cleaning, grinding, sieving, and packaging equipment designed for amaranth processing (Boukid 2021).

11.10.3 Buckwheat Noodle Production Setup

This case study examines the machinery and equipment required for a buckwheat noodle production setup, covering processes such as milling, mixing, extrusion, drying, and packaging specific to buckwheat noodle production (Obadi and Xu 2021).

11.11 Future Trends and Developments in Pseudocereal Machinery

In the future, the development of pseudocereal machinery is expected to focus on several key trends and advancements. Automation will play a significant role, with the integration of robotics, artificial intelligence, and machine learning technologies. This will lead to greater process control, precision, and efficiency while reducing the reliance on manual labor. Energy efficiency will also be a priority, with manufacturers working towards designs and technologies that minimize energy consumption, reduce waste, and optimize resource utilization. Customization and flexibility will be emphasized, with machinery being designed to adapt to different pseudocereal varieties, processing requirements, and end-product specifications. Improved productivity and throughput will be achieved through the development of faster processing equipment, improved material handling systems, and streamlined production workflows. Digitalization and data analytics will play a crucial role, allowing for real-time monitoring of equipment performance, predictive maintenance, and data-driven decision-making. Enhanced quality control measures will be implemented, incorporating advanced sensors, vision systems, and quality control technologies. Sustainability and waste reduction will also be prioritized, with the integration of recycling and waste management systems, as well as the utilization of renewable energy sources. Ultimately, improved food safety measures will be incorporated, such as advanced sanitation features, automated cleaning systems, and microbial control technologies. These future advances in pseudocereal machinery will enable manufacturers to meet consumer demands, increase productivity, ensure food safety, and contribute to sustainable and efficient processing practices.

11.15 Conclusion

In conclusion, the utilization of machinery and equipment in pseudocereal processing plays a vital role in achieving efficient and reliable production operations.

Throughout this chapter, we have explored various aspects of machinery specific to pseudocereal processing, including cleaning, milling, separation, drying, packaging, and quality control.The chapter has provided a comprehensive overview of the different types of machinery and equipment used in pseudocereal processing, highlighting their functionalities and importance at each stage of the production process. From single-screw extruders and twin-screw extruders to expansion chambers, we have seen how these machines contribute to the creation of a wide range of pseudocereal-based products with desirable textures and forms. Additionally, we have examined the significance of bagging machines, filling machines, and sealing machines in the packaging stage, ensuring accurate filling, sealing, labeling, and preservation of pseudocereal products. These machines enable efficient and standardized packaging operations, facilitating product distribution and consumer satisfaction.Moreover, we have explored the importance of moisture analyzers, particle size analyzers, and texture analyzers in assessing and controlling critical product attributes, such as moisture content, particle size distribution, and texture. These analytical tools ensure consistent product quality and stability and enable quality control and product development. Throughout this chapter, we have gained valuable insights into the current state of machinery and equipment in pseudocereal processing. However, it is important to note that the field is continuously evolving, and future advancements are anticipated. In the future, we can expect to witness further developments in pseudocereal machinery, including advanced automation, customization, energy efficiency, and enhanced data-driven optimization. These advances will contribute to increased productivity, sustainability, and improved food safety measures in pseudo processing operations. In conclusion, the appropriate selection and utilization of machinery and equipment are crucial for achieving efficient, consistent, and high-quality pseudocereal processing. By staying abreast of the latest advances and incorporating innovative technologies, manufacturers can meet the growing demand for pseudocereal-based products while ensuring optimal efficiency and sustainability in the industry.

References

Aarnisalo, Kaarina, Kaija Tallavaara, Gun Wirtanen, Riitta Maijala, and Laura Raaska. 2006. "The Hygienic Working Practices of Maintenance Personnel and Equipment Hygiene in the Finnish Food Industry." *Food Control* 17 (12): 1001–11.

Aghbashlo, Mortaza, Rahmat Sotudeh-Gharebagh, Reza Zarghami, Arun S. Mujumdar, and Navid Mostoufi. 2014. "Measurement Techniques to Monitor and Control Fluidization Quality in Fluidized Bed Dryers: A Review." *Drying Technology* 32 (9): 1005–51.

Akpinar, E., A. Midilli, and Y. Bicer. 2003. "Single Layer Drying Behaviour of Potato Slices in a Convective Cyclone Dryer and Mathematical Modeling." *Energy Conversion and Management* 44 (10): 1689–1705.

Alam, Hasin, and S. Hasan Saeed. 2013. "Modern Applications of Electronic Nose: A Review." *International Journal of Electrical and Computer Engineering (IJECE)* 3 (1). https://doi.org/10.11591/ijece.v3i1.1226.

Altun, N., Chuangfu Emre, and Jiann-Yang Xiao. 2009. "Separation of Unburned Carbon from Fly Ash Using a Concurrent Flotation Column." *Fuel Processing Technology* 90 (12): 1464–70.

Amankwah, E.Y.A., 2019. *Drying of yam with solar adsorption system.* (Doctoral dissertation, Wageningen University and Research).

Amir, Andrew M., and Michael B. McDonagh. 2012. "Zinc Oxide Particles: Synthesis, Properties and Applications." *Chemical Engineering Journal* 185: 1–22.

Amyotte, Paul R., and Rolf K. Eckhoff. 2010. "Dust Explosion Causation, Prevention and Mitigation: An Overview." *Journal of Chemical Health and Safety* 17 (1): 15–28.

Andreotti, Bruno, Yoël Forterre, and Olivier Pouliquen. 2013. *Granular Media: Between Fluid and Solid.* Cambridge University Press.

Angeli, Viktória, Pedro Miguel Silva, Danilo Crispim Massuela, Muhammad Waleed Khan, Alicia Hamar, Forough Khajehei, Simone Graeff-Hönninger, and Cinzia Piatti. 2020. "Quinoa (*Chenopodium quinoa* Willd.): An Overview of the Potentials of the 'Golden Grain' and Socio-economic and Environmental Aspects of Its Cultivation and Marketization." *Foods (Basel, Switzerland)* 9 (2): 216.

Arendt, Elke K., and Emanuele Zannini. 2013. *Cereal Grains for the Food and Beverage Industries.* Elsevier.

Aryafar, Ahmad, Reza Mikaeil, Sina Shaffiee Haghshenas, and Sami Shaffiee Haghshenas. 2018. "Application of Metaheuristic Algorithms to Optimal Clustering of Sawing Machine Vibration." *Measurement: Journal of the International Measurement Confederation* 124 (August): 20–31.

Bai, Xinjian, Tao Tao, Linyue Gao, Cheng Tao, and Yongqian Liu. 2023. "Wind Turbine Blade Icing Diagnosis Using RFECV-TSVM Pseudo-Sample Processing." *Renewable Energy* 211 (July): 412–19.

Baiano, A..2021. Craft beer: An overview. *Comprehensive reviews in food science and food safety.* 20(2):1829–1856.

Benettayeb, Asmaa, Muhammad Usman, Coffee Calvin Tinashe, Traore Adam, and Boumediene Haddou. 2022. "A Critical Review with Emphasis on Recent Pieces of Evidence of *Moringa oleifera* Biosorption in Water and Wastewater Treatment." *Environmental Science and Pollution Research International* 29 (32): 48185–209.

Bhowmik, Ankita, Shantanu Bhunia, Anupam Debsarkar, Rambilash Mallick, Malancha Roy, and Joydeep Mukherjee. 2021. "Development of a Novel Helical-Ribbon Mixer Dryer for Conversion of Rural Slaughterhouse Wastes to an Organic Fertilizer and Implications in the Rural Circular Economy." *Sustainability* 13 (16): 9455.

Black, J., and Ronald A. Temple. 2017. *DeGarmo's Materials and Processes in Manufacturing.* John Wiley & Sons.

Boukid, Fatma. 2021. "Cereal-Derived Foodstuffs from North African-Mediterranean: From Tradition to Innovation." In *Cereal-Based Foodstuffs: The Backbone of Mediterranean Cuisine*, 117–50. Springer International Publishing.

Boukid, Fatma. n.d. *Cereal-Based Foodstuffs: The Backbone of Mediterranean Cuisine.* Springer Nature.

Brown, H. M., and H. E. Arrowsmith. 2015. "Sampling for Food Allergens." In *Handbook of Food Allergen Detection and Control*, 181–97. Elsevier.

Byrn, Stephen, Maricio Futran, Hayden Thomas, Eric Jayjock, Nicola Maron, Robert F. Meyer, Allan S. Myerson, Michael P. Thien, and Bernhardt L. Trout. 2015. "Achieving Continuous Manufacturing for Final Dosage Formation: Challenges and How to Meet Them. May 20–21, 2014 Continuous Manufacturing Symposium." *Journal of Pharmaceutical Sciences* 104 (3): 792–802.

Cappelli, A., N. Oliva, and E. Cini. 2020. "A Systematic Review of Gluten-Free Dough and Bread: Dough Rheology, Bread Characteristics, and Improvement Strategies." *Applied Sciences* 10 (18):6559.

Chen, Zhiquan, Xin Tong, and Zhanfu Li. 2020. "Numerical Investigation on the Sieving Performance of Elliptical Vibrating Screen." *Processes (Basel, Switzerland)* 8 (9): 1151.

Choton, Skarma, Neeraj Gupta, Julie D. Bandral, Nadira Anjum, and Ankita Choudary. 2020. "Extrusion Technology and Its Application in Food Processing: A Review." *The Pharmaceutical Innovation* 9 (2): 162–68.

Coles, Richard, Derek Mcdowell, and Mark J. Kirwan. 2003. *Food Packaging Technology*. Vol. 5. CRC Press.

Coțovanu, Ionica, Silviu-Gabriel Stroe, Florin Ursachi, and Silvia Mironeasa. 2022. "Addition of Amaranth Flour of Different Particle Sizes at Established Doses in Wheat Flour to Achieve a Nutritional Improved Wheat Bread." *Foods (Basel, Switzerland)* 12 (1): 133.

Courtois, F., M. Abud Archila, C. Bonazzi, J. M. Meot, and G. Trystram. 2001. "Modeling and Control of a Mixed-Flow Rice Dryer with Emphasis on Breakage Quality." *Journal of Food Engineering* 49 (4): 303–9.

Da Costa, Sara, D. Vítor, and Luísa Alves. 2014. "5 Dry Milling." In *Engineering Aspects of Cereal and Cereal-Based Products*. p.97, CRC Press.

Daehn, Katrin E., André Cabrera Serrenho, and Julian Allwood. 2019. "Finding the Most Efficient Way to Remove Residual Copper from Steel Scrap." *Metallurgical and Materials Transactions B* 50 (3): 1225–40.

Denton, E. J., and J. B. Gilpin-Brown. 1961. "The Distribution of Gas and Liquid within the Cuttlebone." *Journal of the Marine Biological Association of the United Kingdom* 41 (2): 365–81.

Devahastin, Sakamon, and Arun Mujumdar. 2003. "Applications for Fluidized Bed Drying." In *Handbook of Fluidization and Fluid-Particle Systems*. 469–484, CRC Press.

Dhull, S. B., M. K. Kidwai, R. Noor, P. Chawla, and P. K. Rose. 2022a. "A Review of Nutritional Profile and Processing of Faba Bean (*Vicia faba* L.). *Legume Science* 4 (3): e129.

Dhull, S. B., M. K. Kidwai, M. Siddiq, and J. S. Sidhu. 2022b. "Faba (Broad) Bean Production, Processing, and Nutritional Profile." In *Dry Beans and Pulses*. https://doi.org/10.1002/9781119776802.ch14.

Dhull, S. B., J. Kinabo, and M. A. Uebersax. 2023. "Nutrient Profile and Effect of Processing Methods on the Composition and Functional Properties of Lentils (*Lens culinaris* Medik): A Review." *Legume Science* 5 (1): e156.

Dzedzickis, Andrius, Jurga Subačiūtė-Žemaitienė, Ernestas Šutinys, Urtė Samukaitė-Bubnienė, and Vytautas Bučinskas. 2021. "Advanced Applications of Industrial Robotics: New Trends and Possibilities." *Applied Sciences (Basel, Switzerland)* 12 (1): 135.

El-Emam, Mahmoud A., Ling Zhou, Weidong Shi, Chen Han, Ling Bai, and Ramesh Agarwal. 2021. "Theories and Applications of CFD–DEM Coupling Approach for Granular Flow: A Review." *Archives of Computational Methods in Engineering. State of the Art Reviews* 28 (7): 4979–5020.

Elgeti, Dana, Mario Jekle, and Thomas Becker. 2015. "Strategies for the Aeration of Gluten-Free Bread-A Review." *Trends in Food Science and Technology* 46 (1): 75–84.

Enrique, Daisy Valle, Érico Marcon, Fernando Charrua-Santos, and Alejandro G. Frank. 2022. "Industry 4.0 Enabling Manufacturing Flexibility: Technology Contributions to Individual Resource and Shop Floor Flexibility." *Journal of Manufacturing Technology Management* 33 (5): 853–75.

Farkas, Dóra, Lajos Madarász, Zsombor K. Nagy, István Antal, and Nikolett Kállai-Szabó. 2021. "Image Analysis: A Versatile Tool in the Manufacturing and Quality Control of Pharmaceutical Dosage Forms." *Pharmaceutics* 13 (5): 685.

Filli, Kalep, I. O. Afam, and Victoria A. Jideani. 2014. "Extrusion Bolsters Food Security in Africa." *Food Technology* 68 (8): 45–55.

Fiorentini, Martina, Amanda J. Kinchla, and Alissa A. Nolden. 2020. "Role of Sensory Evaluation in Consumer Acceptance of Plant-Based Meat Analogs and Meat Extenders: A Scoping Review." *Foods (Basel, Switzerland)* 9 (9): 1334.

Fu, Bin Xiao. 2008. "Asian Noodles: History, Classification, Raw Materials, and Processing." *Food Research International (Ottawa, Ont.)* 41 (9): 888–902.

Gawenda, Tomasz. 2021. "Production Methods for Regular Aggregates and Innovative Developments in Poland." *Minerals (Basel, Switzerland)* 11 (12): 1429.

Grubben, G. J. H., and T. Soetjipto Partohardjono. 1997. "Plant Resources of South East Asia." *Botanical Journal of the Linnean Society* 123 (3), p.261.

Gullichsen, Johan, Hannu Paulapuro, and Esa Lehtinen. n.d. *Pigment Coating and Surface Sizing of Paper Series Editors This Word Document Was Downloaded From/Please Remain This Link Information When You Reproduce.*

Gunasekaran, Sundaram. 1999. "Pulsed Microwave-Vacuum Drying of Food Materials." *Drying Technology* 17 (3): 395–412.

Hadde, Enrico K., and Jianshe Chen. 2021. "Texture and Texture Assessment of Thickened Fluids and Texture-Modified Food for Dysphagia Management." *Journal of Texture Studies* 52 (1): 4–15.

Han, Xiao-Miao, Jun-Jie Xing, Cong Han, Xiao-Na Guo, and Ke-Xue Zhu. 2021. "The Effects of Extruded Endogenous Starch on the Processing Properties of Gluten-Free Tartary Buckwheat Noodles." *Carbohydrate Polymers* 267: 118170.

Hany, S., I. Ahmed, Zicheng El-Seesy, and Yang Hu. 2022. "Recent Developments in Solar Drying Technology of Food and Agricultural Products: A Review." *Renewable and Sustainable Energy Reviews* 157:112070.

Harle, K. J., S. M. Howden, L. P. Hunt, and M. Dunlop. 2007. "The Potential Impact of Climate Change on the Australian Wool Industry by 2030." *Agricultural Systems* 93 (1–3): 61–89.

Hart, Ashlee B. 2019. *Convening Cultures in Ancient Thrace: An Evaluation of Interaction on Ceramic Technological Choice From Iron Age Bulgaria.* (Doctoral dissertation, State University of New York at Buffalo).

Hemlata, Roshan V., and Michael A. Tiwari. 2016. "Hot-Melt Extrusion: From Theory to Application in Pharmaceutical Formulation." *AAPS PharmSciTech* 17 (1): 20–42.

Hernández-Martínez, Alejandro, Sergio Giraldo, Xavier Alcobé, Ignacio Becerril-Romero, Marcel Placidi, Víctor Izquierdo-Roca, Paul Pistor, Alejandro Pérez-Rodríguez, Edgardo Saucedo, and Matías Valdés. 2022. "Kinetics and Phase Analysis of Kesterite Compounds: Influence of Chalcogen Availability in the Reaction Pathway." *Materialia* 24: 101509.

Hovmand, Svend. 2020. "Fluidized Bed Drying." In *Handbook of Industrial Drying*, 195–248. CRC Press.

Humphries, W., and J. H. Vincent. 1976. "An Experimental Investigation of the Detention of Airborne Smoke in the Wake Bubble behind a Disk." *Journal of Fluid Mechanics* 73 (3): 453–64.

Iranshahi, Kamran, Donato Rubinetti, Daniel I. Onwude, Marios Psarianos, Oliver K. Schlüter, and Thijs Defraeye. 2023. "Electrohydrodynamic Drying versus Conventional Drying Methods: A Comparison of Key Performance Indicators." *Energy Conversion and Management* 279: 116661.

Jankovic, A. 2015. "Developments in Iron Ore Comminution and Classification Technologies." In *Iron Ore*, 251–82. Elsevier.

Jayachandran, M., Ch Rami Reddy, Sanjeevikumar Padmanaban, and A. H. Milyani. 2021. "Operational Planning Steps in Smart Electric Power Delivery System." *Scientific Reports* 11 (1): 17250.

Kaleemullah, S., M. Raveendra Reddyand, and B. Prabhakar. 2023. *Performance Evaluation of Minor Millet Processing Machines Suitable for Small and Medium Scale Industries*. AkiNik Publications.

Kaushal, Pragati, and Navneet Kumar. 2022. "Processing of Cereals." In *Agro-Processing and Food Engineering: Operational and Application Aspects*, 415–54. Springer.

Kaye, Brian H. 1997. "Mixing of Powders." In *Handbook of Powder Science & Technology*, 568–85. Springer US.

Kear, B. H., and L. E. McCandlish. 1993. "Chemical Processing and Properties of Nanostructured WC-Co Materials." *Nanostructured Materials* 3 (1–6): 19–30.

Khairuddin, Muhammad Arif Najmi, and Ola Lasekan. 2021. "Gluten-Free Cereal Products and Beverages: A Review of Their Health Benefits in the Last Five Years." *Foods (Basel, Switzerland)* 10 (11): 2523.

Khanpit, Vishal V., P. Sonali, and Sachin A. Tajane. 2022. "Extrusion for Soluble Dietary Fiber Concentrate: Critical Overview on Effect of Process Parameters on Physicochemical, Nutritional, and Biological Properties." *Food Reviews International*. 39 (9): 1–22.

Khosravi, A., A. Martinez, and J. T. DeJong. 2020. "Discrete Element Model (DEM) Simulations of Cone Penetration Test (CPT) Measurements and Soil Classification." *Canadian Geotechnical Journal* 57 (9): 1369–87.

Kim, J. G., A. E. Yousef, and S. Dave. 1999. "Application of Ozone for Enhancing the Microbiological Safety and Quality of Foods: A Review." *Journal of Food Protection* 62 (9): 1071–87.

KuShaari, Kuzilati, Preetanshu Pandey, Yongxin Song, and Richard Turton. 2006. "Monte Carlo Simulations to Determine Coating Uniformity in a Wurster Fluidized Bed Coating Process." *Powder Technology* 166 (2): 8190.

Kushwaha, Amanendra K., Merbin John, Manoranjan Misra, and Pradeep L. Menezes. 2021. "Nanocrystalline Materials: Synthesis, Characterization, Properties, and Applications." *Crystals* 11 (11): 1317.

Lacroix, Monique. 2009. "Mechanical and Permeability Properties of Edible Films and Coatings for Food and Pharmaceutical Applications." In *Edible Films and Coatings for Food Applications*, 347–66. Springer.

Li, Yang, Lizhang Xu, Liya Lv, Yan Shi, and Yu Xun. 2022. "Study on Modeling Method of a Multi-parameter Control System for Threshing and Cleaning Devices in the Grain Combine Harvester." *Agriculture* 12 (9): 1483.

Lippmann, Richard P. 1990. "Review of Neural Networks for Speech Recognition." In *Readings in Speech Recognition*, 374–92. Elsevier.

Liu, Guanghua, Jiangtao Li, and Kexin Chen. 2013. "Combustion Synthesis of Refractory and Hard Materials: A Review." *International Journal of Refractory and Hard Metals* 39 (July): 90–102.

Liu, Mingjun, Yadong Gong, Jingyu Sun, Benjia Tang, Yao Sun, Xinpeng Zu, and Jibin Zhao. 2023. "The Accuracy Losing Phenomenon in Abrasive Tool Condition Monitoring and a Noval WMMC-JDA Based Data-Driven Method Considered Tool Stochastic Surface Morphology." *Mechanical Systems and Signal Processing* 198: 110410

Liu, Xuwei, Carine Le Bourvellec, Yu Jiahao, Lei Zhao, Kai Wang, Yang Tao, Catherine M. G. C. Renard, and Zhuoyan Hu. 2022. "Trends and Challenges on Fruit and Vegetable Processing: Insights into Sustainable, Traceable, Precise, Healthy, Intelligent, Personalized and Local Innovative Food Products." *Trends in Food Science and Technology* 125 (July): 12–25.

Liu, Zengran, Guangyi Zhang, and Xiangmei Zhang. 2014. "Urban Street Foods in Shijiazhuang City, China: Current Status, Safety Practices and Risk Mitigating Strategies." *Food Control* 41 (July): 212–18.

Madarász, Lajos, Lilla Alexandra Mészáros, Ákos Köte, Attila Farkas, and Zsombor Kristóf Nagy. 2023. "AI-Based Analysis of In-Line Process Endoscope Images for Real-Time Particle Size Measurement in a Continuous Pharmaceutical Milling Process." *International Journal of Pharmaceutics* 641: 123060.

Malik, M., R. Sindhu, S. B. Dhull, C. Bou-Mitri, Y. Singh, S. Panwar, and B. S. Khatkar. 2023. "Nutritional Composition, Functionality, and Processing Technologies for Amaranth." *Journal of Food Processing and Preservation* 2023: 1753029. https://doi.org/10.1155/2023/1753029.

Majumder, Prasanta, Bachu Deb, Rajat Gupta, and Shyam S. Sablani. 2022. "A Comprehensive Review of Fluidized Bed Drying: Sustainable Design Approaches, Hydrodynamic and Thermodynamic Performance Characteristics, and Product Quality." *Sustainable Energy Technologies and Assessments* 53: 102643.

Mariotti, M., C. Alamprese, M. A. Pagani, and M. Lucisano. 2006. "Effect of Puffing on Ultrastructure and Physical Characteristics of Cereal Grains and Flours." *Journal of Cereal Science* 43 (1): 47–56.

Marriott, Norman G., M. Wes Schilling, and Robert B. Gravani. 2018. "Meat and Poultry Plant Sanitation." In *Principles of Food Sanitation*, 311–40. Springer International Publishing.

Martin, Charlie. 2016. "Twin Screw Extruders as Continuous Mixers for Thermal Processing: A Technical and Historical Perspective." *AAPS PharmSciTech* 17 (1): 3–19.

Meng, Xiangchen, Yongxian Huang, Jian Cao, Junjun Shen, and Jorge F. dos Santos. 2021. "Recent Progress on Control Strategies for Inherent Issues in Friction Stir Welding." *Progress in Materials Science* 115: 100706.

Moes, Johannes J., Marco M. Ruijken, Erik Gout, Henderik W. Frijlink, and Michael I. Ugwoke. 2008. "Application of Process Analytical Technology in Tablet Process Development Using NIR Spectroscopy: Blend Uniformity, Content Uniformity and Coating Thickness Measurements." *International Journal of Pharmaceutics* 357 (1–2): 108–18.

Mourtzis, Dimitris, John Angelopoulos, and Nikos Panopoulos. 2022. "Operator 5.0: A Survey on Enabling Technologies and a Framework for Digital Manufacturing Based on Extended Reality." *Journal of Machine Engineering* 22 (1): 43–69.

Mujica, A., Angel Rubio, A. Muñoz, and R. J. Needs. 2003. "High-Pressure Phases of Group-IV, III–V, and II–VI Compounds." *Reviews of Modern Physics* 75 (3): 863–912.

Mujumdar, Arun S. 2004. "Research and Development in Drying: Recent Trends and Future Prospects." *Drying Technology* 22 (1–2): 1–26.

Muzzio, Fernando J., Albert Alexander, Chris Goodridge, Elizabeth Shen, Troy Shinbrot, Konanur Manjunath, Shrikant Dhodapkar, and Karl Jacob. 2004. "Solids Mixing." In *Handbook of Industrial Mixing*, 887–985. John Wiley & Sons, Inc.

Neal, Christopher T., and Tyler L. Wright. 2013. "Optimizing Hammer Mill Performance Through Screen Selection and Hammer Design." *Biofuels* 4 (1): 85–94.

Oates, Joseph. 2008. *Lime and Limestone: Chemistry and Technology, Production and Uses.* John Wiley & Sons.

Obadi, Mohammed, and Bin Xu. 2021. "Review on the Physicochemical Properties, Modifications, and Applications of Starches and Its Common Modified Forms Used in Noodle Products." *Food Hydrocolloids* 112: 106286.

Offiah, Vivian, Vassilis Kontogiorgos, and Kolawole O. Falade. 2019. "Extrusion Processing of Raw Food Materials and By-Products: A Review." *Critical Reviews in Food Science and Nutrition* 59 (18): 2979–98.

Olakanmi, Sunday J., Digvir S. Jayas, and Jitendra Paliwal. 2022. "Implications of Blending Pulse and Wheat Flours on Rheology and Quality Characteristics of Baked Goods: A Review." *Foods (Basel, Switzerland)* 11 (20): 3287.

Onyinge, George. 2019. *Development and Validation of a Drying Model for Rastrineobola argentea Fish in an Indirect Forced Convection Solar Dryer.* (Doctoral dissertation, Maseno University).

Orrego, Carlos, Natalia Eduardo, and Luisa Fernanda Salgado. 2023. "Freeze Drying and Vacuum Drying." In *Drying Technology in Food Processing*, 203–40. WP Publication.

Pagani, M., Mara Lucisano Ambrogina, and Manuela Mariotti. 2007. "Traditional Italian Products from Wheat and Other Starchy Flours." In *Handbook of Food Products Manufacturing*, 327–88. John Wiley & Sons, Inc.

Pandey, Vivek, Hao Chen, Jiaxin Ma, and João M. Maia. 2021. "Extension-Dominated Improved Dispersive Mixing in Single-Screw Extrusion. Part 2: Comparative Analysis with Twin-Screw Extruder." *Journal of Applied Polymer Science* 138 (5): 49765.

Parikh, Dilip. 2017. *How to Optimize Fluid Bed Processing Technology: Part of the Expertise in Pharmaceutical Process Technology Series.* Academic Press.

Peariso, D. 2008. *Preventing Foreign Material Contamination of Foods.* John Wiley & Sons.

Pirsa, Sajad, Iraj Karimi Sani, and Sanaz Sadat Mirtalebi. 2022. "Nano-biocomposite Based Color Sensors: Investigation of Structure, Function, and Applications in Intelligent Food Packaging." *Food Packaging and Shelf Life* 31: 100789.

Prabha, Krishna, Abdullah Payel Ghosh, Rosmin M. Joseph, Reshma Krishnan, Sandeep Singh Rana, Rama Chandra Pradhan, and R. C. Pradhan. 2021. "Recent Development, Challenges, and Prospects of Extrusion Technology." *Future Foods: A Dedicated Journal for Sustainability in Food Science* 3: 100019.

Premi, Monica, and Vishal Sharma. 2022. "Mixing and Forming." In *Agro-Processing and Food Engineering: Operational and Application Aspects*, 253–305. Springer.

Provan, David J., David D. Woods, Sidney W. A. Dekker, and Andrew J. Rae. 2020. "Safety II Professionals: How Resilience Engineering Can Transform Safety Practice." *Reliability Engineering and System Safety* 195: 106740.

Punia, S., and S. B. Dhull. 2019. "Chia Seed (Salvia hispanica L.) Mucilage (a Heteropolysaccharide): Functional, Thermal, Rheological Behaviour and Its Utilization." *International Journal of Biological Macromolecules* 140: 1084–1090.

Qi, Qingbin, Wei Wang, Yu Wang, and Yu Dan. 2020. "Robust Light-Driven Interfacial Water Evaporator by Electrospinning SiO2/MWCNTs-COOH/PAN Photothermal Fiber Membrane." *Separation and Purification Technology* 239: 116595.

Rani, Rekha, Payal Karmakar, Neha Singh, and Ronit Mandal. 2023. "Encapsulation Methods in the Food Industry: Mechanisms and Applications." In *Advances in Food Process Engineering: Novel Processing, Preservation, and Decontamination of Foods*, 93. CRC Press.

Sayyeda, Zeenat A., Isabelle Kamm, Tina Nguyen, Jackson D. Pellett, and Archana Kumar. 2021. "Loss on Drying Using Halogen Moisture Analyzer: An Orthogonal Technique for Monitoring Volatile Content for In-Process Control Samples During Pharmaceutical Manufacturing." *Organic Process Research and Development* 25 (2): 300–307.

Razykov, T. M., C. S. Ferekides, D. Morel, E. Stefanakos, H. S. Ullal, and H. M. Upadhyaya. 2011. "Solar Photovoltaic Electricity: Current Status and Future Prospects." *Solar Energy (Phoenix, Ariz.)* 85 (8): 1580–1608.

Ribeiro, Tânia Bragança, Glenise Bierhalz Voss, Marta Correia Coelho, and Manuela Estevez Pintado. 2022. "Food Waste and By-Product Valorization as an Integrated Approach with Zero Waste: Future Challenges." In *Future Foods*, 569–96. Elsevier.

Robinson, Martin, and Paul W. Cleary. 2012. "Flow and Mixing Performance in Helical Ribbon Mixers." *Chemical Engineering Science* 84 (December): 382–98.

Rogers, G. G., and Leonardo Bottaci. 1997. "Modular Production Systems: A New Manufacturing Paradigm." *Journal of Intelligent Manufacturing* 8 (2): 147–56.

Ronnenberg, Herman. 2016. "Material Culture of Breweries" In *Material Culture of Breweries* 2.

Roughan, Matthew, Subhabrata Sen, Oliver Spatscheck, and Nick Duffield. 2004. "Class-of-Service Mapping for QoS: A Statistical Signature-Based Approach to IP Traffic Classification." In *Proceedings of the 4th ACM SIGCOMM Conference on Internet Measurement*, 135–48.

Saini, Praveen, Nitin Kumar, Sunil Kumar, and Anil Panghal. 2022. "Fluidized Bed Drying: Recent Developments and Applications." In *Thermal Food Engineering Operations*, 197–219. Wiley.

Sanjeev, K., and M. N. Ramesh. 2006. "Low Oxygen and Inert Gas Processing of Foods." *Critical Reviews in Food Science and Nutrition* 46 (5): 423–51.

Sanna, Aimaro, Sujing Li, Rob Linforth, Katherine A. Smart, and John M. Andrésen. 2011. "Bio-oil and Bio-char from Low Temperature Pyrolysis of Spent Grains Using Activated Alumina." *Bioresource Technology* 102 (22): 10695–703.

Santos, Kamila. 2021. *Use of Different Hydrocoloids for the Production of Mixed Structure of Açaí, Banana, Peanut, and Guarana Syrup*. ibict.

Sarkar, Ashok, and Bin Xiao Fu. 2022. "Impact of Quality Improvement and Milling Innovations on Durum Wheat and End Products." *Foods (Basel, Switzerland)* 11 (12): 1796.

Sasthri, Vijaykrishnaraj, Nivedha Muthugopal, and Pichan Krishnakumar. 2020. "Advances in Conventional Cereal and Pseudocereal Processing." *Innovative Processing Technologies for Healthy Grains* 1: 61–81.

Scanlon, M. G., S. Thakur, R. T. Tyler, A. Milani, T. Der, and J. Paliwal. 2018. "The Critical Role of Milling in Pulse Ingredient Functionality." *Cereal Foods World* 63 (5): 201–6.

Schlem, Roman, Christine Friederike Burmeister, Peter Michalowski, Saneyuki Ohno, Georg F. Dewald, Arno Kwade, and Wolfgang G. Zeier. 2021. "Energy Storage Materials for Solid-State Batteries: Design by Mechanochemistry." *Advanced Energy Materials* 11 (30): 2101022.

Schoeman, Letitia, and Marena Manley. 2019. "Oven and Forced Convection Continuous Tumble (FCCT) Roasting: Effect on Physicochemical, Structural and Functional Properties of Wheat Grain." *Food and Bioprocess Technology* 12 (1): 166–82.

Schuler, Steve F., Robert K. Bacon, Patrick L. Finney, and Edward E. Gbur. 1995. "Relationship of Test Weight and Kernel Properties to Milling and Baking Quality in Soft Red Winter Wheat." *Crop Science* 35 (4): 949–53.

Sendek, Austin D., Brandi Ransom, Ekin D. Cubuk, Lenson A. Pellouchoud, Jagjit Nanda, and Evan J. Reed. 2022. "Machine Learning Modeling for Accelerated Battery Materials Design in the Small Data Regime." *Advanced Energy Materials* 12 (31): 2200553.

Shah, Manzoor, Ahmad Shabir, and B. N. Mir. 2021. "Advances in Extrusion Technologies." In *Food Formulation: Novel Ingredients and Processing Techniques*, 147–63. John Wiley & Sons.

Shahidi, Fereidoon. 2005. *Bailey's Industrial Oil and Fat Products, Edible Oil and Fat Products: Processing Technologies*. Vol. 5. John Wiley & Sons.

Shalko, Y. 2022. *Analysis of Production Data Monitoring and Visualization Systems for Cyber-Physical Production Systems (Doctoral dissertation)*.

Shapiro, M., and V. Galperin. 2005. "Air Classification of Solid Particles: A Review." *Genie Des Procedes [Chemical Engineering and Processing]* 44 (2): 279–85.

Sharma, Atul, C. R. Chen, and Nguyen Vu Lan. 2009. *Renewable and Sustainable Energy Reviews* 13: 1185–1210.

Sharma, Harish, and Navneet Kumar, eds. n.d. *Agro-Processing and Food Engineering: Operational and Application Aspects*. Springer Nature.

Shekunov, Boris Y., Pratibhash Chattopadhyay, Henry H. Y. Tong, and Albert H. L. Chow. 2007. "Particle Size Analysis in Pharmaceutics: Principles, Methods and Applications." *Pharmaceutical Research* 24 (2): 203–27.

Shen, Rongxi, Hongru Li, Enyuan Wang, Tongqing Chen, Taixun Li, He Tian, and Zhenhai Hou. 2020. "Infrared Radiation Characteristics and Fracture Precursor Information Extraction of Loaded Sandstone Samples with Varying Moisture Contents." *International Journal of Rock Mechanics and Mining Sciences Oxford, England: 1997)* 130: 104344.

Shun-Ichiro, Mervyn S., John D. Paterson. 1986. "Rheology of Synthetic Olivine Aggregates: Influence of Grain Size and Water." *Journal of Geophysical Research: Solid Earth* 91 (B8): 8151–76.

Siegmann, E., S. Enzinger, P. Toson, P. Doshi, J. Khinast, and D. Jajcevic. 2021. "Massively Speeding up DEM Simulations of Continuous Processes Using a DEM Extrapolation." *Powder Technology* 390 (September): 442–55.

Sieniutycz, Stanislaw. 2015. "A Graphical Approach to Heat and Mass Exchange between Gas and Granular Solid." *Cybernetics and Physics* 4 (4): 116–33.

Singh, Baljit, Chetan Sharma, and Savita Sharma. 2020. Fundamentals of Extrusion Processing. https://doi.org/10.31219/osf.io/xqa5n.

Singh, Rakesh K., and Singh Nepal. 2005. "Quality of Packaged Foods." In *Innovations in Food Packaging*, 24–44. Elsevier.

Sinha, J. P., Ashwani Kumar, and Elmar Weissmann. 2023. "Seed Processing for Quality Upgradation." *Malavika Dadlani*.

Song, S., S. Lu, and A. Lopez-Valdivieso. 2002. "Magnetic Separation of Hematite and Limonite Fines as Hydrophobic Flocs from Iron Ores." *Minerals Engineering* 15 (6): 415–22.

Sozzi, Agustina, Mariana Zambon, Germán Mazza, and Daniela Salvatori. 2021. "Fluidized Bed Drying of Blackberry Wastes: Drying Kinetics, Particle Characterization and Nutritional Value of the Obtained Granular Solids." *Powder Technology* 385 (June): 37–49.

Spinks, Geoffrey M. 2020. "Advanced Actuator Materials Powered by Biomimetic Helical Fiber Topologies." *Advanced Materials (Deerfield Beach, Fla.)* 32 (18): e1904093.

Srinivas, A. 2022. "Millet Milling Technologies." In *Handbook of Millets - Processing, Quality, and Nutrition Status*, 173–203. Springer Nature Singapore.

Stark, C. R. 2012. "Feed Processing to Maximize Feed Efficiency." In *Feed Efficiency in Swine*, 131–51. Wageningen Academic Publishers.

Starý, Michal, František Novotný, Marcel Horák, and Marie Stará. 2020. "Sampling Robot for Primary Circuit Pipelines of Decommissioned Nuclear Facilities." *Automation in Construction* 119: 103303.

Steven, David B., Scot Bennett, Patrick W. Cheu, Donald B. Conley, Steven Guzek, and John Gray 2005. "EXUBERA®: Pharmaceutical Development of a Novel Product for Pulmonary Delivery of Insulin." *Diabetes Technology and Therapeutics* 7 (6): 896–906.

Stojanovic, Mirjana D., V. Slavica Bostjancic-Rakas, and D. Jasna Markovic-Petrovic. 2019. "Scada Systems in the Cloud and Fog Environments: Migration Scenarios and Security Issues." *Facta Universitatis - Series Electronics and Energetics* 32 (3): 345–58.

Subramani, Deepak, Sharmila Tamilselvan, Maheswari Murugesan, and Shivaswamy 2020. "Optimization of Sand Puffing Characteristics of Quinoa Using Response Surface Methodology." *Current Research in Nutrition and Food Science Journal* (August),8(2): 504–15.

Sudhakar, Anjali, Kanupriya Choudhary, and Cinu Subir Kumar Chakraborty. 2023. "Extrusion Technology of Food Products: Types and Operation." In *Advances in Food Process Engineering: Novel Processing, Preservation, and Decontamination of Foods*, 45. CRC Press.

Sun, Cong. n.d. *A Visual Tracking System for Honeybee 3D Flight Trajectory Reconstruction and Analysis*.

Svihus, B., K. H. Kløvstad, V. Perez, O. Zimonja, S. Sahlström, R. B. Schüller, W. K. Jeksrud, and E. Prestløkken. 2004. "Physical and Nutritional Effects of Pelleting of Broiler Chicken Diets Made from Wheat Ground to Different Coarsenesses by the Use of Roller Mill and Hammer Mill." *Animal Feed Science and Technology* 117 (3–4): 281–93.

Tambare, Parkash, Chandrashekhar Meshram, Cheng-Chi Lee, Rakesh Jagdish Ramteke, and Agbotiname Lucky Imoize. 2021. "Performance Measurement System and Quality Management in Data-Driven Industry 4.0: A Review." *Sensors (Basel, Switzerland)* 22 (1): 224.

Tariq, Usman, Irfan Ahmed, Ali Kashif Bashir, and Kamran Shaukat. 2023. "A Critical Cybersecurity Analysis and Future Research Directions for the Internet of Things: A Comprehensive Review." *Sensors (Basel, Switzerland)* 23 (8). https://doi.org/10.3390/s23084117.

Teunou, E., and D. Poncelet. 2002. "Batch and Continuous Fluid Bed Coating-Review and State of the Art." *Journal of Food Engineering* 53 (4): 325–40.

Thomas, L., Bhat, A., Cheriyan, H. and Nirmal Babu, K., 2017. Value chain development and technology practices of spices crop in India (cardamom, ginger, turmeric, black pepper and cinnamon). Challenges and opportunities in value chain of spices in south Asia. Dhaka (Bangladesh): SAARC Agriculture Centre, 56-115. Indian Institute of Spices Research.

Tieu, Vinh. 2021. *Creating Programmable Logic Control Program for a Storing Station.*

Valentiuk, N., and G. Stankevych. 2020. *Peculiarities of the Process of Purification of Amarant Grain from Impurities.*

Van Heurck, Henri. n.d. "A Treatise on the Diatomaceae: Containing Introductory Remarks on the Structure, Life History, Collection, Cultivation and Preparation of Diatoms, and a Description and Figure Typical of Every Known Genus, as Well as a Description and Figure of Every Species Found in the North Sea and Countries Bordering It." *Including Great Britain.*

Varzakas, Theodoros, V. Polychniatou, and Constantina Tzia. 2014. "Mixing-Emulsions." In *Food Engineering Handbook*, 181–252. CRC Press.

Venables, H. J., and J. I. Wells. 2001. "Powder Mixing." *Drug Development and Industrial Pharmacy* 27 (7): 599–612.

Verica, Bojana, Ana Balanč, Steva Belščak-Cvitanović, Kata Lević, Ana Trifković, Ivana Kalušević, Draženka Kostić, Branko Komes, and Viktor Bugarski. 2015. "Trends in Encapsulation Technologies for Delivery of Food Bioactive Compounds." *Food Engineering Reviews* 7 (4): 452–90.

Volkhonov, Michael, Anton Abalikhin, Alexander Krupin, and Ivan Maximov. 2020. "Studying the Operational Efficiency of the Centrifugal-Impact Feed Grain Crusher of the New Design." *Eastern-European Journal of Enterprise Technologies* 5 (1–107): 44–51.

Wilkinson, Arthur N., and Anthony J. Ryan. 1998. *Polymer Processing and Structure Development.* Springer Science & Business Media.

Wimberly, James E. 1983. *Technical Handbook for the Paddy Rice Postharvest Industry in Developing Countries.* International Rice Research Institute.

Xu, Jingwen, Yiqin Zhang, Weiqun Wang, and Yonghui Li. 2020. "Advanced Properties of Gluten-Free Cookies, Cakes, and Crackers: A Review." *Trends in Food Science and Technology* 103 (September): 200–213.

Xue, Fei, and Fei Gao. 2022. "Experimental Investigation of Energy Efficiency of an Air Classifier Mill Pulverizing a Raw Material of Aquafeed." *Particulate Science and Technology* 40 (1): 104–12.

Yadav, Mohini, Vagish Dwibedi, Swati Sharma, and Nancy George. 2022. "Biogenic Silica Nanoparticles from Agro-Waste: Properties, Mechanism of Extraction and Applications in Environmental Sustainability." *Journal of Environmental Chemical Engineering* 10 (6): 108550.

Yang, Kai, Haining Wang, Haining Wang, and Limin Sun. 2022. "An Effective Intrusion-Resilient Mechanism for Programmable Logic Controllers against Data Tampering Attacks." *Computers in Industry* 138: 103613.

Yazdani-Asrami, Mohammad, Wenjuan Song, Antonio Morandi, Giovanni De Carne, Joao Murta-Pina, Anabela Pronto, Roberto Oliveira, et al. 2023. "Roadmap on Artificial Intelligence and Big Data Techniques for Superconductivity." *Superconductor Science and Technology* 36 (4): 043501.

Zhang, Hao, Wanbing Qiao, Xizhong An, Ye Xinglian, and Jiang Chen. 2022. "CFD-DEM Study on Fluidization Characteristics of Gas-Solid Fluidized Bed Reactor Containing Ternary Mixture." *Powder Technology* 401: 117354.

Zhao, Donghua, and Weizhong Guo. 2020. "Shape and Performance Controlled Advanced Design for Additive Manufacturing: A Review of Slicing and Path Planning." *Journal of Manufacturing Science and Engineering* 142 (1): 1–87.

Chapter 12

Traditional Foods Based on Pseudocereals

Praveen Nayak, Atul Anand Mishra,
Rama Nath Shukla, Rohit Biswas

12.1 Introduction

The best-known representatives of pseudocereals are buckwheat, amaranth, quinoa, chia, and canihua, the latter being less well known. All these pseudocereals have very good nutritional value with high concentrations of essential fats, amino acids, minerals, and some vitamins.

Traditional cuisine embodies the essence of a particular culture, its rich history, and the way of life it represents. While globalization has undoubtedly connected the world, it is important to acknowledge that distinct dietary patterns continue to prevail among different countries, as Slimani et al. (2002) have reported. Exploring traditional foods provides valuable insights into the evolution of dietary patterns over time and how they have been influenced and shaped.

The issue of food insecurity is a worldwide concern that revolves around the inadequate availability of nourishing food options. Food insecurity is a complex problem influenced by various factors, including economic, agricultural, environmental, and social changes. Climate change, in particular, poses a significant threat to food security due to its profound effects on the agricultural industry. The ongoing rise in global temperatures and the subsequent decrease in available water and cultivable land are among the crucial factors contributing to a decline in crop yields. The food supply chain and access to healthy foods are disrupted by these adverse effects on the agricultural sector.

In spite of advances in agricultural technology that have been able to maximize the production of staples, a shift towards the use of environmentally viable crops,

 DOI: 10.1201/9781003325277-12

with minimal impact on the environment, is needed if adverse effects from climate change are to be avoided. In order to avoid malnutrition, diversity of diet is crucial as consumption of staple crops itself frequently leads to insufficient essential nutrients. The development of sustainable crops which are resilient to climate change and the production of food products that deliver fundamental nutrients for maintaining a healthy diet should therefore be given utmost importance in combating food insecurity (Biswas et al., 2020; Punia & Dhull, 2019).

Both of these criteria are met by amaranth, quinoa, and millet grains, all discussed in this chapter, which would contribute to food security. They have physiological properties which allow them to cope with drought, high salt soils, and high temperatures; they are also able to thrive in favourable conditions of low fertility. Though these crops have been a part of local culture in developing countries across Asia and South and Central America, the lack of awareness of their benefits has led to reduced consumption. By providing education and raising awareness among farmers about the potential for cultivating these grains with reduced reliance on natural resources, we can promote the widespread adoption of these crops within the agricultural community. This would lead to faster integration and implementation of sustainable farming practices (Balakrishnan et al. 2022)

Traditionally, amaranth seeds were roasted, cooked, flaked, or popped for human consumption (Malik et al., 2023). The above-ground parts of the plant have been served after cooking, similar to how raw spinach is prepared for eating as a vegetable. The popping process holds significant interest due to the anatomical structure of amaranth seeds. In modern times, popped and extruded amaranth grains have gained popularity as the primary ingredient in crunchy bars, breakfast cereals, and snacks (Ramos-Diaz et al., 2013; Singh et al., 2023).

A promising technology to process amaranth is extrusion cooking. Extrusion cooking has been found to significantly enhance the water solubility index of amaranth, increasing it from 11% to 61% (Robin et al., 2015). The highest expansion index was achieved using the addition of 20% amaranth to maize, in comparison with indices obtained when quinoa, canihua, and 100% maize were used (Ramos-Diaz et al., 2013).

12.2 Quinoa

Quinoa is regarded as a highly nutritious food due to it being rich in high-quality protein, lipids, minerals, starch, and vitamins, including vitamins E and C, as noted by Nisar et al. (2018). Quinoa is a valuable resource of essential amino acids, including methionine (0.4–1.0%) and lysine (5.1–6.4%), which are commonly deficient in other grains and legumes. Furthermore, this grain is recognized as a gluten-free option due to its low or negligible levels of prolamin. Quinoa is classified as a functional food that has been shown to lower the risk of various diseases, such as celiac disease, while also promoting overall health, as highlighted by Abdellatif (2018).

Some anti-nutritional compounds, such as saponins, which are bitter and toxic, are mainly found in the hull of the quinoa grains (Dhull et al., 2022a, 2022b, 2023). As a result, it is necessary to remove the hull of quinoa through a process known as dehulling or polishing and washing before consuming it (Sharma & Lakhawat (2017).

The Food and Agriculture Organization (FAO) in Rome has recognized quinoa as a crop that has the potential to make a significant contribution to global food security in the twenty-first century. This is because quinoa is resilient and can tolerate difficult conditions, such as drought, stress and salinity, and can be grown in marginal regions. In acknowledgment of its promising qualities, the United Nations (UN) designated the year 2013 as the International Year of Quinoa. The primary objective behind declaring 2013 as the International Year of Quinoa was to raise worldwide awareness regarding the issues of food security, nutrition, and poverty eradication as highlighted by Sharma and Lakhawat (2017).

Various products have been developed by incorporating quinoa through the utilization of different cooking methods.

12.2.1 Boiling Method

Under the boiling method, kesaribath and upma underwent standardization at different ratios to optimize the sensory attributes and texture of the prepared product. Upma is a savory dish made with semolina, vegetables, and spices, whereas Kesaribath is a sweet dish made with semolina, sugar, and saffron.

12.2.1.1 Upma

Quinoa upma is a South Indian dish made with quinoa, a nutritious and protein-rich seed. To prepare quinoa upma, quinoa is cooked in water or vegetable broth and then tempered with mustard seeds, curry leaves, and other spices like cumin and turmeric. Chopped vegetables, like onion, tomato, carrot, and green chili, can be added for additional flavor and nutrition. It is a healthy and tasty alternative to traditional upma, which is typically made with semolina or cream of wheat.

For upma, the optimal ratio of semolina to water was found to be 1:2, and the addition of vegetables, such as onions, tomatoes, and green chilies, was found to enhance sensory appeal and nutritional value of the dish (Figure 12.1).

Figure 12.1 Process flow chart for preparing upma.

12.2.1.2 Kesaribath

Quinoa kesaribath is an Indian dessert made with quinoa. To make quinoa kesarib-ath, quinoa is cooked in milk and flavored with saffron, cardamom, and sugar. The result is a creamy, fragrant dessert that is similar to rice pudding. It is often served as a special treat for festivals or other celebrations. Quinoa kesaribath is a healthy twist on the traditional Indian dessert.

For kesaribath, the optimal ratio of semolina to water was found to be 1:3 (Supriya Kavali et al., 2019), and the addition of saffron and cardamom was found to enhance the aroma and flavor of the dish. Additionally, the use of ghee (clarified butter) instead of oil was found to improve the texture and mouthfeel of the final product (Figure 12.2).

Overall, the boiling method was found to be an effective way to standardize the preparation of these popular South Indian dishes, ensuring consistent quality and sensory attributes across different batches.

12.2.2 Frying Method

12.2.2.1 Chakli

Quinoa chakli is a popular savory snack from India, typically made during festivals or special occasions. It is a type of crispy spiral-shaped snack, similar to a pretzel or a crispy noodle. To make quinoa chakli, quinoa flour is mixed with other flours, such as rice flour, gram flour, and tapioca flour, along with spices such as cumin, chili powder, and salt. The dough is then rolled into thin spiral shapes using a chakli maker or a cookie press and deep-fried until crispy

For chakli, the optimal ratio of quinoa to rice flour was found to be 1:1, and the addition of ajwain and chilli powder was found to enhance the flavor and nutri-tional value of the dish (Supriya Kavali et al., 2019) (Figure 12.3).

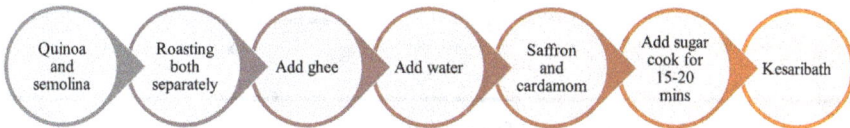

Figure 12.2   Process flow chart for kesaribath.

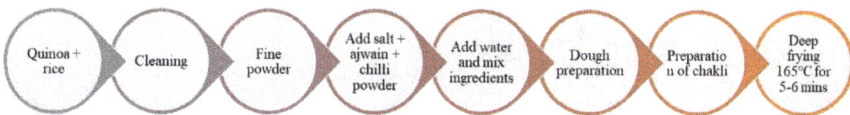

Figure 12.3   Process flow chart for chakli.

12.2.2.2 Nippattu

Quinoa nippattu is a popular snack in South India that is made using quinoa flour along with other ingredients, such as rice flour, roasted peanuts, roasted gram, curry leaves, chili powder, and salt. The mixture is kneaded with water and oil to form a dough, which is then flattened into thin circles and deep-fried in oil until crispy. Nippattu is a crunchy, savory snack that is often enjoyed with a cup of tea or coffee. Quinoa, being a nutritious and gluten-free grain, adds an extra healthy element to this already delicious snack

For nippattu, the optimal ratio of quinoa to rice flour was found to be 1:1, and the addition of peanuts, chilli powder, and curry leaves was found to enhance the flavor and nutritional value of the dish. Additionally, the incorporation of pulses enhances the mouthfeel and overall texture of the product (Supriya Kavali et al., 2019) (Figure 12.4).

12.2.3 Roasting Method

12.2.3.1 Laddu

To make quinoa laddu, the quinoa is first cooked and then mixed with sugar or jaggery (unrefined cane sugar), clarified butter, and nuts such as almonds and cashews. The mixture is then shaped into small balls or laddus. Quinoa laddu is a healthy alternative to traditional Indian desserts, as quinoa is high in protein, fiber, and essential nutrients. It is often served during festivals and celebrations in India and is a popular snack among health-conscious individuals due to its nutritional value.

For laddu, the optimal ratio of quinoa to gram flour was found to be 3:2, and the addition of sugar powder with cardamom was found to enhance the flavor and overall acceptability of the dish (Sharma 2017) (Figure 12.5).

Figure 12.4 Process flow chart for nipattu.

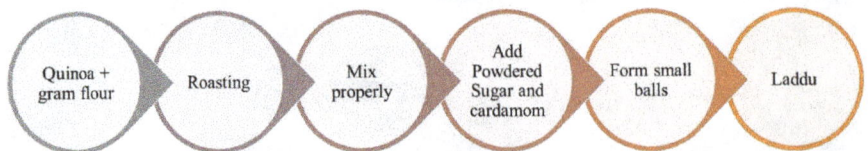

Figure 12.5 Process flow chart for laddoos.

12.2.3.2 Chikki

Quinoa chikki is a popular snack in India made from quinoa seeds, jaggery, and some-times peanuts. To make the chikki, the quinoa seeds are roasted in a pan until they start to pop, then they are mixed with melted jaggery and formed into a thin, flat layer. The mixture is then cut into small squares or rectangles once it cools and hardens. Quinoa chikki is also known for its crunchy texture, nut-infused flavor, and sweetness.

For chikki, addition of groundnut and flax seed in the recipe enhances the flavor and overall acceptability of the dish (Jain & Grover, 2017) (Figure 12.6).

Amaranth, chia, and quinoa are among the most extensively researched American food products, according to Orona-Tamayo et al. (2018). The level of fascination in amaranth is supported by numerous international publications, including Marcone and Kakuda (1999), Rivera et al. (2010), Venskutonis & Kraujalis (2013), De Beer et al. (2016), Kristbergsoon & Ötles (2016), Porras et al. (2016), García-Caldera & Velázquez-Contreras (2017), Machado et al. (2015), and Rios-Hoyo et al. (2017).

12.3 Amaranth

Amaranth, a seed of Mexican origin that has been consumed since the Pre-Columbian era, is often eaten as a snack called alegría prepared from popped amaranth mixed with either honey or sugar, as noted by Porras et al. (2016) and Rios-Hoyo et al. (2017). It has high protein content, being particularly rich in essential amino acids like lysine, and is also rich in dietary fiber, calcium, and iron, with up to 20 times more of these minerals than other seeds, according to Marcone & Kakuda (1999) and Venskutonis & Kraujalis (2013). In addition, Rivera et al. (2010) suggest that amaranth offers benefits beyond basic nutrition and could be used to prepare food products for people with non-communicable diseases (NCDs), a sentiment echoed by Machado et al. (2015), Porras et al. (2016), and García-Caldera & Velázquez-Contreras (2017).

Because of its high content of both macro- and micronutrients, amaranth has been used to create new products, as noted by De Beer et al. (2016). The interest in this food has been growing in recent years, with an increase in the area of land allocated for its cultivation in Mexico (SIAP, 2015), and its production and consumption spreading to other countries, according to Porras et al. (2016).

Products developed by incorporation of amaranth are listed below.

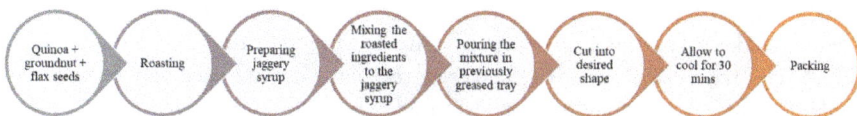

Figure 12.6   Process flow chart for chikki.

12.3.1 Alegría

Alegría is a type of Mexican origin **candy** that is made using amaranth seeds and sugar and honey. It is mainly produced in the town of Santiago Tulyehualco, located in the Xochimilco borough of Mexico City. The candy has been known as alegría, which translates to "joy" in English, since the sixteenth century. In September 2016, the alegría of Tulyehualco was officially declared Patrimonio Cultural Intangible de la Ciudad de México, meaning that it is an intangible part of the cultural heritage of Mexico City.

Amaranth-based snacks known as "alegrías" have gained widespread popularity as the preferred method of enjoying amaranth. To prepare this candy, the amaranth seeds are puffed in a hot pan in absence of oil, a process that only takes a few seconds to start. Once the popping has finished, the seeds are mixed with honey or sugar syrup, and occasionally with additional ingredients such as sizzled seeds (e.g., peanuts or pumpkin) or shredded dried fruits. The mixture prepared is then molded into various shapes and packed for the markets (Porras et al., 2016).

12.3.2 Atole

Atole is a traditional hot beverage of Mexican origin, made from maize and masa. Its name is believed to come from the Nahuatl word "ātōlli" or be of Mayan origin. Atole is also prepared with amaranth that can be flavored with vanilla, cinnamon, guava, or chocolate, which is known as champurrado or atole. It is often presented alongside tamales and is particularly favored during the Day of the Dead and Las Posadas festival period (Beliaev et al., 2009).

12.3.3 Kiwicha

Kiwicha, also known as *Amaranthus caudatus* L., is a highly valued crop that grows in Peru and is considered one of the oldest crops in North and South America. According to scientific studies, it was cultivated by the Incan culture 4,000 years ago (Loaiza et al., 2016). Kiwicha is widely regarded as one of the most precious food items in the world.

One common preparation method is to pop the seeds and use them as a topping for salads, yogurt, or oatmeal. The popped seeds can also be mixed with honey, agave syrup, or molasses to create a sweet treat similar to alegría. Another way to enjoy kiwicha is by cooking it as a porridge with milk or water and adding flavors such as cinnamon, vanilla, or fruit. In Peru, kiwicha is also used to make a type of bread called "pan de kiwicha" and a fermented drink called "chicha de kiwicha." The leaves of the plant can be used raw in salads or cooked as a leafy green vegetable. With its versatility and nutrient-rich profile, kiwicha is a valued crop and food source in Peru and other parts of the world (Villanueva et al., 2007).

12.3.4 Laddoos

Amaranth flour can be used to make a variety of Indian sweets, including laddoos. To make laddu with amaranth flour, the flour is first roasted in a pan until it turns slightly brown and emits a nutty aroma. Then, a mixture of ghee (clarified butter), jaggery (unrefined cane sugar), cardamom powder, and chopped nuts such as almonds and cashews is added to the flour and blended well. The mixture is then shaped into small balls, or laddus, and allowed to cool and set. This sweet and nutty treat is often served during festivals and celebrations in India (Patil et al., 2020).

12.3.5 Sattoo

Sattoo is a traditional Indian beverage made from roasted grains or seeds, including amaranth. To make sattoo from amaranth, the seeds are first washed, dried, and then roasted in a dry pan until they turn golden brown. The roasted seeds are then ground into a fine powder and mixed with water or milk to make a nutritious beverage. Some variations of sattoo may involve adding other ingredients such as jaggery (unrefined sugar), cardamom, or nuts for added flavor and nutrition. Sattoo is known for its high protein and fiber content, and is often consumed as a breakfast drink in some parts of India and Nepal (Joshi et al., 2008).

12.4 Buckwheat

Buckwheat, also known as *Fagopyrum esculentum* Moench, is a pseudocereal crop that has been domesticated in China since at least 1000 BC. Later, buckwheat was introduced to North America by colonists. Buckwheat is a hardy plant that is capable of growing in harsh environments and has a short growing period (Cai et al., 2016).

Buckwheat is known for its superior nutritional composition. It is a plentiful source of polyunsaturated essential fatty acids, which are important for maintaining good health. It also contains elevated levels of minerals such as magnesium and copper, and vitamins like niacin (Zhu et.al., 2016). In addition, buckwheat is a source of resistant starch, dietary fiber, and antioxidant compounds, which act as free radical quenchers (Zhou et al., 2015).

One of the unique features of buckwheat is its high-quality protein, and represents a nutrient powerhouse when compared with other cereal grains like wheat and rice. Buckwheat protein contains all the essential amino acids in sufficient amounts, making it a valuable source of protein for vegetarians and vegans (Jin et al., 2022).

Overall, buckwheat is a nutritious and versatile crop that can provide a range of health benefits (Bobkov, 2016). Its nutritional profile, combined with its capacity to withstand unfavourable conditions and its short cultivation period, makes it a promising crop for sustainable agriculture and food security

For centuries, people have created a diverse range of traditional foods using buckwheat. Buckwheat is a versatile ingredient and can be used to prepare various dishes, which can be broadly categorized into two types: dishes made from buckwheat groats and those made from buckwheat flour.

12.4.1 Buckwheat Groats

Buckwheat groats are a versatile and nutritious food that has been consumed for centuries in various societies. They are a type of grain that is high in protein, fiber, and vital nutrients, such as magnesium (Mg), zinc (Zn), and iron (Fe), and vitamin B_6 (Yilmaz et al., 2020). Buckwheat groats are also naturally free from gluten, making them a perfect choice for people with gluten intolerance or gluten-sensitive celiac disease (Bobkov, 2016).

One of the significant benefits of buckwheat groats consumption is their high fiber content, which promotes digestive health and helps lower cholesterol levels. Additionally, buckwheat groats are a good source of plant-based protein, making them an ideal food for vegetarians and vegans who want to incorporate more protein into their diets. Buckwheat groats are also relatively low in calories, making them an excellent option for people trying to lose weight or maintain a healthy weight (Sofi et al., 2022).

Buckwheat groats can be cooked and used in various dishes such as salads, soups, and stir-fries. In eastern European and Asian cuisines, they are often used as a substitute for rice or other grains. They are widely accessible in health food shops and specialty markets and could be enjoyed as a healthy and delicious addition to any diet.

Several foods have been developed using buckwheat groats or flour.

12.4.1.1 Kasha

Kasha is a traditional eastern European dish made from roasted buckwheat groats, which are the hulled seeds of the buckwheat plant.

Roasting the groats not only enhances their nutty flavor but also increases the bioavailability of some of the nutrients. For example, the process of roasting can enhance the availability of certain minerals, such as zinc (Zn) and iron (Fe), and may also improve the digestibility of the protein.

Kasha is a low-glycemic-index food (Chung et al., 2008), meaning it has a minimal impact on blood sugar levels. This is due to the ubiquity of resistant starch, a type of carbohydrate that resists digestive process and acts like dietary fiber in the body.

Kasha sometimes served as a side serving dish or used as a base for other recipes. It can be seasoned with herbs and spices, mixed with sautéed vegetables, or used as a filling for stuffed peppers or cabbage rolls. In Russia, kasha is often served with milk and honey for breakfast.

Overall, kasha is a healthy and flavorful dish that is easy to prepare and versatile. It can be enjoyed as a comforting meal on a cold day or as a nutritious addition to any meal.

12.4.1.2 Memil Muk

Muk is a Korean food that has been a favorite for a long time. It can be made from various grains like buckwheat, mung bean, or acorn, and is known for its unique texture that is soft and chewy. The process of making muk involves obtaining starch from the grains, boiling it in water until it thickens, and then stiffening it to form a jelly-like substance (Kim et al., 2011).

Memil muk is traditionally made using buckwheat flour and water, but some variations may include additional ingredients such as sugar or salt to enhance the flavor. The blend is heated over mild heat and stirred constantly until it reaches a thick and smooth consistency. This process can take up to 30 minutes, and the mixture needs to be continuously stirred to prevent lumps from forming.

After the mixture thickens, it is strained through a fine-mesh sieve or cheesecloth to remove any remaining lumps or impurities. The strained liquid is then transferred to a container and left to cool and set in the refrigerator for several hours until it solidifies into a jelly-like texture.

Memil muk is a low-calorie and low-fat dessert that is rich in fiber, vitamins, and minerals (Zhu et al., 2016). Buckwheat flour provides a considerable amount of dietary fiber, which helps regulate blood sugar levels and promotes digestive health (Kim et al., 1997). It also contains vital minerals such as zinc, iron, and magnesium (Raguindina et al., 2021), as well as antioxidants that help protect against cell damage and inflammation.

The jelly-like texture of memil muk is soft and smooth, making it easy to digest and suitable for individuals with sensitive stomachs (Zhu et al., 2016). The addition of toppings like honey, sugar syrup, or fruit and nuts can provide additional flavor and nutritional benefits. For example, adding fruits like blueberries or strawberries can increase the antioxidant content of the dessert (Huang et al., 2012), while adding nuts like almonds or walnuts can provide healthy fats and proteins (Ros, 2010).

12.4.1.3 Soba

Soba is a thin noodle of buckwheat flour produced in Japan.. It is a staple food in Japanese cuisine and is commonly served hot or cold. Soba can be enjoyed in various forms, including in soups, stir-fries, salads, and as a side dish.

The word "soba" itself comes from the Japanese characters 蕎 (soba) and 麦 (mugi), which mean buckwheat and wheat, respectively (Ikeda et al., 2000). From 1603 until 1868 in the Tokugawa period, also known as the Edo period, a tradition of eating soba was developed and traditional soba noodles are prepared primarily from buckwheat flour, which is a gluten-free whole grain that contains significant levels of protein, fiber, and various minerals, including magnesium, iron, and zinc (Yalcin, 2021). Buckwheat is also a valuable source of antioxidants, which can help mitigate the harmful effects of oxidative stress and inflammation in the body (Yilmaz et al., 2020).

In terms of preparation, it is worth noting that soba noodles, in addition to buckwheat flour, can contain some wheat flour, while wheat flour is only added in small quantities to help bind the dough, which means that they may not be entirely gluten-free (Giménez-Bastida et al., 2015). It's important to check the label or ask the manufacturer if you have gluten intolerance or celiac disease. Additionally, overcooking soba noodles can cause them to lose some of their nutritional value and become mushy in texture, so it is recommended to cook them for a shorter amount of time and rinse them thoroughly with cold water to stop the cooking process (Lui et al., 2016). Soba is known for its nutty and earthy flavor and is often praised for its health benefits.

12.4.1.4 Buckwheat Tea

Tea prepared using buckwheat is called memil-cha in Korea, soba-cha in Japan, and kuqiao-cha in China, and is prepared by roasting buckwheat and brewing it as a tea. Similar to other customary teas of Korean origin, memil-cha could be consumed hot or cold and is often served as a substitute for water (Guo et al., 2017).

Buckwheat tea is not classified as a conventional tea, such as green, black, or aromatic tea. Rather, it's a form of roasted rice tea that is derived from Tartary buckwheat (*Fagopyrum tataricum*) seeds, which are sometimes referred to as Tartary buckwheat rice or TBT for short. To create TBT, these seeds are ground, mixed with water, extruded, dehydrated, dried, and roasted. TBT has gained increasing popularity in Asia as well as in Europe due to its exceptional malty aroma, wellness benefits, and user-friendliness.

Buckwheat tea is a popular drink in various regions of the world, especially in Japan, Korea, and China (Xu et al., 2019). It is often consumed as a healthy alternative to coffee or black tea due to its low caffeine content and high nutritional value. Buckwheat tea is enriched with antioxidants, minerals, vitamins, and phytochemicals, including rutin, a flavonoid that has been shown to improve blood circulation and control blood pressure (Peng et al., 2015).

In addition to its health advantages, buckwheat tea has a delightful nutty and earthy taste that is refreshing and comforting. It can be savored either hot or cold and is frequently complemented with a sweetener like honey or sugar to counterbalance its slightly bitter taste. Furthermore, buckwheat tea can be mixed with other herbs or spices, such as ginger, cinnamon, or lemon, to produce a more intricate and flavorful infusion.

12.5 Summary and Conclusion

Pseudocereals encompass a collection of plant species that, despite not being true grasses, are commonly categorized and utilized in a similar manner to cereals owing to their nutritional characteristics and adaptability in cooking. Quinoa, amaranth, and buckwheat are among the notable pseudocereals that have been part of traditional diets across diverse cultures throughout history. Lately, there has been an increasing fascination with these pseudocereals due to their significant nutritional content and potential advantages for overall well being.

Quinoa, which originates from the Andean region of South America, has become widely favored as a substitute for grains containing gluten. It offers a considerable amount of protein, fiber, and vital nutrients, such as iron and magnesium. Quinoa's versatility allows for its incorporation into numerous dishes, ranging from salads and stir-fries to breakfast cereals.

Amaranth, which originated in Central and South America, is an additional pseudocereal that has been cultivated for centuries. It is renowned for its significant protein content and contains all the essential amino acids, making it a valuable plant-based protein source. Amaranth can be cooked in a similar manner to rice, popped like popcorn for a crunchy snack, or ground into flour to be utilized in baking..

Despite its name, buckwheat is unrelated to wheat and is, in fact, gluten free. It possesses a unique nutty taste and is frequently featured in traditional cuisine in countries such as Russia, Japan, and China. Buckwheat exhibits versatility and can be enjoyed in various forms, including groats, flour, or noodles, thereby providing a diverse range of culinary opportunities.

In addition to their nutritional qualities, these pseudocereals have garnered recognition for their sustainable cultivation methods. They are renowned for being resilient crops that require minimal water and can thrive in diverse climates. Consequently, they are frequently regarded as environmentally friendly alternatives to conventional cereal grains.

In conclusion, pseudocereal based traditional foods like quinoa, amaranth, and buckwheat have been enjoyed in diverse cultures across the globe for a significant period of time. These unique grains not only provide distinct flavors but also offer considerable nutritional advantages, making them versatile ingredients for a wide range of culinary creations. With the increasing interest in promoting health and sustainability through dietary choices, these pseudocereals have gained recognition as valuable and enriching additions to culinary traditions worldwide.

References

Abdellatif, A.S.A. (2018). Chemical and technological evaluation of quinoa (*Chenopodium quinoa* Willd.) cultivated in Egypt. *Acta Scientific Nutritional Health*, 2, 115–125.

Balakrishnan, G., Schneider, R.G. (2022 Aug 13). The role of amaranth, quinoa, and millets for the development of healthy, sustainable food products - A concise review. *Foods*, 11(16), 2442. doi: 10.3390/foods11162442.

Beliaev, D., Davletshin, A., Tokovinine, A. (2009). Sweet cacao and sour atole: Mixed drinks on classic maya ceramic vases. *Pre-Columbian Foodways*, 257–272. doi: 10.1007/978-1-4419-0471-3_10.

Biswas, R., Singh, N., Mishra, A.A., Chawla, P. (2020). Mathematical models and kinetic studies for the assessment of antimicrobial properties of metal nanoparticles. In Sanju Bala Dhull, Prince Chawla, Ravinder Kaushik (Eds.), *Nanotechnological Approaches in Food Microbiology*, pp. 1–29, Boca Raton:CRC Press.

Bobkov, S. (2016). Biochemical and technological properties of buckwheat grains. In M. Zhou (Ed.), *Molecular Breeding and Nutritional Aspects of Buckwheat*, pp. 423–440, London:Academic Press.

Cai, Y.Z., Corke, H., Wang, D., Li, W.D. (2016), *Buckwheat. Reference Module in Food Science*. doi: 10.1016/b978-0-08-100596-5.00034-2.

Chung, H.J., Liu, Q., Pauls, K.P., Fan, M.Z., Yada, R. (2008). *In vitro* starch digestibility, expected glycemic index and some physicochemical properties of starch and *flour from common bean (Phaseolus vulgaris L.)* varieties grown in Canada. *Food Research International*, 41(9), 869–875. doi: 10.1016/j.foodres.2008.03.013.

De Beer, H., Mielmann, A., Coetzee, L. (2016). Exploring the acceptability of amaranth-enriched bread to support household food security. *British Food Journal*, 118(11), 2632–2646.

Dhull, S.B., Kidwai, M.K., Noor, R., Chawla, P., Rose, P.K. (2022a). A review of nutritional profile and processing of faba bean (*Vicia faba* L.). *Legume Science*, 4(3), e129.

Dhull, S.B., Kidwai, M.K., Siddiq, M., Sidhu, J.S. (2022b). Faba (broad) bean production, processing, and nutritional profile. In M. Siddiq, M.A. Uebersax (Eds.), *Dry Beans and Pulses*. doi: 10.1002/9781119776802.ch14.

Dhull, S.B., Kinabo, J., Uebersax, M.A. (2023). Nutrient profile and effect of processing methods on the composition and functional properties of lentils (*Lens culinaris* Medik): A review. *Legume Science*, 5(1), e156.

García-Caldera, N., Velázquez-Contreras, F. (2017). Amaranth pasta in Mexico: A celiac overview. *Journal of Culinary Science and Technology*, 17, 1–9. doi: 10.1080/15428052.2017.1405862.

Giménez-Bastida, J.A., Piskula, M.K., Zieliński, H. (2015). Recent advances in processing and development of buckwheat derived bakery and non-bakery products-a review. *Polish Journal of Food and Nutrition Sciences*, 65(1), 9–20.

Guo, H., Yang, X., Zhou, H., Luo, X., Qin, P., Li, J., Ren, G. (2017). Comparison of nutritional composition, aroma compounds, and biological activities of two kinds of Tartary buckwheat tea. *Journal of Food Science*, 82(7), 1735–1741. doi: 10.1111/1750-3841.13772.

Huang, W.Y., Zhang, H.C., Liu, W.X., Li, C.Y. (2012). Survey of antioxidant capacity and phenolic composition of blueberry, blackberry, and strawberry in Nanjing. *Journal of Zhejiang University Science Part B*, 13(2), 94–102. doi: 10.1631/jzus.B1100137; PMID: 22302422; PMCID: PMC3274736.

Ikeda, K., Asami, Y. (2000). Mechanical characteristics of buckwheat noodles. *Fagopyrum*, 17, 67–72.

Jain, T., Grover, K. (2017). Nutritional evaluation of garden cress Chikki. *Agricultural Research and Technology Open Access Journal*, 4(2), 1–5.

Jin, J., Ohanenye, I.C., Udenigwe, C.C. (2022). Buckwheat proteins: Functionality, safety, bioactivity, and prospects as alternative plant-based proteins in the food industry. *Critical Reviews in Food Science and Nutrition*, 62(7), 1752–1764.

Joshi, B.K. (2008). Buckwheat genetic resources: Status and prospects in Nepal. *Agriculture Development Journal*, 5, 13–30.

Kim, A.J., Han, M.R., Rho, J.O. (2011). Quality characteristics of Cheongpomook prepared with different levels of mungbean powder. Korean. *Korean Journal of Human Ecology*, 20(6), 1229–1237. doi: 10.5934/KJHE.2011.20.6.1229.

Kim, H.S., Ahn, S.Y. (1997). Effect Amylose and Amylopectin on the Texture of mook. *Korean Journal of Human Ecology*, 6, 157–166.

Kristbergsoon, K., Ötles, S. (2016), *Functional Properties of Traditional Foods*, Springer, New York.

Liu, R., Wei, Y.M., Xing, Y.N., Wang, J., Zhang, Y.Q., Zhang, B. (2016). Sensory quality and physico-chemical properties of three types of commercial dried Chinese noodles. *Advance Journal of Food Science and Technology*, 10(4), 262–269.

Loaiza, M.A.P.P., López-Malo, A., Jiménez-Munguía, M.T. (2016). Nutraceutical properties of amaranth and chia seeds. *Functional Properties of Traditional Foods*, 189–198. doi: 10.1007/978-1-4899-7662-8_13.

Machado, N.M., Joy, C., Dutra, I., Carvalho, E., Andre, H.M. (2015). Addition of quinoa and amaranth flour in gluten-free breads: Temporal profile and instrumental analysis. *LWT – Food Science and Technology*, 62(2), 1011–1018.

Malik, M., Sindhu, R., Dhull, S.B., Bou-Mitri, C., Singh, Y., Panwar, S., Khatkar, B.S. (2023). Nutritional composition, functionality, and processing technologies for amaranth. *Journal of Food Processing and Preservation*, 2023, 1753029. doi: 10.1155/2023/1753029.

Marcone, M.F., Kakuda, Y. (1999). A comparative study of the functional properties of amaranth and soybean globulin isolates. *Food/Nahrung*, 43(6), 368–373.

Nisar, M., More, D.R., Zubair, S., Sawate, A.R., Hashmi, S.I. (2018). Studies on Development of Technology for Preparation of Cookies incorporated with Quinoa seed flour and its nutritional and sensory quality evaluation. *International Journal of Chemical Studies*, 6(2), 3380–3384.

Orona-Tamayo, D., Valverde, M.E., Paredes-López, O. (2018). Bioactive peptides from selected Latin American food crops – A nutraceutical and molecular approach. *Critical Reviews in Food Science and Nutrition*, 58(1), 1–27. doi: 10.1080/10408398.2018.1434480.

Patil, S.B., Jena, S. (2020). Utilization of underrated pseudo-cereals of North East India: A systematic review. *Nutrition and Food Science*, 50(6), 1229–1240. doi: 10.1108/nfs-11-2019-0339.

Peng, L.X., Zou, L., Wang, J.B., Zhao, J.L., Xiang, D.B., Zhao, G. (2015). Flavonoids, antioxidant activity and aroma compounds analysis from different kinds of Tartary buckwheat tea. *Indian Journal of Pharmaceutical Sciences*, 77(6), 661–667.

Porras, M.A.P., López-Malo, A., Jiménez-Munguía, M.T. (2016). Nutraceutical properties of amaranth and chia seeds. In K. Kritsbergsson and S. Otles (Eds.), *Functional Properties of Traditional Foods*, Springer, New York, pp. 189–198.

Punia, S., Dhull, S.B. (2019). Chia seed (*Salvia hispanica* L.) mucilage (a heteropolysaccharide): Functional, thermal, rheological behaviour and its utilization. *International Journal of Biological Macromolecules*, 140, 1084–1090.

Raguindina, P.F., Itodoa, O.A., Stoyanov, J., Dejanovic G.M., Gamba M., Asllana E., Minder B., Bussler W., Metzger B., Muka T., et al. (2021). A Systematic review of phytochemicals in oat and buckwheat. *Food Chemistry*, 338, 127982. doi: 10.1016/j.foodchem.2020.127982.

Ramos Diaz, J.M., Kirjoranta, S., Tenitz, S., Penttilä, P.A., Serimaa, R., Lampi, A.-M., Jouppila, K. (2013). Use of Amaranth, Quinoa and Kañiwa in extruded corn-based snacks. *Journal of Cereal Science*. doi: 10.1016/j.jcs.2013.04.003.

Rios-Hoyo, A., Romo-Araiza, A., Meneses-Mayo, M., Gutierrez-Salmeán, G. (2017). Prehispanic functional foods and nutraceuticals in the treatment of dyslipidemia associated to cardiovascular disease: A mini-review. *International Journal of for Vitamin and Nutrition Research*, 1–14. doi: 10.1024/0300-9831/a000290.

Rivera, G., Bocanegra-García, V., Monge, A. (2010). Traditional plants as source of functional foods. *CyTA: Journal of Food*, 8(2), 159–167.

Robin, F., Théoduloz, C., Srichuwong, S. (2015). Properties of extruded whole grain cereals and pseudocereal flours. *International Journal of Food Science and Technology*, 50(10), 2152–2159. doi: 10.1111/ijfs.12893.

Ros, E. (2010, Jul). Health benefits of nut consumption. *Nutrients*, 2(7), 652–682. doi: 10.3390/nu2070652. Epub 2010 Jun 24. PMID: 22254047; PMCID: PMC3257681.

Sharma, G., Lakhawat, S. (2017). Nutrition facts & functional potential of quinoa (*Chenopodium quinoa*), an ancient Andean grain: A review. *Journal of Pharmacognosy and Phytochestry*, 6(4), 1488–1489.

SIAP (2015). Avance de siembras y cosechas. Resumen nacional por estado. *Servicio de Información Agroalimentaria y Pesquera, Available at*: http://infosiap.siap.gob.mx:8080/agricola_siap_ gobmx/AvanceNacionalCultivo.do (accessed 20 February 2018).

Singh, N., Biswas, R., Banerjee, M. (2023). A systematic review to identify obstacles in the agricultural supply chain and future directions. *Journal of Agribusiness in Developing and Emerging Economies*, Vol. ahead-of-print No. ahead-of-print. doi: 10.1108/JADEE-12-2022-0262.

Slimani, N., Fahey, M., Welch, A.A., et al. (2002). Diversity of dietary patterns observed in the European Prospective Investigation into Cancer and Nutrition (EPIC) project. *Public Health Nutrition*, 5(6B), 1311–1328.

Sofi, S.A., Ahmed, N., Farooq, A., Rafiq, S., Zargar, S.M., Kamran, F., Dar, T.A., Mir, S.A., Dar, B.N., Mousavi Khaneghah, A. (2022). Nutritional and bioactive characteristics of buckwheat, and its potential for developing gluten-free products: An updated overview. *Food Science and Nutrition*, 1–21. doi: 10.1002/ fsn3.3166.

Kavali, S., Shobha, D., Shekhara Naik, R., Brundha, A.R. (2019). Development of value added products from quinoa using different cooking methods. *The Pharma Innovation Journal*, 8(7), 548–554.

Venskutonis, P.R., Kraujalis, P. (2013). Nutritional components of amaranth seeds and vegetables: A review on composition, properties, and uses. *Comprehensive Reviews in Food Science and Food Safety*, 12(4), 381–412.

Villanueva, O., Arnao, I. (2007). Purifi cación de una proteína de 35kDa rica en lisina, de la fracción albúmina de *Amaranthus caudatus* (kiwicha). *Anales de la Facultad de Medicina*, Universidad Nacional Mayor de San Marcos, 68(4), 344–350.

Xu, Q., Wang, L., Li, W., Xing, Y., Zhang, P., Wang, Q., Li, H., Liu, H., Yang, H., Liu, X., Ma, Y. (2019 Nov 29). Scented Tartary buckwheat tea: Aroma components and antioxidant activity. *Molecules*, 24(23): 4368. doi: 10.3390/molecules24234368.

Yalcin, S. (2021). Quality characteristics, mineral contents and phenolic compounds of gluten free buckwheat noodles. *Journal of Food Science and Technology*, 58(7), 2661–2669.

Yilmaz, H.Ö., Ayhan, N.Y., Meriç, Ç.S. (2020). Buckwheat: A useful food and its effects on human health. *Current Nutrition and Food Science*, 16(1), 29–34.

Zhou, X., Hao, T., Zhou, Y., Tang, W., Xiao, Y., Meng, X., Fang, X. (2015). Relationships between antioxidant compounds and antioxidant activities of Tartary buckwheat during germination. *Journal of Food Science and Technology*, 52(4), 2458–2463.

Zhu, F. (2016). Chemical composition and health effects of Tartary buckwheat. *Food Chemistry*, 203, 231–245.

Chapter 13

Recent Trends in Pseudocereal-based Foods

Debapam Saha, Abhishek Pradhan,
Ayan Sarkar, Punyadarshini Punam Tripathy

13.1 Introduction

For centuries, cereals have constituted the primary constituent of human dietary staples. Cereals provide a significant proportion (up to 50% globally) of the daily dietary energy requirement. However, a few cereal proteins may cause hypersensitivity responses in sensitive people, leading to chronic sickness. With economic expansion, urbanisation, and lifestyle and dietary changes, allergies and food intolerances have increased globally in recent decades. Globally, the prevalence of gluten intolerance is on the rise. Three frequent conditions linked to gluten intake are wheat allergy, coeliac disease, and gluten sensitivity or non-coeliac gluten sensitivity.

In the twenty-first century, bakers and cereal scientists face a significant challenge in creating food products that won't harm these individuals. The utilisation of different cereals and flours necessitates the search for alternatives to gluten, such as other flour constituents, preparation of composite flours, and dough treatment, or a change in baking technique. Alternatives to traditional bakery products will be possible as a result of modern technological breakthroughs, the use of gluten-free flours, carbohydrates, hydrocolloids, and novel culinary ingredients (see Figure 13.1). There exist several types of edible pseudocereals that are fit for human consumption, such as teff, buckwheat, amaranth, quinoa, and wild rice. In fact, researchers worldwide are increasingly focussing their efforts on exploring the

DOI: 10.1201/9781003325277-13

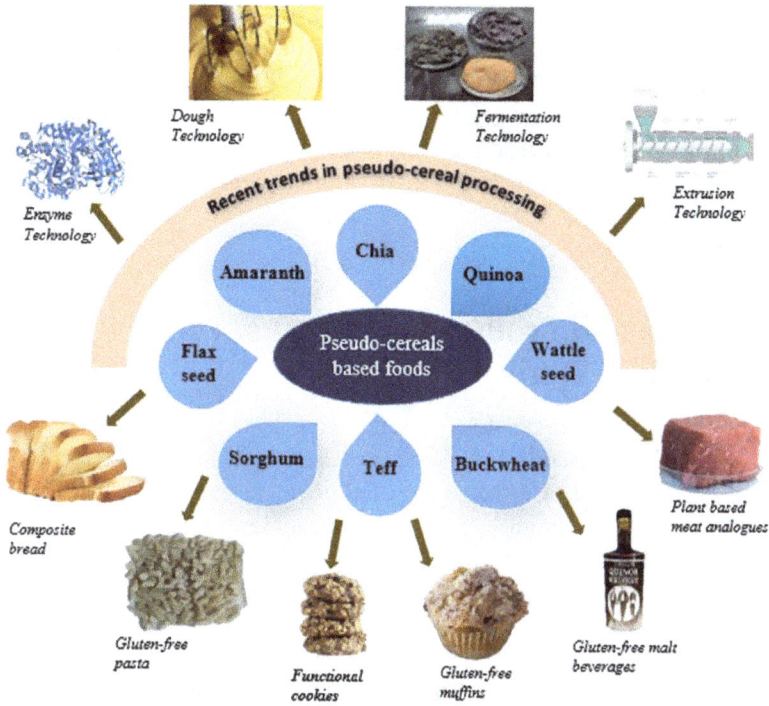

Figure 13.1   Recent advances in processing technologies for pseudocereal-based product development.

untapped potential of pseudocereal-based products, unravelling their nutritional composition, and health benefits as demonstrated in Figure 13.2. Additionally, there are cereal grains that lack gluten, such as millet, maize, sorghum, fibre and rice. Amaranth has been found to contain substantial quantities of fibre, zinc, iron, calcium, and magnesium (Gambus et al., 2002). Similarly, thiamin, riboflavin, and

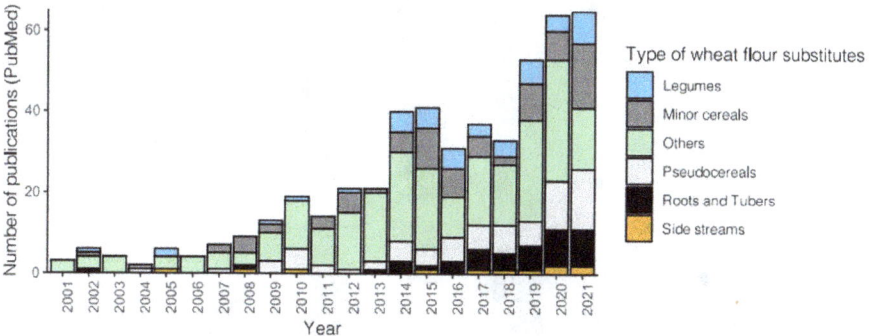

Figure 13.2   Number of research publications per year in the PubMed database related to composite bread over the past 20 years (2001–2021) (Wang & Jian, 2022).

niacin are all abundant in buckwheat, making it an excellent source of B vitamins (Lebiedzińska & Szefer, 2006). Quinoa is rich in protein, dietary fibre, magnesium, calcium, iron, and B vitamins (Vega-Gálvez et al., 2010). Millets are rich in essential nutrients such as thiamin, riboflavin, potassium, and calcium (Lebiedzińska & Szefer, 2006). Other than cereal grains, edible seeds and pseudocereals may be considered feasible options for inclusion in gluten-free diets owing to the nutritional qualities they possess.

13.2 Origin and Production of Pseudocereals and their Derived Products

Pseudocereals refer to the dicotyledonous species that have higher starch content and beneficial physicochemical characteristics that closely resemble those of authentic cereals. The three most widely recognised pseudocereals in the food industries are buckwheat, amaranth, and quinoa. The nutritional profile and the proximate compositions of these pseudocereals, along with some minor cereals, according to the data of the USDA (Agregán et al., 2023), given in Figure 13.3, reveal that they are an important source of carbohydrates, high-quality protein, fibre, and lipids. The origins of quinoa cultivation were mainly in the region of the Andes in South America, dating as far back as 5000–3000 BC. The production of quinoa

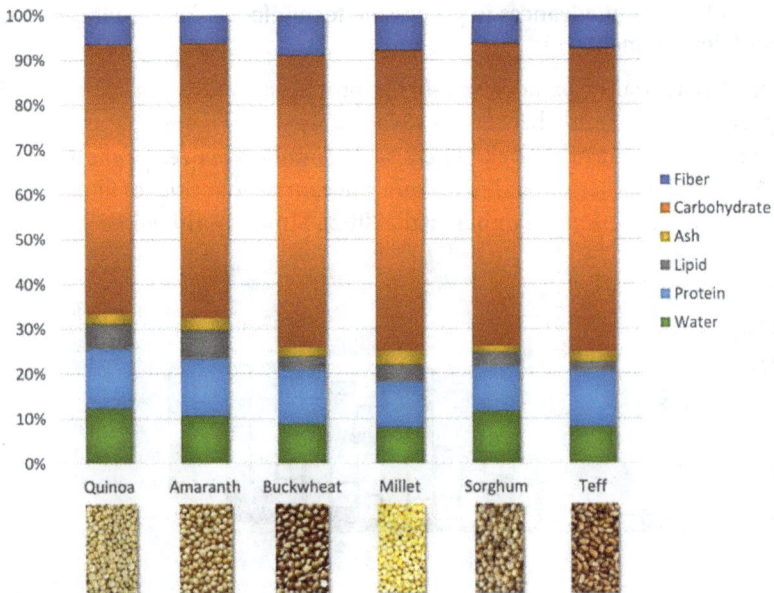

Figure 13.3 **Proximate composition of pseudocereals and minor cereals. Data from the USDA food composition database (Agregán et al., 2023).**

has exhibited a consistent increasing trend in recent decades, particularly after 2013, with both production and consumption experiencing exponential growth, as reported by Hunt et al. (2018). The cultivation of quinoa in 2019 covered 184,585 hectares, primarily in Bolivia, Peru, and Ecuador. The production of quinoa during this period amounted to 161,415 metric tonnes (Hunt et al., 2018).

The origin of buckwheat can be traced back to the mid-sixth millennium BC in the southwestern region of China. The global cultivation of buckwheat in 2019 covered an area of 1,673,478 hectares, resulting in a production output of roughly 2,042,401 metric tonnes (World Health Organization, 2020). The primary producers of buckwheat are the Russian Federation, accounting for 46.72%, and China, accounting for 29.73%. Following these two countries are the USA, Ukraine, Kazakhstan, and Japan, accounting for 4.77%, 4.13%, 4.03%, and 3.90%, respectively.

The *Amaranthus* species are indigenous to Central and South America, except for certain types like *Amaranthus spinosus* L., which thrive in the tropical and subtropical areas of India. The Aztecs and Mayans of Central America relied heavily on amaranth as a primary food source. However, with the arrival of Europlean colonisers, its consumption drastically declined to insignificant levels (Graziano et al., 2022). Today, amaranth production is also prevalent in various South American nations, along with Russia, India, China, and Kenya.

13.3 Different Pseudocereals Used for Making Modern Gluten-free Products

Pseudocereal-based gluten-free products (GFP), ranging from gluten-free (GF) breads and pasta to snacks and cereals, not only provide a safe alternative for those with dietary restrictions but also deliver a unique flavour profile and a rich source of essential nutrients. With their rising popularity, pseudocereal-based GFPs are revolutionising the way we approach GF diets, offering a tasty and inclusive solution for individuals seeking a diverse and healthy eating experience. However, this gluten replacement poses a severe technological challenge due to gluten's viscoelastic properties, that are largely essential for the dough's gas- and water-holding capacity during fermentation (Hager et al., 2012). Consequently, significant efforts are needed to enhance the sensory quality, functional, and techno-functional attributes of pseudocereal-based GFPs, some of which are described in Table 13.1.

13.3.1 Quinoa-based Products

Quinoa is a nutrient-dense pseudocereal which provides a variety of health benefits due to its high protein, fibre, and mineral contents. The significance of quinoa in food processing originates from its versatility as a constituent, which enables the creation of a wide variety of gluten-free, nutrient-rich products that appeal to a variety of dietary requirements. The incorporation of quinoa into food processing

Table 13.1   Application of Pseudocereals in Various Food Formulations

Pseudo-cereals	Products/ formulations	Findings/ Characteristics of developed products	References
Buckwheat	Bread (Rice + Buckwheat)	• Acceptable organoleptic properties • Decreased starch-retro-gradation with improved anti-staling properties.	(Torbica et al., 2012)
	Bread (Buckwheat + corn starch)	• Increased antioxidant and reducing capacity • Improved functional properties.	(Wronkowska et al., 2010)
	Bread (Buckwheat + exopolysaccharide)	• Improved baking quality • Evenness in surface morphology	(Rühmkorf et al., 2012)
	Spaghetti (30% buckwheat)	• Hydration level increased • Decreased cooked firmness	(Yalla & Manthey, 2006)
	Beer (100% Buckwheat malt)	• Lower viscosity of worts • Higher amylase activity of fermentable extracts	(Phiarais et al., 2010)
Flaxseed	Ice creams (0.5% flaxseed protein concentrates)	• Potential substitute for gelatine • Improved meltdown time and overrun percentage	(Dev & Quensel, 1989)
	Pasta (15% flaxseed addition)	• Longer shelf-life with antifungal properties • Reduced firmness after cooking	(Manthey et al., 2008)
	Muffins (0–40% raw and roasted ground flaxseed)	• Enhanced crumb hardening • Higher water absorption • Lesser dough development time	(Sudha et al., 2010)
	Pasta (5% or 10% flaxseed + semolina)	• Weakening effect on dough	(Mercier et al., 2014)
Teff	Bread (30% Teff (white variety) incorporation in wheat flour)	• 2-fold increase in Fe content • In-vitro antioxidant activity increased. • Increased dough development time, degree of softening, bitterness and textural hardness of crumb.	(Alaunyte et al., 2012)
	Bread (40% Teff (white and brown variety)+ wheat flour)	• 5-10-fold increase in Fe content • Mn, Cu, Zn, Mg content significantly enhanced. • Decreased crumb resilience and loaf volume.	(Ronda et al., 2015)

(Continued)

Table 13.1 (Continued) Application of Pseudocereals in Various Food Formulations

Pseudo-cereals	Products/ formulations	Findings/ Characteristics of developed products	References
	Sourdough bread (Teff + hetero-fermentation by lactic acid bacteria *L. plantarum FST 1.7*)	• Decreased the dough strength without affecting the staling rate. • Increased the crumb porosity and cell volume • Inferior aroma than wheat-based bread.	(Hager, Lauck, et al., 2012)
	Injera (Teff + white sorghum in 1:1 ratio)	• Fermentation reduced phytic acid content. • Improved in-vitro bio-accessibility of Fe.	(Baye et al., 2014)
	Pasta (Tagliatelle produced from teff + 40% common bean flour)	• Increased protein and dietary fibre content. • GI reduced from 60 to 39.	(Giuberti et al., 2016)
	Lactic acid beverage (Teff malt wort + *L. amylolyticus*)	• Acceptable sensory profile • Improved free amino nitrogen in the fermented wort.	(Gebremariam et al., 2013)
Quinoa	Bread (non-sourdough) (10,20% quinoa + wheat flour)	• Water absorption decreased by 2 percent; • Dough development time increased by 7 and 14 %. • Bread weight decreased by 2 percent. • Specific volume decreased by 11 and 30 %, respectively.	(Rodriguez-Sandoval et al., 2012)
	Extruded snack (Quinoa (100%) + whey protein concentrates (12.5%) + cashew pulp (12.5%)	• WAI decreased by 16 %, WSI increased by 50 %, and peak viscosity decreased by 41 %. • MC deceased 45 %, bulk density increased by 3 %, and volume expansion deceased by 39 %.	(Onwulata et al., 2010)
	Cookies (Quinoa+ peanut butter + chocolate chip)	• Significantly sweeter than the wheat-based cookies • Acceptable sensory profile to that of wheat-based cookies	(Harra et al., 2011)
	Spaghetti (quinoa flour + chickpea flour + broad bean flour + maize flour + soy flour)	• Increased shear viscosity and elongation • Increased the bulkiness and adhesiveness • Increased the firmness of quinoa dough • Sensory attributes did not improve	(Mastromatteo et al., 2012)

(Continued)

Table 13.1 (Continued)   Application of Pseudocereals in Various Food Formulations

Pseudo-cereals	Products/ formulations	Findings/ Characteristics of developed products	References
	Breakfast cereals (quinoa + wheat flour + rice flour)	• GA and GABA levels increased during extrusion and flaking.	(Joye et al., 2011)
	Edible film (quinoa protein + CH)	• Smooth, continuous as well as compact structure. • Increased WVP by 147 % and EAB by 271 % • Increased TS by 5 % and thickness by 163 %. • Decreased water activity a by 40 % and decomposition temperature by 56 %.	(Abugoch et al., 2011)
Amaranth	Bread (amaranth incorporation in wheat: 5%, 10%, 15%, 20% and 30%)	• Significant nutrient improvement with increasing incorporation levels. • Up to 10% incorporation, exhibited enhanced organoleptic characteristics and rheological properties.	(Emire & Arega, 2012)
	Biscuits (up to 35% amaranth substitution in wheat flour)	• Improved organoleptic properties (colour, taste, flavour and appearance)	(Sindhuja et al., 2005)
	Functional beverage (amaranth-chia-based beverage in 70:30 ratio)	• Higher protein and calorie content • High overall acceptance of the developed beverage among the panellists.	(Argüelles-López et al., 2018)
	Amaranth oil (squalene)	• It contains greater antioxidant content and relieves from free-radical induced oxidative injuries of the skin.	(Huang et al., 2009)
Sorghum	Sourdough bread	• Highly nutritious with greater biological activity • Suitable for gluten-intolerant consumers • Improved rheological and textural properties.	(Olojede, Sanni, & Banwo, 2020b)
	Porridge (Ogi-baba) (Sorghum + soybean)	• Smooth, creamy, free-flowing thin porridge • Has variable sensorial and nutritional qualities. • Good source of bioactive compounds (Probiotic foods)	(Adelekan & Oyewole, 2010)

(Continued)

Table 13.1 (Continued)   Application of Pseudocereals in Various Food Formulations

Pseudo-cereals	Products/ formulations	Findings/ Characteristics of developed products	References
	Pasta	• Mild flavour and chewy texture with good source of bioactive compounds. • Rich in minerals like manganese, zinc, iron, and magnesium.	(Palavecino et al., 2019)
	Pito (fermented alcoholic beverage of sorghum)	• Contained unfermented sugars, amino acids, and lactic acid. • Good source of vitamin B complexes.	(Ezekiel et al., 2015)

(TEAC: Trolox equivalent antioxidant capacity; GI: Glycaemic index; WSI: Water solubility index; WAI: Water absorption index; MC: Moisture content; GA: Glutamic acid; GABA: Gamma-aminobutyric acid; EAB: Elongation at break; WVP: Water vapour permeability and TS: tensile strength)

has facilitated the production of a diverse range of food formulations, including pasta, breads, and snacks. This approach has been adopted to cater to the growing demand for plant-based, health-conscious options that do not compromise on taste or nutritional content. Several researchers have conducted studies on the advancement of food products based on quinoa and have documented alterations in diverse quality characteristics.

Chlopicka et al. (2012) developed a composite bread consisting of wheat and incorporating 15% or 30% of quinoa flour (QF). The findings demonstrated a 36% rise in the total flavonoid content (TFC), an 11% increase in the total phenolic content (TPC), and a remarkable 47% rise in antioxidant capacity (AC) of the bread. Similarly, Alvarez-Jubete et al. (2010) found that the addition of 50% QF could increase the TFC, TPC, and AC values of bread by 24%, 5%, and 13%, respectively. In a study performed by Hager et al. (2012), based on bread staling, the authors replaced wheat flour with 100% quinoa flour and found that the staling rate of bread dropped significantly by around 95%. Diaz et al. (2013) observed that quinoa/maize-based snacks contained more dietary fibre than amaranth/maize-based snacks. Additionally, the hardness of quinoa snacks was found to be less than that of the amaranth snack.

In other products, such as pasta, the dough's quality attributes could be defined by its elasticity, viscosity, water absorption capacity, and cohesiveness. The optimal pasta products should be robust, malleable, and have a uniform surface (Fuad & Prabhasankar, 2010). Incorporating quinoa flour into the dough mixture resulted in diminished elasticity and strength, decreased the tensile strength of the dried pasta, reduced the cooking losses, and achieved a softer texture of the cooked pasta as described by Sissons et al. (2007). One method for mitigating these detrimental effects involved using structuring agents, such as sodium carboxymethyl cellulose (CMC), casein (CAS), whey protein isolates (WPI), chitosan (CH), and pre-gelatinised starches of legume, oat, quinoa (15, 10 and 5%) and emulsifiers, as reported by Schoenlechner et al. (2010). Chillo et al. (2009) reported that incorporation of CMC at a concentration of 0.1%, 0.2%, or 0.3%, or PS at concentrations of 10%, 20%, or 30% had no discernible effect on the dough's breaking strength or the organoleptic qualities of the dry and cooked quinoa spaghetti.

The potential for quinoa flour to exhibit compatibility with a variety of food ingredients in the production of snacks has been reported. According to the comparative analysis by Diaz et al. (2013), the addition of 20% QF in maize snacks resulted in reduction in their hardness compared with snacks made solely from maize. The observed phenomenon could potentially be attributed to the coexistence of additional constituents, thereby diminishing the intermolecular associations between starch and proteins, as suggested by Diaz et al. (2013). Furthermore, according to Harra et al. (2011), substitution of WF with 50% or 100% quinoa flour resulted in peanut butter cookies with enhanced sweetness and chewiness.

13.3.2 Buckwheat-based Products

Buckwheat is a pseudocereal that is gluten-free and a member of the Polygonaceae family. The two most commonly cultivated types are common buckwheat (*Fagopyrum esculentum*) and Tartary buckwheat (*Fagopyrum tataricum*). Recent interest in buckwheat as a potential functional food has increased due to its abundance of starch and proteins, antioxidants, dietary fibre and trace elements. Buckwheat is a suitable dietary option for individuals who must stick to a diet free of gluten, owing to its exceptional nutritional composition (Saturni et al., 2010). The elimination of gluten poses considerable challenges for manufacturers and bakers. Presently, numerous GFPs in the market are of substandard quality, characterised by unsatisfactory flavour and mouthfeel (Giménez-Bastida et al., 2015).

Buckwheat addition into a GF experimental formulation improved the bread's physicochemical properties and enhanced the proteins, folates and minerals, particularly Cu and Mg, in the end product, with predominantly oleic and linoleic acids (Giménez-Bastida et al., 2015). Also, according to a study conducted by Alvarez-Jubete et al. (2010), it was found that buckwheat gluten-free bread exhibited excellent baking characteristics, including increased volume and a softer crumb, when compared to the gluten-free control sample. Buckwheat-based products have been found to improve the texture, sensory characteristics, and shelf life of gluten-free bakery items through the incorporation of various components, such as starches, emulsifiers, hydrocolloids, non-gluten proteins, prebiotics, and their combinations. These components serve as substitutes for gluten and facilitate improved protein cross-linking. For instance, Smerdel et al. (2012) reported improved baking qualities and macroscopic structure of buckwheat flour with the addition of trans-glutaminase or proteins from different sources (caseinate, soya isolate, egg-white powder). Incorporation of 20% buckwheat into rice-based GF cookies showed appreciable sensory scores among panellists, with acceptable physicochemical and textural properties expressed in terms of cross section, shape, rupture, and appearance of the overall structure (Torbica et al., 2012). Many researchers claimed that buckwheat had been successfully incorporated in durum wheat semolina for the manufacture of GF-pasta. Rice-based noodles enriched with corn starch and 20% buckwheat with xanthan gum (XG) addition exhibited higher ash content (K, Mg and P content) along with decreased levels of anti-nutritional factors, colour values, and cooking loss. However, it was observed that the production of pure buckwheat noodles was quite difficult due to the lack of sufficient cohesive protein in the dough (Giménez-Bastida et al., 2015). In addition to the aforementioned GF products, buckwheat has demonstrated exceptional suitability for the production of gluten-free beer with acceptable fizziness and bitterness (Giménez-Bastida et al., 2015). Malting of buckwheat grains includes several processing stages like steeping (commonly done in dilute NaOH solution), germination (at 20°C), and the kilning process (High temperature and short time, HTST process) causing physicochemical changes in the pseudocereal grain. Thus, the time-temperature combinations play a vital role in the enzymatic regime in the GF-beer production line.

13.3.3 Amaranth-based Products

Amaranth, a gluten-free pseudocereal, is renowned for its exceptional nutritional qualities, which make it a valuable dietary addition. It contains high protein, fibre, vitamins, and ash content with several health benefits, like reduction in cholesterol and enhanced digestion. Its importance in food processing originates from its potential as a versatile and nutrient-dense ingredient that can be used in a variety of GFPs, from baked goods to snacking, to satisfy the rising demand for healthier and plant-based alternatives.

The abundant nutritional value of amaranth grain makes it highly versatile for various baking purposes (Lorenz, 1981). According to Sanz-Penella et al. (2013), adding amaranth flour in bread preparation considerably increased the quantity of protein, fat, ash, dietary fibre, and mineral content of the bread. However, the incorporation of amaranth was also accompanied by greater levels of phytates, which may have a negative impact on the mineral's bioavailability. According to Zebdewos et al. (2015), porridge made with 70% amaranth and 30% chickpea contained more iron, zinc, calcium, protein, and lower phytate than whole maize or a combination of the three basic components (maize, chickpea, and amaranth). In a study conducted by Sindhuja et al. (2005), focussing on cookies made from wheat and amaranth flour, it was discovered that the addition of amaranth flour significantly improved the cookies, even when replacing only 25–30% of the wheat flour. The hardness of the cookies decreased as the amount of amaranth increased.

Moreover, the cookies exhibited enhanced organoleptic characteristics, including improvements in colour, taste, flavour, and appearance (Sindhuja et al., 2005).

In today's world, there has been an increasing recognition of the importance of consumables in the mitigation and management of illnesses. The utilisation of these substances is a commendable approach for dispensing essential nutrients and bioactive compounds to diverse population demographics. Argüelles-López et al. (2018) developed a beverage using a combination of amaranth and chai flour in a 70:30 ratio using extrusion and germination techniques. According to Espino-González et al. (2018), the preparation of the amaranth beverage involved the concoction of 88.8 g of amaranth grains, 0.35 g of salt, 41.3 g of sugar, and one litre of water, along with orange or grape flavour. The resulting beverage was found to contain 0.35 g/L electrolytes, 1.5% protein, and 10% carbohydrates. The differences in cycling performance and hydration status of cyclists over a time trial were attributed to the amaranth beverage's greater caloric content (52.48 kcal/100 mL) than industrial beverages (24 kcal/100 mL).

13.3.4 Flaxseed-based Products

Common flax (*Linum usitatissimum* L.), has been utilised as a crop for producing both fibre and oilseed since ancient times. Throughout history, flaxseed has been utilised within the industrial sector for its applications as a linen textile or a drying oil in various products such as paints, varnishes, and linoleum. Flaxseed oil is considered a desirable food item for its elevated levels of α-linolenic acid, dietary fibre, and superior protein content, rendering it a promising option for nutritional applications (Rabetafika et al., 2011). In fact, approximately 70% of global flaxseed oil production is intended for technical applications in the paint, varnish, linoleum, and PVC plastic industries. Additionally, flaxseed oil finds application in the formulation of inks and personal care commodities. The flaxseed fibre is utilised in numerous industries, including textile, paper, and bio-composite production.

Presently, flaxseed has new opportunities in the food industry due to rising consumer demand for functional foods. The food industry is considering the incorporation of flaxseed into the traditional food products in order to provide consumers with the opportunity to reap its health benefits. According to Mercier et al. (2014), α-linolenic acid (ALA) constitutes a significant proportion of approximately 50–62% of flaxseed oil with a nutrient-dense composition that exerts biological remedial properties such as anticancer, antioxidative, and anti-oestrogenic activities. Some of the probable applications of flaxseed in tailoring GFPs are discussed in Table 13.1.

White bread enriched with 15% (wt) pulverised flaxseed resulted in more than 2 logs relative increase in omega 3 fatty acids as compared to flaxseed-free bread (Menteş et al., 2008). In case of flaxseed-enriched macaroni and pasta, a study conducted by Lee et al. (2004) reported a rise in the conjugated diene levels subsequent to the hydration and mixing process of pasta dough, both with and without flaxseed enrichment, particularly while processing at a higher temperature of around 90°C. The dough that lacked flaxseed had a greater proportion of conjugated diene content compared to the dough that contained 15% ground whole flaxseed or 15% ground flaxseed hull. This reduction in conjugated dienes suggests the existence of potent antioxidant compounds in flaxseed, such as lignans, which effectively inhibit lipid oxidation. Also, the enriched pasta had a 6% lower cooked weight than the traditional pasta. Furthermore, increasing the flaxseed content in the pasta from 15% to 30% led to a reduction in the swelling and diameter of the pasta after cooking (Villeneuve et al., 2013). The findings indicate that the inclusion of flaxseed leads to a reduction in water absorption during the cooking process, potentially due to the dilution of starch.

Interestingly, it had been found that cookies and muffins enriched with flaxseed were a little darker in appearance with improved firmness. It was found that, with flaxseed enrichment, there was a reduction in the water activity from 0.461 to 0.356 in the cookies and their moisture content decreased from 6.5% to 4.8% at an incorporation rate of 18%, which helped to slow down microbial development and biological degradation of the final product (Mercier et al., 2014). However, an acceptable sensory scores were obtained up to only 30% flaxseed incorporation (Mercier et al., 2014). Moreover, it was claimed by Ahmed (1999) that flaxseed caused a lubricating effect on maize-based cereal bars as suggested by the decreasing torque of the extruder from 15.8 to 10.8 Nm in flaxseed-free bars, associated with less power consumption and economical production of the same. From a sensory aspect, an incorporation of flaxseed up to 12% was considered to be acceptable for the commercial production of cereal bars, whereas more than 36% flaxseed incorporation imparted a distinct bitter taste to the bars.

13.3.5 Sorghum-based Products

Sorghum is a versatile and drought-tolerant cereal that is of critical importance in processing industries. It's high in nutritional value, and it contains significant amounts of fibre, antioxidants, and important minerals, all of which contribute to its impact. Sorghum is used in a variety of processing industries, including brewing, baking, snack manufacture, and animal feed, providing healthy and interesting choices for the food and beverage processing industries. Sorghum is a key ingredient in boosting the sustainability and creativity of the processing industries because of its versatility, nutritional value, and economic advantages.

Numerous studies in the past have focussed on formulating and producing sorghum-based food products for enhancing health benefits, consumer acceptability, and economic value. Sorghum has the potential to produce various types of bread such as kissra bread, flatbread, khamir bread, sourdough bread (SDB), and fry bread. However, sorghum flour or dough has poor viscoelasticity and texture because it lacks gluten protein (Olojede et al., 2020). The use of starter cultures in the fermentation process of sorghum-based sourdough resulted in the production of bread that exhibited improved rheological properties, texture, protein digestibility, antioxidant characteristics, and consumer acceptance (Olojede et al. 2020a, 2020b). Sorghum bread can be improved through the incorporation of natural stabilising agents such as xanthan gum. The inclusion of xanthan gum to the dough mixture improved the elasticity and firmness of the dough. The potential cause of this phenomenon may be attributed to the stabilising, gelling, or thickening attributes of xanthan gum (Jafari et al., 2018).

Sorghum-derived extruded food items, such as noodles, pasta, crisps, and snacks, exhibit potential as viable products, particularly for individuals with coeliac disease (Pezzali et al., 2020). According to the study conducted by Rashwan et al. (2021), the production of noodles using red and white Egyptian sorghum resulted in a greater overall acceptability rating of 70.1–80.15%. Benhur et al. (2015) reported that the pasta produced from sorghum flour displayed a notable protein content of approximately 170 g/kg, a dietary fibre content of approximately 80 g/kg, and a polyphenol content of 2.6 g gallic acid-equivalents (GAE)/kg. According to Pezzali et al. (2020), the production of sorghum crisps by the extrusion of white and red sorghum flour has shown the ability to increase the utilisation of sorghum-based foods in the market.

Popular traditional beverages are fermented cereal-based drinks produced using grains like sorghum and maize. Kunu-zaki is a type of non-alcoholic beverage that is widely consumed as a nutritious sorghum-based fermented drink. The methodology for preparing this beverage is cost-effective and it is readily available in nearby markets (Ndulaka et al., 2014). Also, mahewu is a breakfast beverage made from sorghum that is frequently utilised as a weaning food for children. It is also suitable for consumption by adults due to its high nutritional value (Blandino et al., 2003). In brief, fermented sorghum-based drinks are among the popular beverages in many nations as sorghum contains phenolic components (tannins, flavonoids, phenol, anthocyanins) and dietary fibres with great nutritional value (Coulibaly et al., 2020).

13.3.6 Wattle Seed-based Products

Since many centuries, the seeds of the legume acacia (wattle) have been part of Indigenous Australian diet. The native population of Western Australia consumes species like *Acacia microbotrya* and *Acacia cyclops* that are native to that area. Wattle seeds make up a significant part of the diets of Australian Indigenous peoples, although the general public does not consider them to be a food source because acacia trees are more commonly used to make high-quality hardwood. The wattle seeds were shown to have higher levels of protein, dietary fibre, zinc, and potassium than several popular annual legumes like lentils and chickpeas. Thus, the wattle seed may bring additional health advantages for consumers when added to food items.

Incorporation of flaxseed powders (*L. usitatissimum*) in the development of GF breads were found to be beneficial. According to Krishna Kumar et al. (2019) addition of 1% wattle seed powders increased the specific loaf volume by 50% while reducing the crumb hardness by 30–65%. These textural changes were correlated with the presence of water-soluble carbohydrates and insoluble fibre, that had the potential to emulsify and absorb water. Upon flaxseed addition, a darker crumb was observed of the GF-crumb surface was observed. Also, a characteristic gel-like, viscoelastic network, connecting protein and starch, was observed. Surprisingly, no significant foaming stability was noted despite the high protein level, probably due to the structure of the specific proteins and their interaction with insoluble fibre.

Composite flour of acacia, cereals and legumes is locally available in Nigeria; when tested in rats at the laboratory scale, the most effective complementation with *Acacia colei* with regard to weight gain, protein efficiency ratio (PER), and animal health was found in red sorghum, brown fonio, and white acacia (Adewusi et al., 2011). In Niger, wattle seed can greatly boost the quantity and quality of protein and increase the levels of several micronutrients in a millet kunu, with 15–20% incorporation (Adewusi et al., 2011). There can be substitutes for acacia seed in conventional infant foods too, made with pulses. Therefore, current research trends primarily focus on the nutritional/toxicological evaluation and assimilation of seeds into regional foods and diets; nonetheless, it is important to keep in mind that this species may become invasive. Engaging potential adopters and stakeholders within the food supply chain is highly recommended for conducting participatory research.

13.3.7 Chia-based Products

Chia-based food products have gained widespread recognition for their exceptional nutritional profile and health benefits. Packed with fibre, antioxidants and omega-3 fatty acids, chia seeds offer a versatile and accessible ingredient for creating innovative and nutritious foods. Their significance lies in providing convenient options that promote wellness and cater to the increasing demand for healthy and functional dietary choices.

Chia seeds contain a significant amount of mucilage and gum, making them a potentially useful resource in food industries (Fernandes & de las Mercedes Salas-Mellado, 2017). The utilisation of chia seeds is predominantly observed in the bakery. This is due to the presence of elevated concentration of carbohydrates present in baked goods in comparison to other vital nutrients, as noted by Romankiewicz et al. (2017). In a study by Coelho & Salas-Mellado (2014), the inclusion of whole chia seeds in breadcrumbs resulted in a softer texture compared with the use of chia flour. In a research study conducted by Romankiewicz et al. (2017), chia seeds were incorporated into wheat flour at a concentration of 4% to 6% for composite bread preparation. Furthermore, the incorporation of chia seeds at a proportion of 6% to 8% in wheat flour has been observed to decrease baking loss. This can be correlated to the significant amount of dietary fibre present in chia seeds, which has the ability to bind with water and hinder evaporation during the bread baking process, as reported by Romankiewicz et al. (2017). According to Iglesias-Puig & Haros (2013), the inclusion of chia seeds in bread resulted in a dark colouration on account of major phenolic contents. . The high content of omega-3 fatty acids, fibre, minerals, vitamins, proteins, phytochemicals in chia seeds have led to its potential applications as nutritional supplements, food additives, and beverage bases.

13.3.8 Teff-based Products

Teff (*Eragrostis teff*) is an annual crop belonging to the Poaceae (grass) family. It is a major food crop indigenous to Eritrea and Ethiopia and is used for the preparation of traditional dishes and beverages, including tella (opaque beer), kitta (unleavened bread), and injera (flatbread). Teff has been rising in popularity in recent years, largely because of its attractive nutritional qualities. In addition to being a grain for human use, teff straw may be utilised as fodder for horses and cattle. The straw can be processed to produce bio-methane and utilised as an adsorbent in wastewater treatment. As a result, teff can be used extensively and sustainably in various food and agri-based formulations as described in Table 13.1.

The integration of teff flour into wheat flour has proven to be effective in the creation of various bakery items such as bread, cake, cookies, and biscuits. In gluten-free formulations, particularly in injera (traditional sourdough flatbread), teff constitutes the principal ingredient; however, the phytate content in teff hinders the bio-availability of essential minerals (like Fe) and proteins (Zhu, 2018). This necessitates pre-treatment of the teff flour by fermentation before incorporation. Fermentation using lactic acid bacteria (LAB) proved to be more efficient in reducing phytate content by up to 70% in injera than the conventional fermentation without inoculation with LAB (Zhu, 2018). When teff flour was used to make gluten-free egg spaghetti, and compared with products made from oat and wheat., it was found that the teff and oat spaghetti were more enriched in dietary fibre and mineral contents, while teff spaghetti had a lower glycaemic index (GI-45) than that of wheat spaghetti (GI-67) (Hager et al., 2012). Sourdough technology has been one of the most recent trends in the development of teff-based products. In fact, the recipe with 20% teff inclusion in rice and buckwheat sourdoughs produced good scores in terms of visual appearance, with a smaller loaf volume for composite bread, although the greater levels of rutin in buckwheat sourdough gave it a slightly bitter taste (Zhu, 2016). The non-sourdough breads of teff were also popular among consumers. According to the comparative *in-vitro* analysis of non-sourdough breads from different flours, teff bread had a predicted glycaemic index (pGI) of 74, which was higher than that of sorghum of 72, oat (pGI) of 71, and commercially available gluten-free breads (pGI) of 69. Teff bread's pGI, however, was lower than that of buckwheat (80) and quinoa (95) breads (Zhu, 2018).

Moreover, teff can be used in the formulation of cookies, extrudate, fat replacer, and weaning food. A blend of teff starch and stearic acid as paste could be utilised as a potential substitute for fat in mayonnaise-like emulsions, with the aim of controlling calorie intake. According to Teklehaimanot et al. (2013), mayonnaise with an 80% teff-emulsion displayed characteristics similar to those of full-fat mayonnaise while having 76% fewer calories. Furthermore, teff can be promisingly used for the production of gluten-free malt. Gebremariam et al. (2013) studied the effects of drying temperature and time (kilning) on the activities of α- and β-amylases in teff malt. According to their findings, the temperature schedule of 18 hours at 30 °C plus 1 hour at 60 °C plus 3 or 5 hours at 65 °C was optimal for preserving enzymes while producing the least amount of dimethyl sulphide. Thus, the minute granules of teff starch exhibit promising potential in various applications, such as gluten-free malt and beverages, fat substitution, the creation of biodegradable films, acting as carriers for functional ingredients in encapsulation processes, and utilisation in industries such as paper, cosmetics, textiles, and photography.

13.4 Advances in the Different Technologies in the Pseudocereal-based Foods

Different technologies have been introduced in the food industry to process pseudocereal-based foods. Among these technologies, enzyme, dough, fermentation, and extrusion technologies are the most effective.

13.4.1 Enzyme Technology

In the food sector, enzymes are frequently used in the preparation of bread, cheese, and infant formula. Specific molecules can be targeted by enzymes, allowing the other nutrients to remain unaltered. Additionally, enzymes release bioactive compounds from the dietary matrix, improving bioavailability and accessibility (Schaffer-Lequart et al., 2017).

The technological attributes of cereals or cereal-derived products are contingent upon the activity of both exogenous and endogenous enzymes. Cereal grains that have not been treated need to go through several pre-treatment processes that could also use enzymes. According to Perdon et al. (2020), enzyme pre-treatment can enhance the technical and nutritional properties of substances by modifying their micro, macro, and molecular structure, as well as the composition and quality of their nutrients, phytochemicals, and potentially harmful components. Mycotoxin reduction can be achieved by using enzymes to bio-transform mycotoxins into harmless metabolites. However, factors such as the source of enzyme, the processing conditions, and its concentration all affect the breakdown of mycotoxins by enzymes, as demonstrated by Loi et al. (2017).

The quantities of free ferulic acid in grains, which can enhance its bioavailability, can be increased by an enzymatic process which is capable of precisely hydrolysing the ester bond implicated in the covalent connection that binds it to other salts or hemicellulose like arabinoxylans. The fermentation of wheat bran, with or without enzymatic treatment, was found to enhance the bio-accessibility of ferulic acid by a factor of five (Anson et al., 2009; Björck et al., 2012).

13.4.2 Dough Technology

Kneading, shaping, and baking are a few of the procedures used in dough technology. There is an additional fermenting process for bread goods. The sensory quality of bread can be enhanced and varied through the use of dough technology, a well-established biotechnological method with great potential for the fermentation of pseudocereals and cereals other than wheat (Schaffer-Lequart et al., 2017). The demand from consumers and the enhanced nutritional profile of wholegrain flour in comparison to refined grains have put much attention on the dough technology of wholegrain flour in foods. Research on dough microbiota of cereals and pseudocereals has shown a significant diversity in their microflora, sparking interest in starting cultures and improving the nutritive and sensory properties of the products (Corsetti & Settanni, 2007). Dough processing technology offers potential in combating non-communicable diseases, such as diabetes, cardiovascular disease, and obesity, by enabling reduction of salt and sugar content in baked goods (Sahin et al., 2019).

In a dough processing system, the LAB species *Leuconostoc oenos* produces erythritol, while *Leuconostoc mesenteroides*, *Lactobacillus sanfranciscensis*, and *Leuconostoc citreum* produce mannitol. In the presence of xylose, the fungus *Candida milleri* also makes xylitol, which increases the sweetness level of the baked goods. The exopolysaccharides generated by yeasts and/or LAB serve as both bulking agents and enhancers of the structure and texture of bread products, effectively substituting for added sugars (Sahin et al., 2019). It is anticipated that dough processing will be employed in the future to enhance existing technology and create foods with specific properties.

13.4.3 Fermentation

Foods based on cereals, such as breads, beverages, porridges, and gruels, involve the well-established method of fermenting whole grains. LAB and yeasts are the predominant food-grade microorganisms used for the fermentation of cereal-based foods. They may be naturally present or added to the flour as starter cultures (Hammes et al., 2005). Fermentation leads to a decrease in pH, resulting in an extended shelf life of the product and reducing the need for preservatives in the final product. The conversion of amino acids, lipids, and carbohydrates during fermentation can have a considerable impact on the flavour of the fermented product (Poutanen et al., 2009; Soetan & Oyewole, 2009). Fermentation boosts the availability of many B vitamins, amino acids, and ash contents of iron, calcium, and zinc, by enzymatically breaking down anti-nutritional components such as phytates and tannins. The creation of non-traditional cereal-based probiotic beverages is linked to innovation in this industry. Due to medical (the prevalence of lactose intolerance and cow's milk allergies) and lifestyle factors, these drinks are growing in popularity. Ideally, fermentation strains would quickly acidify the substrate, shielding it from contamination while improving it at the same time. Drinks made from fermented cereals function as probiotics, prebiotics, and synbiotics, and may help with digestive health (Basinskiene & Cizeikiene, 2020).

13.4.4 Extrusion

Extrusion cooking, a versatile method commonly employed in the food industry, enables the production of aerated cereal-based products with a wide array of shapes and textures. Nowadays, the majority of extruded goods (cereals for breakfast or salty snacks) are made from refined flour. The presence of a substantial amount of insoluble dietary fibre in bran contributes to wholegrain cereals having a reduced water holding capacity compared to starch, making them less physicochemically compatible with other grain components (Robin et al., 2011). Moreover, the texture of extruded wheat grain flours is inferior to that of refined flours due to the reduced elasticity of the starchy melt during the heating process involved in extrusion. High shear forces during extrusion lead to thermal cleavage and depolymerisation of the starch molecules. Because of this, straight chains are created, which are subjected to retrogradation. The enhancement of the expansion volume of extruded products is achieved by targeting the solubility of insoluble fibre present in bran. For instance, the ferulic acid ester cross-links in maize bran were hydrolysed after being efficiently solubilised using an alkaline solution (Blake, 2006). Studies have shown that extrusion holds the potential to greatly improve the nutritional value of food, particularly in the case of quinoa and amaranth. The extrusion process enhances the nutritional content of these grains. Interestingly, single-screw extruders were used to create gluten-free extruded items from the pure grains of amaranth and quinoa (Hemalatha et al., 2016). A comprehensive examination of the impact of extrusion processing parameters on quinoa flour demonstrated moderate expansion properties in comparison to widely utilised cereal grains. The findings suggest that relying solely on quinoa may not be optimal for the production of directly expanded goods. However, in products where straight expansion is not a critical textural requirement, incorporating quinoa can be more beneficial (Kowalski et al., 2016).

13.5 Addition of Special Additives

Sensory analysis of pseudocereal-based food formulations revealed lower acceptability in comparison with market-available wheat-based products. Due to gluten dilution or the absence of gluten protein, enhancers are being added to imitate the viscoelastic behaviour of wheat gluten and to enhance the baking and other techno-functional properties of pseudocereal-based foods. The utilisation of several additives like microbial transglutaminase (TGase) could be employed in an effort to improve the quality of bread, as demonstrated by Renzetti et al. (2008).

Nonetheless, the application of the aforementioned additive did not have any significant impact on the bread-making qualities and batter rheology of teff, oat, and sorghum, but those of buckwheat and brown rice were improved. The observed variation could potentially be attributed to dissimilarities in the protein composition, non-protein constituents, and quantity of TGase incorporation. In a study conducted by Hager and Arendt (2013), hydroxypropylmethylcellulose (HPMC) and/or xanthan gum were utilised as additives to enhance the bread-making properties of gluten-free pseudocereal flours. The inclusion of xanthan gum linearly decreased the volume of all kinds of bread. Teff and maize bread volumes increased with the addition of HPMC, whereas rice bread volumes decreased linearly with HPMC concentration. The incorporation of HPMC did not have an influence on the volume of buckwheat bread. According to Hager and Arendt (2013), incorporating xanthan gum enhanced the crumb hardness of buckwheat and teff breads and lowered the hardness of maize bread. Consequently, the impact of incorporating hydrocolloids into the texture of gluten-free bread is contingent upon the specific hydrocolloid and flour types utilised, as well as the particular textural parameters under examination. The physicochemical characteristics of flour influenced by the addition of HPMC and xanthan gum are subject to alterations based on the ionic strength and pH of the bread matrix, as well as the application of heat and shearing forces. There is notable variation in the chemical composition of various flours, and their interactions with hydrocolloids therefore also vary considerably (Hager & Arendt, 2013). Moreover, an interesting interaction between flaxseed concentration (0–50%) and sweetener content (45–55%) was also observed in flaxseed-containing cereal bars by Mridula et al. (2013). They reported a significant decrease in textural hardness with a rise in sweetener content from 45% to 55%, whereas the opposite response was observed for the control sample.

In recent food formulations, some pseudocereals are often used as additives themselves. Acacia gum, commonly referred to as gum arabic, derived from *Acacia* spp., is utilised as a binding and emulsifying agent within the baking industry to enhance the rheological properties of wheat. Krishna Kumar et al. (2019) demonstrated favourable outcomes through the integration of gum arabic into wheat bread. The inclusion of gum arabic at a concentration of 0.3% resulted in elevated gas production and retention, ultimately contributing to greater bread softness.

Therefore, the application of additives such as hydrocolloids, emulsifiers, and enzymes contribute to enhancing the texture, structure, and overall acceptability of foods. They help in mitigating the inherent challenges faced in gluten-free formulations, such as poor dough elasticity, reduced volume, and dryness. Additionally, additives can extend the shelf life of these products through prevention of microbial growth and oxidative deterioration. However, achieving a balance between the utilisation of additives and maintaining transparent labelling is crucial to ensure both consumer safety and satisfaction. By leveraging the potential of additives, the food industries can continue to innovate and formulate delicious and nutritive pseudocereal-based foods.

13.6 Consumer Needs

Consumer needs and suggestions regarding the recent trends in pseudocereal-based food products particularly revolve around enhancing the overall quality, accessibility, and sustainability of these products. Consumers are seeking continuous improvement in terms of taste, texture, and variety, as they need pseudocereal-based food options that are not only nutritious but also delicious and enjoyable to consume. Additionally, there is a growing demand for greater accessibility to these products, both in terms of availability in stores and affordability, so that a wider range of consumers can access and incorporate them into their diets. Consumers also express a desire for sustainability, including considerations such as responsible sourcing, environmental friendly packaging, ethical production practices, and longevity in storage. By addressing these consumer needs, manufacturers and suppliers can drive further growth in the market and establish a strong consumer base that values high-quality, accessible, and sustainable pseudocereal-based food products.

13.7 Human Health Aspects

Children and teenagers who regularly consume whole grains are likely to lead healthy lives. Additionally, they exhibit an exceptional range of physical and cognitive activity proficiency (Williams, 2014). The determination of a precise connection between eating whole grains and changes in bacterial flora is quite difficult. The majority of research focusses on how the gut flora changes to show the beneficial qualities of whole grains. The pH, length of intestinal transit, and the weight of faeces, and the concentration of short-chain fatty acids were the focus of other research (Koecher et al., 2019). Wholegrain cereals are typically used to manage weight because they lower the cholesterol level and hunger (Kim, 2018). Cereals lessen the risk of developing cancer, diabetes, and inflammatory diseases because they are high in bioactive components such polyphenols, phytosterols, and β-glucans. Due to their high concentration of such beneficial substances, cereals lower the risk of developing cancer, diabetes, and inflammatory illnesses (J. Fu et al., 2020; M. Fu et al., 2020). The consumption of pseudocereal-based diets is likely to enhance multiple mechanisms, including improved gut transit (McIntosh et al., 2003; Ross et al., 2011), modifications in gut microbiota metabolism, microbial populations with potential prebiotic effects (Carvalho-Wells et al., 2010; Costabile et al., 2008; McIntosh et al., 2003; Ross et al., 2011), and improved methyl donor (betaine) status (Ross et al., 2011). The health benefits imparted by pseudocereals, owing to their nutrient-dense bioactive and phytochemical compositions, are shown in Figure 13.4.

Enhancing the nutritional value of cereals necessitates the removal of harmful components. Various techniques can be employed to eliminate or reduce antinutrients and create functional foods. It is widely recognised that utilising

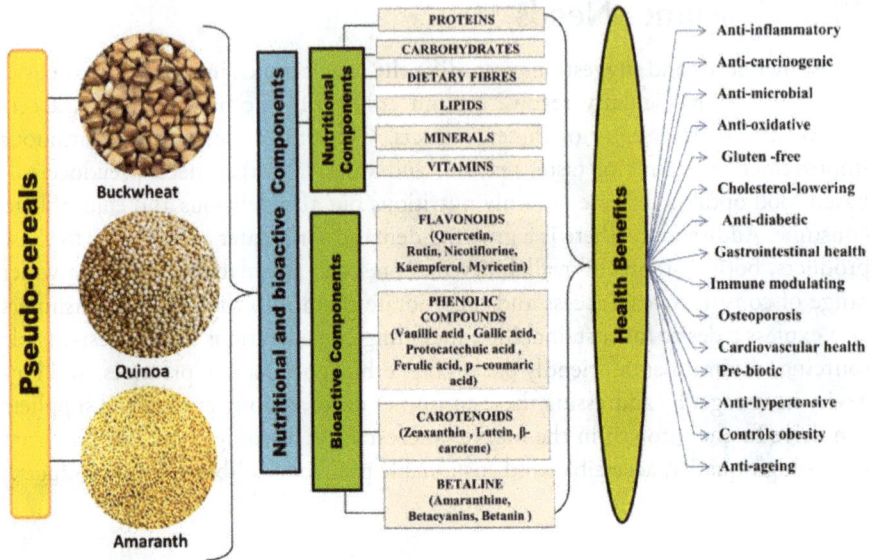

Figure 13.4   Nutritional and bioactive components present in pseudo-cereals and their health benefits (Thakur et al., 2021).

straightforward and cost-effective conventional processing methods is an efficient strategy to process grains more efficiently. They increase the health benefits of grains and increase the benefits to the body.

13.8 Conclusions and Future Aspects

The current developments in the field of pseudocereal-derived food products have demonstrated their significant potential in the dietary domain, as well as their capacity to meet the increasing need for varied, health-promoting, and eco-friendly food alternatives. The inclusion of pseudocereals in diverse food formulations has furnished customers with substitutes that are free from gluten and characterised by distinct flavours. Recent trends in pseudocereal-based products with the application of additives have revolutionised the food industry by offering an array of innovative, nutritious, and appealing options to consumers. The strategic use of additives has improved the sensory attributes, functional properties, and nutritional value of these products, aligning them with the growing demand for healthier and more convenient food choices. With advances in food science and technology, manufacturers have been able to create gluten-free bread, pasta, snacks, and breakfast cereals that closely mimic the taste and texture of their wheat-based counterparts. Additionally, the potential for pseudocereal-based food products in the future is encouraging, as current research is dedicated to enhancing processing methods and sensory characteristics, and investigating new uses in functional foods and inventive food endeavours. The food industry is likely to be influenced by the changing consumer preferences towards healthier and more sustainable options. Pseudocereal-based foods are expected to have a significant impact in this regard, owing to their nutritional advantages and potential for diverse uses.

References

Abugoch, L. E., Tapia, C., Villamán, M. C., Yazdani-Pedram, M., & Díaz-Dosque, M. (2011). Characterization of quinoa protein–chitosan blend edible films. *Food Hydrocolloids*, *25*(5), 879–886.

Adelekan, A., & Oyewole, O. (2010). Production of ogi from germinated sorghum supplemented with soybeans. *African Journal of Biotechnology*, *9*(42), 7114–7121.

Adewusi, S., Falade, O., & Harwood, C. (2011). Wattle as food: Nutritional and toxic considerations. *WATTLE We Eat for Dinner*.

Agregán, R., Guzel, N., Guzel, M., Bangar, S. P., Zengin, G., Kumar, M., & Lorenzo, J. M. (2023). The Effects of processing technologies on nutritional and anti-nutritional properties of pseudocereals and minor cereal. *Food and Bioprocess Technology*, *16*(5), 961-986.

Ahmed, Z. S. (1999). Physico-chemical, structural and sensory quality of corn-based flax-snack. *Food/Nahrung*, *43*(4), 253–258.

Alaunyte, I., Stojceska, V., Plunkett, A., Ainsworth, P., & Derbyshire, E. (2012). Improving the quality of nutrient-rich Teff (*Eragrostis tef*) breads by combination of enzymes in straight dough and sourdough breadmaking. *Journal of Cereal Science*, *55*(1), 22–30. https://doi.org/10.1016/j.jcs.2011.09.005.

Alvarez-Jubete, L., Auty, M., Arendt, E. K., & Gallagher, E. (2010). Baking properties and microstructure of pseudocereal flours in gluten-free bread formulations. *European Food Research and Technology*, *230*(3), 437–445.

Alvarez-Jubete, L., Wijngaard, H., Arendt, E. K., & Gallagher, E. (2010). Polyphenol composition and in vitro antioxidant activity of amaranth, quinoa, buckwheat and wheat as affected by sprouting and baking. *Food Chemistry*, *119*(2), 770–778.

Anson, N. M., Selinheimo, E., Havenaar, R., Aura, A.-M., Mattila, I., Lehtinen, P., Bast, A., Poutanen, K., & Haenen, G. R. (2009). Bioprocessing of wheat bran improves *in vitro* bioaccessibility and colonic metabolism of phenolic compounds. *Journal of Agricultural and Food Chemistry*, *57*(14), 6148–6155.

Argüelles-López, O. D., Reyes-Moreno, C., Gutiérrez-Dorado, R., Sánchez-Osuna, M. F., López-Cervantes, J., Cuevas-Rodríguez, E. O., Milán-Carrillo, J., & Perales-Sánchez, J. X. K. (2018). Functional beverages elaborated from amaranth and chia flours processed by germination and extrusion. *Biotecnia*, *20*(3), 135–145.

Basinskiene, L., & Cizeikiene, D. (2020). Cereal-based nonalcoholic beverages. In *Trends in Non-Alcoholic Beverages*, 63–99.

Baye, K., Mouquet-Rivier, C., Icard-Vernière, C., Picq, C., & Guyot, J. P. (2014). Changes in mineral absorption inhibitors consequent to fermentation of Ethiopian injera: Implications for predicted iron bioavailability and bioaccessibility. *International Journal of Food Science and Technology*, *49*(1), 174–180.

Benhur, D. R., Bhargavi, G., Kalpana, K., Vishala, A., Ganapathy, K., & Patil, J. (2015). Development and standardization of sorghum pasta using extrusion technology. *Journal of Food Science and Technology, 52*(10), 6828–6833.

Björck, I., Östman, E., Kristensen, M., Anson, N. M., Price, R. K., Haenen, G. R., Havenaar, R., Knudsen, K. E. B., Frid, A., & Mykkänen, H. (2012). Cereal grains for nutrition and health benefits: Overview of results from in vitro, animal and human studies in the HEALTHGRAIN project. *Trends in Food Science and Technology, 25*(2), 87–100.

Blake, O. A. (2006). Effect of molecular and supramolecular characteristics of select dietary fibers on extrusion expansion.

Blandino, A., Al-Aseeri, M., Pandiella, S., Cantero, D., & Webb, C. (2003). Cereal-based fermented foods and beverages. *Food Research International, 36*(6), 527–543.

Carvalho-Wells, A. L., Helmolz, K., Nodet, C., Molzer, C., Leonard, C., McKevith, B., Thielecke, F., Jackson, K. G., & Tuohy, K. M. (2010). Determination of the *in vivo* prebiotic potential of a maize-based whole grain breakfast cereal: A human feeding study. *British Journal of Nutrition, 104*(9), 1353–1356.

Chillo, S., Civica, V., Iannetti, M., Suriano, N., Mastromatteo, M., & Del Nobile, M. A. (2009). Properties of quinoa and oat spaghetti loaded with carboxymethylcellulose sodium salt and pregelatinized starch as structuring agents. *Carbohydrate Polymers, 78*(4), 932–937.

Chlopicka, J., Pasko, P., Gorinstein, S., Jedryas, A., & Zagrodzki, P. (2012). Total phenolic and total flavonoid content, antioxidant activity and sensory evaluation of pseudocereal breads. *LWT – Food Science and Technology, 46*(2), 548–555.

Coelho, M. S., & Salas-Mellado, M. d. l. M. (2014). Chemical characterization of chia (*Salvia hispanica* L.) for use in food products. *Journal of Food and Nutrition Research, 2*(5), 263–269.

Corsetti, A., & Settanni, L. (2007). Lactobacilli in sourdough fermentation. *Food Research International, 40*(5), 539–558.

Costabile, A., Klinder, A., Fava, F., Napolitano, A., Fogliano, V., Leonard, C., Gibson, G. R., & Tuohy, K. M. (2008). Whole-grain wheat breakfast cereal has a prebiotic effect on the human gut microbiota: A double-blind, placebo-controlled, crossover study. *British Journal of Nutrition, 99*(1), 110–120.

Coulibaly, W. H., Bouatenin, K. M. J.-P., Boli, Z. B. I. A., Alfred, K. K., Bi, Y. C. T., N'sa, K. M. C., Cot, M., Djameh, C., & Djè, K. M. (2020). Influence of yeasts on bioactive compounds content of traditional sorghum beer (tchapalo) produced in Côte d'Ivoire. *Current Research in Food Science, 3*, 195–200.

Dev, D., & Quensel, E. (1989). Functional properties of linseed protein products containing different levels of mucilage in selected food systems. *Journal of Food Science, 54*(1), 183–186.

Diaz, J. M. R., Kirjoranta, S., Tenitz, S., Penttilä, P. A., Serimaa, R., Lampi, A.-M., & Jouppila, K. (2013). Use of amaranth, quinoa and kañiwa in extruded corn-based snacks. *Journal of Cereal Science, 58*(1), 59–67.

Emire, S. A., & Arega, M. (2012). Value added product development and quality characterization of amaranth (*Amaranthus caudatus* L.) grown in East Africa. *African Journal of Food Science and Technology, 3*(6), 129–141.

Espino-González, E., Muñoz-Daw, M. J., Rivera-Sosa, J. M., María, L., Cano-Olivas, G. E., De Lara-Gallegos, J. C., & Enríquez-Leal, M. C. (2018). The influence of an amaranth-based beverage on cycling performance: A pilot study. *Biotecnia, 20*(2), 31–36.

Ezekiel, C. N., Abia, W. A., Ogara, I. M., Sulyok, M., Warth, B., & Krska, R. (2015). Fate of mycotoxins in two popular traditional cereal-based beverages (kunu-zaki and pito) from rural Nigeria. *LWT – Food Science and Technology, 60*(1), 137–141.

Fernandes, S. S., & de las Mercedes Salas-Mellado, M. (2017). Addition of chia seed mucilage for reduction of fat content in bread and cakes. *Food Chemistry, 227,* 237–244.

Fu, J., Zhang, Y., Hu, Y., Zhao, G., Tang, Y., & Zou, L. (2020). Concise review: Coarse cereals exert multiple beneficial effects on human health. *Food Chemistry, 325,* 126761.

Fu, M., Sun, X., Wu, D., Meng, L., Feng, X., Cheng, W., Gao, C., Yang, Y., Shen, X., & Tang, X. (2020). Effect of partial substitution of buckwheat on cooking characteristics, nutritional composition, and in vitro starch digestibility of extruded gluten-free rice noodles. *LWT, 126,* 109332.

Fuad, T., & Prabhasankar, P. (2010). Role of ingredients in pasta product quality: A review on recent developments. *Critical Reviews in Food Science and Nutrition, 50*(8), 787–798.

Gambus, H., Gambus, F., & Sabat, R. (2002). The research on quality improvement of gluten-free bread by amaranthus flour addition. *Zywnosc, 9*(2), 99–112.

Gebremariam, M. M., Zarnkow, M., & Becker, T. (2013). Effect of drying temperature and time on alpha-amylase, beta-amylase, limit dextrinase activities and dimethyl sulphide level of teff (Eragrostis tef) malt. *Food and Bioprocess Technology, 6*(12), 3462–3472.

Giménez-Bastida, J. A., Piskuła, M., & Zieliński, H. (2015). Recent advances in development of gluten-free buckwheat products. *Trends in Food Science and Technology, 44*(1), 58–65.

Giuberti, G., Gallo, A., Fiorentini, L., Fortunati, P., & Masoero, F. (2016). In vitro starch digestibility and quality attributes of gluten free 'tagliatelle' prepared with teff flour and increasing levels of a new developed bean cultivar. *Starch-Stärke, 68*(3–4), 374–378.

Graziano, S., Agrimonti, C., Marmiroli, N., & Gullì, M. (2022). Utilisation and limitations of pseudocereals (quinoa, amaranth, and buckwheat) in food production: A review. *Trends in Food Science and Technology 125,* 154–165.

Hager, A.-S., & Arendt, E. K. (2013). Influence of hydroxypropylmethylcellulose (HPMC), xanthan gum and their combination on loaf specific volume, crumb hardness and crumb grain characteristics of gluten-free breads based on rice, maize, teff and buckwheat. *Food Hydrocolloids, 32*(1), 195–203.

Hager, A.-S., Lauck, F., Zannini, E., & Arendt, E. K. (2012). Development of gluten-free fresh egg pasta based on oat and teff flour. *European Food Research and Technology, 235*(5), 861–871.

Hager, A.-S., Wolter, A., Czerny, M., Bez, J., Zannini, E., Arendt, E. K., & Czerny, M. (2012). Investigation of product quality, sensory profile and ultrastructure of breads made from a range of commercial gluten-free flours compared to their wheat counterparts. *European Food Research and Technology, 235*(2), 333–344.

Hammes, W. P., Brandt, M. J., Francis, K. L., Rosenheim, J., Seitter, M. F., & Vogelmann, S. A. (2005). Microbial ecology of cereal fermentations. *Trends in Food Science and Technology, 16*(1–3), 4–11.

Harra, N., Lemm, T., Smith, C., & Gee, D. (2011). Quinoa flour is an acceptable replacement for all purpose flour in a peanut butter cookie. *Journal of the American Dietetic Association, 9*(111), A45.

Hemalatha, P., Bomzan, D. P., Rao, B. S., & Sreerama, Y. N. (2016). Distribution of phenolic antioxidants in whole and milled fractions of quinoa and their inhibitory effects on α-amylase and α-glucosidase activities. *Food Chemistry, 199*, 330–338.

Huang, Z.-R., Lin, Y.-K., & Fang, J.-Y. (2009). Biological and pharmacological activities of squalene and related compounds: Potential uses in cosmetic dermatology. *Molecules, 14*(1), 540–554.

Hunt, H. V., Shang, X., & Jones, M. K. (2018). Buckwheat: A crop from outside the major Chinese domestication centres? A review of the archaeobotanical, palynological and genetic evidence. *Vegetation History and Archaeobotany, 27*(3), 493–506.

Iglesias-Puig, E., & Haros, M. (2013). Evaluation of performance of dough and bread incorporating chia (Salvia hispanica L.). *European Food Research and Technology, 237*(6), 865–874.

Jafari, M., Koocheki, A., & Milani, E. (2018). Functional effects of xanthan gum on quality attributes and microstructure of extruded sorghum-wheat composite dough and bread. *LWT, 89*, 551–558.

Joye, I. J., Lamberts, L., Brijs, K., & Delcour, J. A. (2011). In situ production of γ-aminobutyric acid in breakfast cereals. *Food Chemistry, 129*(2), 395–401.

Kim, C. H. (2018). Microbiota or short-chain fatty acids: Which regulates diabetes? *Cellular and Molecular Immunology, 15*(2), 88–91.

Koecher, K. J., McKeown, N. M., Sawicki, C. M., Menon, R. S., & Slavin, J. L. (2019). Effect of whole-grain consumption on changes in fecal microbiota: A review of human intervention trials. *Nutrition Reviews, 77*(7), 487–497.

Kowalski, R. J., Medina-Meza, I. G., Thapa, B. B., Murphy, K. M., & Ganjyal, G. M. (2016). Extrusion processing characteristics of quinoa (*Chenopodium quinoa* Willd.) var. Cherry Vanilla. *Journal of Cereal Science, 70*, 91–98.

Krishna Kumar, R., Bejkar, M., Du, S., & Serventi, L. (2019). Flax and wattle seed powders enhance volume and softness of gluten-free bread. *Food Science and Technology International, 25*(1), 66–75.

Lebiedzińska, A., & Szefer, P. (2006). Vitamins B in grain and cereal–grain food, soy-products and seeds. *Food Chemistry, 95*(1), 116–122.

Lee, R. E., Manthey, F. A., & Hall III, C. A. (2004). Content and stability of hexane extractable lipid at various steps of producing macaroni containing ground flaxseed. *Journal of Food Processing and Preservation, 28*(2), 133–144.

Loi, M., Fanelli, F., Liuzzi, V. C., Logrieco, A. F., & Mulè, G. (2017). Mycotoxin biotransformation by native and commercial enzymes: Present and future perspectives. *Toxins, 9*(4), 111.

Lorenz, K. (1981). *Amaranthus hypochondriacus* – Characteristics of the starch and baking potential of the flour. *Starch-Stärke, 33*(5), 149–153.

Manthey, F. A., Sinha, S., Wolf-Hall, C. E., & Hall III, C. A. (2008). Effect of flaxseed flour and packaging on shelf life of refrigerated pasta. *Journal of Food Processing and Preservation, 32*(1), 75–87.

Mastromatteo, M., Chillo, S., Civica, V., Iannetti, M., Suriano, N., & Del Nobile, M. A. (2012). A multistep optimization approach for the production of healthful pasta based on nonconventional flours. *Journal of Food Process Engineering, 35*(4), 601–621.

McIntosh, G. H., Noakes, M., Royle, P. J., & Foster, P. R. (2003). Whole-grain rye and wheat foods and markers of bowel health in overweight middle-aged men. *The American Journal of Clinical Nutrition, 77*(4), 967–974.

Menteş, Ö., Bakkalbaşi, E., & Ercan, R. (2008). Effect of the use of ground flaxseed on quality and chemical composition of bread. *Food Science and Technology International, 14*(4), 299–306.

Mercier, S., Villeneuve, S., Moresoli, C., Mondor, M., Marcos, B., & Power, K. A. (2014). Flaxseed-enriched cereal-based products: A review of the impact of processing conditions. *Comprehensive Reviews in Food Science and Food Safety, 13*(4), 400–412.

Mridula, D., Singh, K., & Barnwal, P. (2013). Development of omega-3 rich energy bar with flaxseed. *Journal of Food Science and Technology, 50*(5), 950–957.

Ndulaka, J., Obasi, N., & Omeire, G. (2014). Production and evaluation of reconstitutable Kunun-Zaki. *Nigerian Food Journal, 32*(2), 66–72.

Olojede, A., Sanni, A., & Banwo, K. (2020a). Rheological, textural and nutritional properties of gluten-free sourdough made with functionally important lactic acid bacteria and yeast from Nigerian sorghum. *LWT, 120*, 108875.

Olojede, A., Sanni, A., Banwo, K., & Adesulu-Dahunsi, A. (2020b). Sensory and antioxidant properties and *in-vitro* digestibility of gluten-free sourdough made with selected starter cultures. *LWT, 129*, 109576.

Onwulata, C., Thomas, A., Cooke, P., Phillips, J., Carvalho, C., Ascheri, J., & Tomasula, P. (2010). Glycemic potential of extruded barley, cassava, corn, and quinoa enriched with whey proteins and cashew pulp. *International Journal of Food Properties, 13*(2), 338–359.

Palavecino, P. M., Ribotta, P. D., León, A. E., & Bustos, M. C. (2019). Gluten-free sorghum pasta: Starch digestibility and antioxidant capacity compared with commercial products. *Journal of the Science of Food and Agriculture, 99*(3), 1351–1357.

Perdon, A. A., Schonauer, S. L., & Poutanen, K. (2020). *Breakfast Cereals and How They Are Made: Raw Materials, Processing, and Production.* Elsevier.

Pezzali, J. G., Suprabha-Raj, A., Siliveru, K., & Aldrich, C. G. (2020). Characterization of white and red sorghum flour and their potential use for production of extrudate crisps. *PLOS ONE, 15*(6), e0234940.

Phiarais, B. P. N., Mauch, A., Schehl, B. D., Zarnkow, M., Gastl, M., Herrmann, M., Zannini, E., & Arendt, E. K. (2010). Processing of a top fermented beer brewed from 100% buckwheat malt with sensory and analytical characterisation. *Journal of the Institute of Brewing, 116*(3), 265–274.

Poutanen, K., Flander, L., & Katina, K. (2009). Sourdough and cereal fermentation in a nutritional perspective. *Food Microbiology, 26*(7), 693–699.

Rabetafika, H. N., Van Remoortel, V., Danthine, S., Paquot, M., & Blecker, C. (2011). Flaxseed proteins: Food uses and health benefits. *International Journal of Food Science and Technology, 46*(2), 221–228.

Rashwan, A. K., Yones, H. A., Karim, N., Taha, E. M., & Chen, W. (2021). Potential processing technologies for developing sorghum-based food products: An update and comprehensive review. *Trends in Food Science and Technology, 110*, 168–182.

Renzetti, S., Dal Bello, F., & Arendt, E. K. (2008). Microstructure, fundamental rheology and baking characteristics of batters and breads from different gluten-free flours treated with a microbial transglutaminase. *Journal of Cereal Science, 48*(1), 33–45.

Robin, F., Dubois, C., Pineau, N., Schuchmann, H. P., & Palzer, S. (2011). Expansion mechanism of extruded foams supplemented with wheat bran. *Journal of Food Engineering, 107*(1), 80–89.

Rodriguez-Sandoval, E., Sandoval, G., & Cortes-Rodríguez, M. (2012). Effect of quinoa and potato flours on the thermomechanical and breadmaking properties of wheat flour. *Brazilian Journal of Chemical Engineering, 29*(3), 503–510.

Romankiewicz, D., Hassoon, W. H., Cacak-Pietrzak, G., Sobczyk, M., Wirkowska-Wojdyła, M., Ceglińska, A., & Dziki, D. (2017). The effect of chia seeds (*Salvia hispanica* L.) addition on quality and nutritional value of wheat bread. *Journal of Food Quality* Volume 2017 Article ID 7352631.

Ronda, F., Abebe, W., Pérez-Quirce, S., & Collar, C. (2015). Suitability of tef varieties in mixed wheat flour bread matrices: A physico-chemical and nutritional approach. *Journal of Cereal Science, 64*, 139–146. https://doi.org/10.1016/j.jcs.2015.05.009.

Ross, A. B., Bruce, S. J., Blondel-Lubrano, A., Oguey-Araymon, S., Beaumont, M., Bourgeois, A., Nielsen-Moennoz, C., Vigo, M., Fay, L.-B., & Kochhar, S. (2011). A whole-grain cereal-rich diet increases plasma betaine, and tends to decrease total and LDL-cholesterol compared with a refined-grain diet in healthy subjects. *British Journal of Nutrition, 105*(10), 1492–1502.

Rühmkorf, C., Rübsam, H., Becker, T., Bork, C., Voiges, K., Mischnick, P., Brandt, M. J., & Vogel, R. F. (2012). Effect of structurally different microbial homoexopolysaccharides on the quality of gluten-free bread. *European Food Research and Technology, 235*(1), 139–146.

Sahin, A. W., Zannini, E., Coffey, A., & Arendt, E. K. (2019). Sugar reduction in bakery products: Current strategies and sourdough technology as a potential novel approach. *Food Research International, 126*, 108583.

Sanz-Penella, J. M., Wronkowska, M., Soral-Smietana, M., & Haros, M. (2013). Effect of whole amaranth flour on bread properties and nutritive value. *LWT – Food Science and Technology, 50*(2), 679–685.

Saturni, L., Ferretti, G., & Bacchetti, T. (2010). The gluten-free diet: Safety and nutritional quality. *Nutrients, 2*(1), 00016–00034.

Schaffer-Lequart, C., Lehmann, U., Ross, A. B., Roger, O., Eldridge, A. L., Ananta, E., Bietry, M.-F., King, L. R., Moroni, A. V., & Srichuwong, S. (2017). Whole grain in manufactured foods: Current use, challenges and the way forward. *Critical Reviews in Food Science and Nutrition, 57*(8), 1562–1568.

Schoenlechner, R., Drausinger, J., Ottenschlaeger, V., Jurackova, K., & Berghofer, E. (2010). Functional properties of gluten-free pasta produced from amaranth, quinoa and buckwheat. *Plant Foods for Human Nutrition, 65*(4), 339–349.

Sindhuja, A., Sudha, M., & Rahim, A. (2005). Effect of incorporation of amaranth flour on the quality of cookies. *European Food Research and Technology, 221*(5), 597–601.

Sissons, M. J., Soh, H. N., & Turner, M. A. (2007). Role of gluten and its components in influencing durum wheat dough properties and spaghetti cooking quality. *Journal of the Science of Food and Agriculture, 87*(10), 1874–1885.

Smerdel, B., Pollak, L., Novotni, D., Čukelj, N., Benković, M., Lušić, D., & Ćurić, D. (2012). Improvement of gluten-free bread quality using transglutaminase, various extruded flours and protein isolates. *Journal of Food and Nutrition Research, 51*(4), 242–253.

Soetan, K., & Oyewole, O. (2009). The need for adequate processing to reduce the anti-nutritional factors in plants used as human foods and animal feeds: A review. *African Journal of Food Science, 3*(9), 223–232.

Sudha, M., Begum, K., Ramasarma, P. (2010). Nutritional characteristics of linseed/flax-seed (*Linum usitatissimum*) and its application in muffin making. *Journal of Texture Studies, 41*(4), 563–578.

Teklehaimanot, W. H., Duodu, K. G., & Emmambux, M. N. (2013). Maize and teff starches modified with stearic acid as potential fat replacer in low calorie mayonnaise-type emulsions. *Starch-Stärke, 65*(9–10), 773–781.

Thakur, P., Kumar, K., & Dhaliwal, H. S. (2021). Nutritional facts, bio-active components and processing aspects of pseudocereals: A comprehensive review. *Food Bioscience*, *42*, 101170.

Torbica, A., Hadnađev, M., & Hadnađev, T. D. (2012). Rice and buckwheat flour characterisation and its relation to cookie quality. *Food Research International*, *48*(1), 277–283.

Vega-Gálvez, A., Miranda, M., Vergara, J., Uribe, E., Puente, L., & Martínez, E. A. (2010). Nutrition facts and functional potential of quinoa (*Chenopodium quinoa* Willd.), an ancient Andean grain: A review. *Journal of the Science of Food and Agriculture*, *90*(15), 2541–2547.

Villeneuve, S., Des Marchais, L.-P., Gauvreau, V., Mercier, S., Do, C. B., & Arcand, Y. (2013). Effect of flaxseed processing on engineering properties and fatty acids profiles of pasta. *Food and Bioproducts Processing*, *91*(3), 183–191.

Wang, Y., & Jian, C. (2022).Sustainable plant-based ingredients as wheat flour substitutes in bread making. *npj Science of Food*, *6*(1), 49.

Williams, P. G. (2014). The benefits of breakfast cereal consumption: A systematic review of the evidence base. *Advances in Nutrition*, *5*(5), 636S–673S.

World Health Organization. (2020). The state of food security and nutrition in the world 2020: transforming food systems for affordable healthy diets (Vol. 2020). Food & Agriculture Organization.

Wronkowska, M., Zielińska, D., Szawara-Nowak, D., Troszyńska, A., & Soral-Śmietana, M. (2010). Antioxidative and reducing capacity, macroelements content and sensorial properties of buckwheat-enhanced gluten-free bread. *International Journal of Food Science and Technology*, *45*(10), 1993–2000.

Yalla, S. R., & Manthey, F. A. (2006). Effect of semolina and absorption level on extrusion of spaghetti containing non-traditional ingredients. *Journal of the Science of Food and Agriculture*, *86*(5), 841–848.

Zebdewos, A., Singh, P., Birhanu, G., Whiting, S., Henry, C., & Kebebu, A. (2015). Formulation of complementary food using amaranth, chickpea and maize improves iron, calcium and zinc content. *African Journal of Food, Agriculture, Nutrition and Development*, *15*(4), 10290–10304.

Zhu, F. (2016). Chemical composition and health effects of Tartary buckwheat. *Food Chemistry*, *203*, 231–245.

Zhu, F. (2018). Chemical composition and food uses of teff (*Eragrostis tef*). *Food Chemistry*, *239*, 402–415.

Chapter 14

Role of Pseudocereals in the Management of Diseases

Subhankar Das, Manjula Ishwara Kalyani

14.1 Introduction

Nowadays, people are more concerned about the negative impact and risk of developing diseases due to an unhealthy lifestyle. Therefore, there has been a shift towards healthier food consumption and lifestyle changes. A healthy lifestyle not only reduces the chances of disease but also contributes to a sustainable environment. As a result, there is an increased demand for healthy, nutritious food. Pseudocereals provide an excellent substitute for true cereals to fulfill the nutritional requirements of consumers. Pseudocereals belong to the families of dicotyledonous plants but have the characteristics of a true cereal. From a botanical perspective, pseudocereals are not true cereals and belong to non-grass plant families. However, they are considered to be the perfect substitute for cereals and are consumed in the same manner as conventional cereals (1, 2). As a result of the presence of several biologically active compounds, such as phenolic acids, trace elements, flavonoids, vitamins, minerals, essential fatty acids, fibers, etc., the nutritional profile of pseudocereals provides a healthy alternative as compared to conventional cereals. Several factors, such as malnutrition, food shortages, and climate change, have urged scientists to find alternative crops that can ensure food security. Therefore, the need of the hour is to diversify from the traditional crops, such as wheat, rice, and corn that currently meet around 50% of calorie demands globally. Pseudocereals, which are denoted as "sub-exploited" or "under-utilized" foods, are the perfect

DOI: 10.1201/9781003325277-14

candidates for ensuring nutritional security due to their agronomic traits and thera-peutic benefits (3–6). Pseudocereals are considered potential crops for the future because of their genetic variation, which makes them adaptable to different climatic conditions, ranging from tropical to temperate climatic zones (7, 8). Furthermore, the consumption of pseudocereals is considered safe in terms of their toxicological profile due to the presence of fewer anti-nutritional components as well as fewer inherent toxic metabolites. For instance, saponins are considered to have a negative impact on the intestinal tract (9). Some of the popular pseudocereals are *Fagopyrum* spp. (buckwheat), *Chenopodium quinoa* (quinoa), *Amaranthus* spp. (amaranth) and *Salvia hispanica* (chia) (10–12). These gluten-free edible seeds are packed with ben-eficial biologically active components (13). Due to their gluten-free nature, pseu-docereals have gained popularity among individuals with celiac disease. Celiac disease, which is also termed gluten-sensitive enteropathy, is triggered in geneti-cally susceptible people by the intake of gluten-containing grains such as wheat, rye, oats, and barley (14, 15). Apart from celiac disease, pseudocereals with their nutritional composition protect an individual from several other diseases, which include diabetes, cancer, inflammation, microbial infections, and cardiovascular diseases. In addition, pseudocereals serve as prebiotic agents for probiotic microor-ganisms. Probiotic bacteria are beneficial bacteria that exhibit several therapeutic impacts, including modulation of the immune system, reduction of blood cho-lesterol, lactose intolerance, and Crohn's disease, as well as relief from diarrhea (16, 17). Pseudocereals have an astounding ability to manage disease due to their diverse bioactive compounds, which makes them perfect candidates to alleviate various symptoms associated with diseases (8, 10, 18, 19). This chapter describes the important types of pseudocereals and their bioactive compounds. Furthermore, we will discuss the disease-managing ability of the bioactive ingredients present in pseudocereals.

14.2 A Brief Overview of Nutritional Aspects of Pseudocereals

According to UN Food and Agriculture Organization (FAO) statistics, hun-ger have impacted approximately 828 million people globally in 2021. This is an increase of around 46 million over the previous year, 2020. Furthermore, malnutrition has endangered the lives of millions of children and increased their risk of death. It has been anticipated that, by 2030, there will be up to 670 million people suffering from hunger (20). In addition, climate change has also impacted the production of various crops across the globe. Globally, overcoming malnutrition as well as ensuring food and nutritional security have become major concerns. Therefore, the dependency on conventional crops has to be diversified towards newer crops. Pseudocereals, with their nutrient-dense

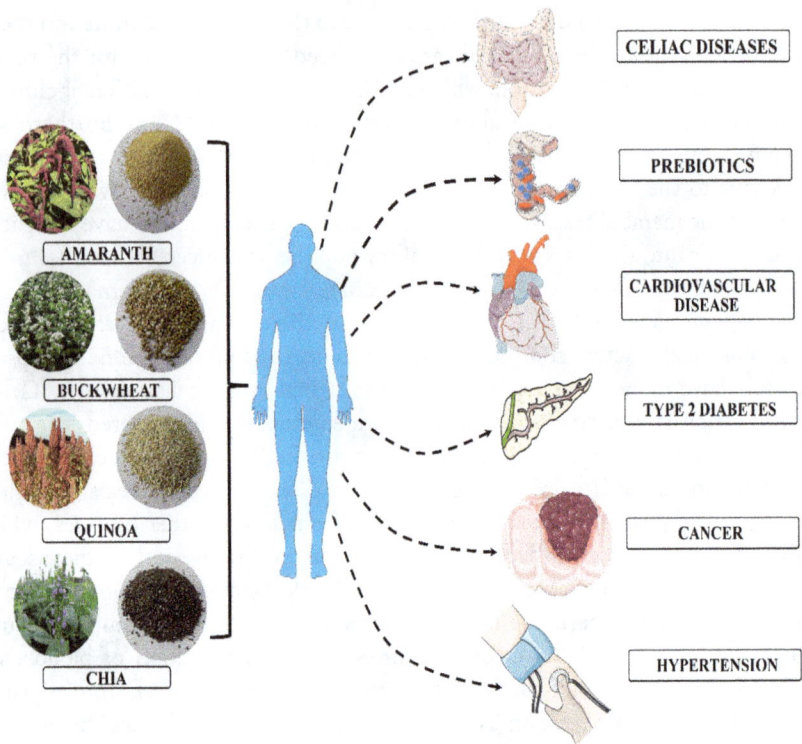

Figure 14.1 **Schematic illustration of disease-managing characteristics of pseudocereals. This Figure was partially built with the aid of Servier Medical Art.**

biocomponents, have emerged as an important substitute for traditional cereals. Furthermore, pseudocereals are combating nutritional challenges and have proven to be a suitable food source to ensure a healthy and nutritional diet. The daily intake of pseudocereals in the diet lowers the chances of getting diseases such as cardiovascular diseases, peptic ulcers, diabetes, various cancers, hypertension, osteoporosis, etc. (as shown in Figure 14.1) (6, 21, 22). Pseudocereals are also known to have a plethora of various biologically active compounds, such as carbohydrates, proteins, vitamins, lipids, minerals, and various other biologically active compounds. In addition, pseudocereals also contain various phytochemicals such as flavonoids, polysterols, phytosteroids, polyphenols, rutin, etc. (23–27). Furthermore, pseudocereals, which are also known as "orphan crops", can withstand adverse environmental conditions, which include drought, extreme temperate zones, and low soil fertility, thereby ensuring food and nutritional security (28–30). Therefore, pseudocereals are denoted as future crops. Here, we will discuss some of the most popular pseudocereals, which include amaranthus, quinoa, buckwheat, and chia, along with their diverse bioactive compounds.

14.2.1 Amaranth

Amaranth is a dicotyledonous plant of the Amaranthaceae family. The name 'Amaranthus' is taken from the Greek word "Anthos" (flower), which denotes "eternal or unfading". Amaranthus are rapidly growing annual pseudocereal grains with lentil-shaped seeds that are around 1 mm in diameter (2, 10, 31). Although amaranth can grow in all types of soils, it grows best in loamy and loamy sandy soils that consist of a high amount of organic matter and have adequate drainage. Despite the crop's ability to flourish on soil with a pH of 8.5, pH 6–7 is regarded as ideal for the growth of amaranth. Nevertheless, research findings suggest that achieving optimal nutrient accumulation in amaranth plants necessitates appropriate mineral levels and pH in the soil (32). Apart from its edible seeds, amaranth leaves can also be consumed as vegetables. The genus *Amaranthus* is one of the oldest cultivated pseudocereals. The most commonly cultivated amaranth species are *Amaranthus caudatus, Amaranthus paniculatus, Amaranthus cruentus,* and *Amaranthus hypochondriacus* (33–38). This could be because *Amaranthus* is a C_4 plant with anatomical features known for their role in an efficient photosynthetic pathway, with efficient CO_2 utilization across a broad temperature range from 25 to 40°C, and low water requirements when compared with true cereals. Because of the continuous challenge of climate change, amaranth has become an ideal alternative to conventional cereals (11, 36). Amaranth is considered to be a super food due to its nutritional as well as therapeutic values. Amaranth is a gluten-free pseudocereal that contains a significant quantity of important amino acids, calcium, tocopherols, vitamins, minerals, fibers, and unsaturated fats such as omega-4, omega-6, etc. (6, 39). Due to the gluten-free nature of amaranth grains, it is an appropriate diet for people who suffer from celiac disease (40). The protein found in the amaranth seed is dispersed predominantly, approximately 65%, in the embryo region, which is likewise high in lysine. Based on the solubility of the proteins present in amaranthus, there are albumins, globulins, glutelins, and prolamins (11). Besides their high protein content, amaranth grains include a variety of important amino acids, including leucine (Leu), arginine (Arg), threonine (Thr), methionine (Met), isoleucine (Ile), tryptophan (Trp), valine (Val), phenylalanine (Phe), histidine (His), and lysine (Lys). Moreover, amaranth grains also contain unsaturated fatty acids such as linoleic acid, also called omega-6 fatty acid, and α-linolenic acid, also known as omega-3 fatty acid, as well as oleic, squalene, palmitic, and stearic acids. Furthermore, ferulic acid, caffeic acid, and *p*-hydroxybenzoic acid are the key phenolic chemicals found in amaranth seed. In addition, amaranth has also been shown to regulate blood lipid levels (35, 41, 42). Amaranth, therefore, appears to be a healthy substitute for conventional cereals, having the potential to satisfy an individual's dietary needs.

14.2.2 Quinoa

Quinoa is known to be one of the oldest grown plant species, having originated 7000 years earlier in the Andean region of South America and being reported to have been consumed by Andean cultures (1, 37, 43). Quinoa is a dicotyledonous C_3 plant, and, like amaranth, the leaves of quinoa can also be consumed along with its grains. Quinoa seeds are small and round and their diameter ranges between 1.5 mm and 2.5 mm, along with seed colors ranging from white to purple, yellow, black, or brown, and the episperm has four layers. Quinoa has remarkable adaptability to adverse environmental conditions, such as extreme temperatures ranging from –4°C to 38°C, relative humidity levels between 44% and 88%, water scarcity, and high levels of salinity and acidity. The nutritional components of quinoa are known to contain oil, protein (13.81 to 21.9%), and dietary fiber (13.4%), among which insoluble fibers represent 11.0% and soluble fibers 2.4%. The fatty acids content of quinoa is mostly made up of linoleic acid, as well as oleic and palmitic acids. It is also high in linoleic acid and the omega-3 fatty acid, linolenic acid, as well as antioxidants such as α and γ tocopherols (11, 32, 33, 43–45). Quinoa is also high in fiber, minerals, which include calcium, potassium, manganese, copper, zinc, iron, phosphorus, and vitamins E, B_6, and C. Quinoa contains an excellent balance of amino acids, including leucine, isoleucine, phenylalanine, histidine, tryptophan, threonine, valine, lysine, and methionine. In addition, quinoa proteins are made up of around 37% 11S globulin, 35% 2S albumin, and 0.5-7% prolamins. Furthermore, it has been reported that quinoa is an excellent source of carotenoids (4.6–18.1 mg/kg dw), which belong to the subfamily of terpenoids (8, 46, 47). The quinoa grains are also rich in other different bioactive compounds, including flavonoids (e.g., kaempferol, quercetin), flavanols, polyphenols, phytosterols, squalene, and fagopyritols (41, 45, 48). Also, quinoa contains saponins, which are regarded as antinutritional compounds. The amount of saponin varies, ranging from 0.01% to 4.65%. Furthermore, the bitter-tasting saponins are mostly found in the seed's outer layer (37, 41). Depending on the saponin level, quinoa comes in both sweet and bitter variants; if the saponin level is less than 0.11% the quinoa will be a sweet variant (38, 49). As a result, eating quinoa can satisfy an individual's nutritional requirements.

14.2.3 Buckwheat

Buckwheat belongs to the Polygonaceae family and the genus *Fagopyrum*. It is a popular food in arid and cold areas around the world. Buckwheat is a dicotyledonous C_3 plant and has triangular-shaped seeds that are approximately 6 to 9 mm in length. Buckwheat embryos are smaller than the seeds of amaranth and quinoa. Moreover, the embryo of buckwheat is located in the distal region of the kernel. Buckwheat can grow in non-fertile soil and can even tolerate acidic soils (as low as pH 5). It is a suitable crop for cold desert regions. However, buckwheat is affected by the frost during its germination (10, 11, 50–52). The most common species of buckwheat used are *Fagopyrum esculentum*, *Fagopyrum tataricum* and *Fagopyrum divobotrys*. Buckwheat fruit is also rich in proteins, lipids, vitamins, fiber, carbohydrates, and minerals, including zinc, copper, calcium, magnesium, selenium, manganese, potassium, zinc, iron, phosphorus, and manganese (11, 14, 50, 51). In addition, buckwheat has a variety of attractive nutritional components, such as antioxidants, dietary fiber, flavonoids, fagopyrins, and thiamin-binding proteins. Moreover, when compared with cereals, buckwheat has greater quantities of lysine, arginine, and aspartic acid. Also, its dietary fiber content is greater than that of amaranth and quinoa. Furthermore, buckwheat contains a variety of phenolic and flavonoid chemicals, including rutin, hyperin, quercetin, and quercitrin. Buckwheat is also known to be a rich source of vitamins B and C, tocopherols, and polyunsaturated fatty acids (1, 10, 50, 53). It is also a suitable alternative to cereals for individuals suffering from celiac disease due to its gluten-free nature and a healthy substitute for cereals (54).

14.2.4 Chia

Chia (*Salvia hispanica* L.) is a member of the Lamiaceae family and an annual plant. The seed of the chia is quasi-oval shaped and ranges in size from 1 mm to 2 mm. The exterior layer of the chia seed is smooth and consists of a testa with a shiny appearance. The outer color of the seeds ranges from brown, grey, to white. Chia consists of mucilage that is situated in the inner portion of the epidermal region of the seed. The seeds, when soaked in water, instantly swell up, causing a rupture in the primary layer of the cell, exposing the epidermis; as a result, the seed is subsequently surrounded by mucilaginous gel (55). Chia seeds are a nutritionally rich superfood that includes high concentrations of omega-3 fatty acids, antioxidants, proteins, amino acids such as arginine, glutamic acid, aspartic acid, dietary fiber, and carbohydrates. Chia seed contains a large number of essential amino acids as well as sulfur-containing amino acids when compared with other pseudocereals (56, 57). Furthermore, chia seeds have a high protein level of around 20%. Therefore, chia seeds are regarded as a superfood and exhibit nutritional superiority over true cereals. The mineral components of chia seeds include sodium, manganese, phosphorus, potassium, and calcium (58, 59). Chia seeds are also an oilseed and have a high known level of α-linolenic fatty acid, making them an important nutritional source of omega-3 fatty acids (60, 61). Furthermore, its gluten-free properties have attracted individuals suffering from celiac disease, helping them to adhere to an appropriate gluten-free diet while, at the same time, providing an individual with all the necessary nutrients (62).

14.3 Bioactive Compounds and Nutritional Components Present in Pseudocereals and Their Effects on Human Diseases

Food is an essential component for the proper growth and development of the human body. Different foods have a varied range of nutritional components, which comprise micronutrients and macronutrients. In addition, a nutritious diet not only promotes appropriate growth and development but also lowers the chances of acquiring chronic diseases, which include diabetes, osteoporosis, cardiovascular disease, various types of cancer, obesity, dental disease, etc. Currently, the food options and nutritional security across the globe have shrunk to very few crops such as maize, rice, and wheat. These crops meet more than 50% of the world's demand for calories (3, 33). According to reports, individuals who consume who-legrains have a decreased risk of diseases and death when compared with people who consume fewer wholegrains. Therefore, the numerous associated benefits of eating nutritious foods have sparked a burgeoning inclination towards adopting a health-conscious way of life. (21, 28). Due to the appealing nutritional content of pseudocereals, they have attracted attention and have also extended the range of available crops that provide for the nutritional requirements of an individual (22). Pseudocereals are dicotyledonous plants that have become an essential part of a balanced diet due to their nutritional and neutraceutical benefits. However, a healthy diet, along with other lifestyle changes, contributes to better management of diseases, reducing symptoms, and boosting quality of life (35). Pseudocereals are edible seeds and the most popular pseudocereals comprise plants such as amaranth (*Amaranthus* spp.), buckwheat (*Fagopyrum* spp.), chia (*Salvia hispanica*) and quinoa (*Chenopodium quinoa*).

14.3.1 Gluten-free Nature of Pseudocereals

Celiac disease, often referred to as "gluten-sensitive enteropathy," is triggered upon consumption of gluten in the diet and occurs in genetically predisposed humans. Celiac disease is reported to be one of the most prevalent diseases with long-lasting symptoms, affecting around 1% of the population. Celiac disease is an autoim-mune disease that is linked to genetic factors (HLADQ2 and HLA-DQ8) that affect the mucosa of the small intestine in individuals who are genetically predis-posed. The immune response gets activated when gluten peptides bind to lym-phocyte antigens HLA-DQ2 and HLA-DQ8 and subsequently present to the T cells in the lamina propria. T cells discover gluten-sensitive CD4+ helper cells, and the immune response begins, followed by various clinical symptoms (10, 18, 63). Celiac disease is known to be associated with a multitude of other diseases, such as enamel hypoplasia, dermatitis herpetiformis, anemia caused by insufficient iron, osteoporosis, arthritis, various types of cancer, autoimmune disorders, and so on (51). Gluten causes continual mucosal inflammation in individuals with celiac disease, eventually resulting in the loss of absorptive properties of the villi and crypt hyperplasia. Gluten is a protein complex comprised of prolamins and gluteins and is generally found in crops such as barley, rye, oats and wheat. The prolamin protein fraction has been given a specific name based on the presence of the protein complexes found in various portions of cereal: in barley (hordeins), in oats (avenins), in wheat (gliadins), and in rye (secalin). Individuals with celiac disease have incomplete digestion of these proteins, resulting in the formation of peptides that produce a class of polypeptides rich in proline and glycine and trigger a T-cell-mediated immunological response. On the contrary, pseudocereals are composed of a high percentage of albumins and globulins, with very low contents of prolamins, which have nutritional consequences for celiac disease patients. Globulins and albumins have lower levels of glutamic acid and proline than prolamins but contain higher levels of important amino acids like lysine. In addition, the consumption of gluten in the dietary regimen also elicits clinical manifestations, including inflammation, atrophy, and hyperplasia in the small intestine among individuals diagnosed with celiac disease. Moreover, gluten can also be found in foods such as fish, milk, and meat (14, 64, 65). The strategy to prevent celiac disease is to maintain a stringent gluten-free diet for the rest of one's life. Furthermore, a strictly gluten-free diet improves clinical symptoms, including intestinal mucosal healing (14). However, following a stringent gluten-free diet puts an individual with celiac disease at risk of nutritional deficiency (40). Studies have also revealed that a stringent gluten-free diet leads to low nutrient absorption of essential nutrients, which include carbohydrates, fats, and proteins. Therefore, pseudocereals serve as an excellent substitute for the gluten diet. Pseudocereals, which are rich in dietary fiber, vitamins, minerals, protein, and starch, fulfill the dietary requirements of individuals suffering from celiac disease. Furthermore, pseudocereals are rich in protein with balanced amino acids, which fortify the gluten-free diet. Quinoa has a protein concentration of >23%, with a high bioavailability of 83% and a balanced amount of all essential amino acids. However, quinoa has been reported to have a low amount of prolamins ($\leq$7%) (66). Quinoa proteins are made up of 37% 11S globulin fraction, 35% 2S albumin fraction, and 0.5–7% of the prolamin fraction (47). The seeds of amaranth also contain a high amount of protein, comprised mainly of 11S and 7S globulins, albumins, globulins, and glutelins. However, amaranth has a low quantity of prolamins (2–3%) (65, 67–69). Furthermore, the amaranth proteins are enhanced with arginine, tryptophan, lysine, and sulfur-containing amino acids in higher amounts than other crops. In addition amaranth grain contains high-quality protein (13% to 19%) that is 90% digestible and has an exceptional amino acid equilibrium, which is superior to that of cereals and certain legumes. Amaranth proteins contain between 4.9–6.1g/100 g of protein, which is comparable to soy and are in low concentration in cereals (70). Studies have further shown that amaranth and quinoa have a higher digestible protein content and nutrient balance than casein (15, 65). Similarly, buckwheat has been identified as an excellent source of protein of high biological value that is devoid of gluten. Also, when compared with cereals, buckwheat contains a proper balance of amino acids, including a high percentage of lysine and arginine (67). Furthermore, germinating or sprouting seeds of buckwheat reduce protease inhibitor activity, leading to high protein digestion ability (33). Buckwheat contains 11–19% protein, with 55% found in the embryo, 35% in the endosperm, and the remainder in the shell. The important protein found in buckwheat is 13S globulin, a unique and rare plant seed storage protein (67).

Chia seeds are also devoid of gluten, which makes them an excellent food for celiac patients. The main protein fraction in chia seeds comprises 52% of globulin, which includes 11S and 7S proteins. In addition, chia seeds have a higher concentration of glutamic acid (Glu), aspartic acid (Asp), arginine (Arg), lysine (Lys), cysteine (Cys), and methionine (Met) than other cereal grains. Buckwheat and chia are pseudocereals that have a high concentration of various important amino acids (56, 57). In addition to being an excellent source of protein, pseudocereals include a variety of other nutritional elements that meet the requirements of celiac patients to follow a proper gluten-free diet. Pseudocereals like amaranth have been demonstrated to be rich in riboflavin, vitamin E, and minerals, including calcium (Ca), magnesium (Mg), and iron (Fe) (15). Furthermore, ferulic acid, caffeic acid, and p-hydroxybenzoic acid are the principal phenolic compounds present in amaranth seeds (35). Furthermore, amaranth seeds are rich in dietary fibers at around 4–8%, compared with cereals at 2% (67). The seed also contains 60 to 65% carbohydrates, of which 57% is starch (33). Similarly, quinoa seeds are rich in bioactive compounds like flavonoids such as kaempferol and quercetin (35). Quinoa contains a high concentration of vitamin E, which ensures the stability of its lipids during storage (47). In addition, quinoa seeds have been reported to contain phenolic compounds that include ecdysteroids, phenolic acids, and various phytohormones (34, 71). Quinoa seeds have been reported to meet up to 55% of a person's daily phosphorus and magnesium requirements (48). Beside this, quinoa seeds also contain calcium, zinc, and iron, in addition to phytochemicals such as saponins, phytosterols, and phytoecdysteroids (72). Quinoa has a high dietary fiber content that includes both soluble and insoluble fibers. It also contains fatty acids, which include linoleic acid as well as linolenic acid (44). Buckwheat is also rich in different nutritionally active compounds like dietary fiber, of which 20–30% are soluble dietary fiber, and also includes vitamin B and vitamin E. The dietary fibers present in the buckwheat grains are higher in concentration than in amaranth and quinoa. Buckwheat serves as an important source of the flavonoid rutin and has pharmaceutical importance, protecting against gastric damage (10, 67, 73). Buckwheat is also rich in flavonoids, phytosterols, flavones, fagopyrins, and thiamin-binding proteins, which have been shown to protect an individual against a variety of chronic diseases (33). The major phenolic components of buckwheat seeds are quercetin, apigenin, and luteolin (74). Chia also serves as an important part of a celiac individual's dietary fiber requirement, which makes up around 30% of its overall weight. Chia also includes vitamins, fatty acids, and a variety of minerals, such as phosphorus, calcium, sodium, and manganese (59). In terms of fats, chia seeds contain 30 to 40% fat, including around 68% of the omega-3 fatty acid α-linolenic acid (75). Amaranthus, chia, buckwheat, and quinoa are thus ideal options for a gluten-free diet with enhanced nutrition that helps an individual with celiac disease fulfill their need for all essential nutrients.

14.3.2 Prebiotic Role of Pseudocereals

The gut microbiome is made up of beneficial bacteria that play a pivotal role in maintaining proper gut health. The term "probiotic" is obtained from the Greek word meaning "for life" (76). Probiotics are live bacteria that impart health benefits when taken by an individual in sufficient proportions. The beneficial bacteria in the human body colonize the gut microbiome right after birth, and the gut microbiome contains a diverse range of microbes, approximately 1000 species (77, 78). Probiotics further enhance the intestinal microbiome and protect the gastrointestinal tract from any infections, inflammatory bowel disease, lactose intolerance, Crohn's disease, antibiotic-associated diarrhea, and high blood cholesterol, and further boost the immune system (79). Prebiotics, on the other hand, are the nutritional support provided to the gut microbiome. Prebiotics are poorly digestible carbohydrates that include sugar polyols, polysaccharides, oligosaccharides, resistant starches, and dietary fiber that provide nourishment as well as enhance the growth of gut microbiota. There have been several studies showing the beneficial role of prebiotics in human health (76, 80–82). Pseudocereals contain certain carbohydrates that serve as prebiotics and further assist in enhancing the probiotic bacterial strains in the digestive system (16). Furthermore, the combination of fermented pseudocereals with probiotic bacteria increases the nutritional content of the food, along with its sensory value and shelf life. The inclusion of pseudocereal in fermented food serves as a vehicle to deliver functional compounds that include minerals, antioxidants, vitamins, probiotics, etc. (81, 82). In most developed countries, fermented pseudocereals are ingested in the form of beverages, porridges, etc., resulting in the indirect ingestion of probiotics. In the study reported by Kocková et al. (2013) pseudocereal porridges, which included amaranth and buckwheat, were fermented by the probiotic bacterial strain *Lactobacillus rhamnosus* GG, whereby the substrate remained stable for a storage period of 21 days, thus ensuring a suitable substrate for probiotics (17, 82). In addition, adding up to 3% quinoa flour has been reported to not affect the bacterial strains *Bifidobacterium animalis* and *Lactobacillus acidophilus* in terms of the bacterial count or viability of the strains throughout the storage period. According to Soltani et al. (2017), fermented probiotic yogurts, containing the bacterial strain *L. rhamnosus* GR-1 and added with quinoa along with other cereals, have shown satisfactory results and had no negative impact on the development and survival of the bacterial strain *L. rhamnosus* GR-1. So, quinoa can be an excellent additive that can be incorporated as a prebiotic in diets (83). Since it is well known that pseudocereals include dietary fibers that are comprised of soluble and insoluble fibers, the study also suggests that soluble fibers present in the colon selectively encourage the growth of bacterial strains such as *lactobacilli* and *bifidobacteria* (44, 67, 84). Furthermore, studies have shown that quinoa and amaranth are ideal substrates because of the availability of certain biologically active compounds that have probiotic effects on the gut flora (68). The incorporation of 30% quinoa, along with 70% soy, in a beverage showed increased viability of *Lactobacillus casei*. Further incorporation of quinoa produced intriguing results, as the pseudocereal exhibited reduced fermentation time and increased viscosity. Furthermore, it was able to supply enough protein, carbohydrate, and lipid content, in addition to having a reduced calorific value than a beverage containing solely soy extract (85). Subsequent investigation revealed that mucilage along with the soluble protein fractions of chia and flax seed work synergistically to increase the viability rate of the probiotic bacterial strains *Bifidobacterium infantis* and *Lactobacillus plantarum* when spray-dried and stored at 4°C. The probiotic bacterial strains were encapsulated in a solution that included chia seed, flax seed, and maltodextrin, in which the combined effect of mucilage and soluble protein enhanced the performance of the probiotic (86). Furthermore, yogurt supplemented with chia seed extract has been shown to reduce fermentation time while increasing lactic acid bacteria count. It has further been reported that chia seed extract imparts beneficial effects on colon health. It also enhances the physiochemical characteristics of the bacterial cell by increasing water-holding capacity, viscosity, color, and antioxidant capabilities (87). As a result, pseudocereals are an excellent bacterial substrate that can help treat and control a variety of chronic conditions.

14.3.3 Antidiabetic Role of Pseudocereals in Type 2 Diabetes

Diabetes is a chronic disorder that develops when the pancreas is unable to produce the right amount of insulin or when one's body is unable to properly respond to the the insulin that is released. Insulin is an important peptide hormone synthesized by the β-cells of the pancreatic islets of Langerhans and is responsible for keeping blood glucose levels normal (84, 88). Type 2 diabetes is reported to be the most prevalent form of diabetes and accounts for around 95% of the total number of people suffering from diabetes. Moreover, obesity and insufficient physical activity are influencing factors in type 2 diabetes. Furthermore, the release of large amounts of free fatty acids by adipose tissue reduces insulin sensitivity in muscle, liver, and fat, resulting in increased glucose levels, and insulin resistance, in type 2 diabetes (89–92). Maintaining a stringent diet and following a healthy lifestyle are the cornerstones of managing type 2 diabetes. Furthermore, type 2 diabetes is also characterized by a rise in the glycemic index, resistance to insulin, and a decrease in the function of the pancreas which leads to low insulin release. Hence, the adoption of a low-glycemic-index diet has been found to correlate with enhanced insulin resistance. Pseudocereals, including amaranth, buckwheat, quinoa, and chia, are an appropriate nutritious diet for type 2 diabetes management (41, 93–95). However, it is also possible to manage type 2 diabetes in its early stage by preventing the ability of the intestinal region to digest and absorb glucose from complex carbohydrates. Antidiabetic drugs, such as synthetic α-glucosidases and α-amylase, help in decreasing carbohydrate absorption by inhibiting the absorption of glucose, subsequently lowering the glycemic index. Nonetheless, the ability of drugs from the α-glucosidase family to manage type 2 diabetes comes with several side effects, including gastrointestinal symptoms that include abdominal pain, diarrhea, and bloating. Other synthetic hypoglycemic drugs include biguanides and sulfonylurea, which also exhibit various negative effects, such as abnormal colon function. Plant-based glycosidase inhibitors, on the other hand, are more potent and have no side effects. Therefore, consumption of whole cereals is encouraged for diabetic individuals (96–98). Chia seeds, for example, contain a significant amount of fiber that can absorb water 15 times more than their weight. It has further been revealed that chia seeds are an excellent dietary fiber source, containing 40% fiber. Also, chia seeds contain 5% soluble fibers that appear as clear mucilage when they come in contact with water. The increased fiber intake aids type 2 diabetes by slowing digestion and, as a result, the rate of glucose release. Furthermore, it also enhances peristaltic movement in the intestine and also helps in lowering plasma cholesterol (59, 95). Furthermore, studies have shown that cooked quinoa can inhibit α-glucosidases *in vitro* while not affecting pancreatic α-amylase. Thus, quinoa provides minimal glucose absorption in the intestinal region with no negative side effects. The inhibitory effects of quinoa and purple maize might be due to the protocatechuic acid components and derivatives of quercetin (93). However, the report suggests that some quinoa cultivars, such as red- and black-seeded ones, contain a high concentration of anthocyanins such as cyanidin, cyanidin 3-*O*-glucosyl-rutinoside, and cyanidin 3-*O*-sambubioside, and other cyanidin 5-*O*-glucosides. It has also been shown that grains high in anthocyanin compounds can lower starch hydrolysis due to their inhibitory characteristics against carbohydrate-digesting enzymes, and hence aid in managing type 2 diabetes as well as obesity (8). After thermal processing, 2-day-old buckwheat sprouts and 10-day-old seedling extracts showed α-amylase and α-glucosidase inhibitory effects, as well as significantly increased total phenolics and antioxidant activity (98). In addition, buckwheat seeds comprise soluble carbohydrates, such as fagopyritols, which are an important source of D-chiro-inositol compounds that have positive effects on controlling the glycemic level in people with type 2 diabetes (35). Buckwheat has been reported to contain the bioflavonoid rutin, which helps in improving glucose homeostasis as well as recovery of retinal function, and subsequently contributes to lowering diabetic symptoms (50). Methanolic extracts from both raw and processed amaranth, as well as finger millets, show inhibitory effects on key enzymes such as α-amylase and α-glucosidase that are related to type 2 diabetes (99). Amaranth species, as previously reported, have beneficial effects in regulating type 2 diabetes due to their ability to inhibit α-amylase (93, 100). Buckwheat contains a majority of resistant starch with slow digestibility, thereby exhibiting a low glycemic index compared with wheat. The carbohydrates rich in fagopyrin and fagopyritols have been shown to have a positive impact on individuals with type 2 diabetes (14, 33, 67). Quinoa, which contains stigmasterol, β-sitosterol, gallic acid, caffeic acid, and ferulic acid, also has preventive action against diabetes (101). Thus, pseudocereals have a greater potential to be included in normal diets, ultimately assisting in the management of glucose levels and the prevention of type 2 diabetes without the risks associated with synthetic medications.

14.3.4 Anticancer Role of Pseudocereals

Cancer is a complicated disease that involves physiological changes in the cells that eventually lead to malignant tumors. Cancer cells are characterized by the unregulated growth of abnormal cells known as neoplasia. The root cause of mortality in cancer patients is the invasion of tumor cells into the surrounding tissues and various organs (102). Also the combination of diverse environmental stressors and poor lifestyle practises contributes to the oxidative stress and synthesis of free radicals specifically reactive oxygen species (ROS) and reactive nitrogen species (RNS) that are responsible for causing tissue damage in humans. Also, it is widely acknowledged that the occurrence of cancer is often linked to the presence of oxidative stress generated by free radicals (103–105). Flavonoids, phenolic acids, trace minerals, fatty acids, and vitamins are abundant in pseudocereals and have been related to improved human health. Plant products exhibit several properties that decrease the chances of developing a range of chronic illnesses, including atherosclerosis as well as cancer, and thereby their consumption has been greatly encouraged. These positive benefits have been associated with the presence of antioxidants, which play essential roles in the suppression of free radical accumulation and oxidative chain reactions inside tissues and membranes.

For instance, plant polyphenols comprise a diverse group of natural antioxidants and are among the major candidates linked to cancer protection and prevention (28, 35, 93). Buckwheat has been reported to be the richest source of various polyphenols, which include quercetin, apigenin, rutin, isoquercitrin, catechin, myricetin, and luteolin. Similarly, quinoa seeds are rich in phenolic compounds such as glycosides of quercetin, caffeic acid, kaempferol, ferulic acid, and vanillic acid. Amaranth, on the other hand, contains phenolic compounds such as caffeic acid, *p*-hydroxybenzoic acid, ferulic acid, sesamin, tyrosol, and cardol (8, 14, 50, 106). In chia seeds, the polyphenols that are present include chlorogenic acid, caffeic acid, myricetin, quercetin, and kaempferol (95). In addition, polyphenol-containing antioxidants have been shown to have several health advantages. These antioxidants, which are also known as free radical scavengers, protect the cells from any oxidative stress that may otherwise lead to oxidative stress-linked diseases like cancer (6, 107). Furthermore, the polyphenol quercetin has been found to have the ability to inhibit cancer cell proliferation by regulating particular signaling pathways, such as decreasing oncogene expression, triggering apoptosis in malignant cells, blocking angiogenesis, etc. Also, it can strongly suppress HepG2 cell proliferation and induce apoptosis, potentially through the involvement of cyclin D1 regulation (108). On the other hand, buckwheat's positive health benefits are attributable in part to its high concentration of bioflavonoids such as rutin. Rutin imparts beneficial biological as well as physiological functions, such as protecting the stomach from any damage, providing protection against UV radiation and X-rays, and reducing oxidative stress (50, 53, 67). Buckwheat further includes trace elements such as selenium at a concentration of 0.0099–0.1208 mg/g, which protects against cancer and Acquired Immune Deficiency Syndrome (AIDS) (33). Biopeptides have wide functionality that exhibits significant effects on human health as a result of their unique amino acid compositions, different peptides, and their reactions towards the body. Studies have revealed the effective properties of these peptides in various diseases, including antihypertensive, anticholesterolemic, antioxidant, anti-inflammatory, anticancer, antimicrobial, and immunomodulatory. Maldonado-Cervantes et al. (2010) reported in their study that the amaranth-derived peptide lunasin, which is composed of 43 amino acids, has the potential to prevent cancer due to its ability to penetrate into the nucleus of the cell and, subsequently, prevent fibroblast cancer cell transformation. Further, the amaranth-derived lunasin-like peptide has been reported to inhibit the acetylation of histones H3 and H4. Acetylation of histones is one of the epigenetic mechanisms proposed for the lunasin peptide's cancer-preventive capabilities, as well as its significance in chromatin modification in cell cycle regulation and in tumor suppressor ability in carcinogenesis (109, 110). A study conducted by Vilcacundo et al. (2018) revealed the antioxidant activity of a peptide fraction of quinoa of less than 5 kDa, while the anticancer potential of quinoa is shown by the peptide fraction greater than 5 kDa (111). Jayaprakasm et al. (2004) discovered the anticancer activities of lunasin against mammalian cell culture models as well as skin cancer animal models (112, 113). Chia seed, because of the large amount of insoluble fiber present, increases intestinal movement and thereby protects an individual from colon cancer and obesity (59). Furthermore, based on the study reported by Ramzi et al. 2017, whereby oil derived from chia seeds exhibited *in vitro* cancer cytotoxic properties, chia seed oil demonstrated antiproliferative activity against human lymphoblastic leukemic cell lines, HeLa, and MCF-7 cells (114). Squalene is an organic compound found in pseudocereals and is a polyunsaturated lipid present on the skin surface that acts as an emollient and antioxidant, with antitumor potential. Squalene also has various other beneficial roles that include chemopreventive properties, and it also serves as a chemotherapeutic agent (115). As a result, pseudocereals play a vital role in both controlling and suppressing various types of cancers and their clinical symptoms.

14.3.5 Pseudocereal Role in Cardiovascular Disease

Cardiovascular disease is one of the leading causes of mortality worldwide, claiming more lives than all other diseases combined, including cancer and respiratory ailments. Cardiovascular disease killed an estimated 17.9 million people in the year 2019, accounting for 32% of all global mortality. Moreover, 85% of these fatalities were caused by heart attacks or strokes. While age is a major risk factor for developing cardiovascular disease, several other risk factors are also associated with it such as smoking, a sedentary lifestyle, dyslipidemia, hypertension, diabetes, and abdominal obesity (35, 116, 117). Pseudocereals have unique blends of various phytochemicals that serve as cardioprotective agents. Pseudocereals have been reported to contain a high concentration of lipids compared with other cereals. α-linolenic acid is an unsaturated fatty acid that can help prevent cardiovascular disease and is abundantly found in pseudocereals such as chia, buckwheat, amaranth, and quinoa. Linoleic acid makes up 50% of the fatty acids in quinoa and amaranth, and around 35% in buckwheat. Furthermore, chia oil contains a high concentration of polyunsaturated fatty acids, with 68% consisting of the omega-3 fatty acid α-linolenic acid, which imparts health benefits to individuals suffering from cardiovascular diseases. Literature studies have revealed that α-linolenic acid is beneficial for health as consumption of omega-3 fatty acids is reported to reduce biological markers linked with cardiovascular diseases (14, 56, 60, 118). Furthermore, both amaranth seeds and sprouts, as well as quinoa seeds, have shown antioxidant activity, which occurs via radical scavenging potential and thus inhibits lipid peroxidation (18). However, polyphenols, which are important components of pseudocereals, also play a critical role in inhibiting lipid peroxidation via radical scavenging activity (119). Polyphenols, for instance, rutin and nicotiflorin have been reported to be beneficial for managing cardiovascular diseases (120). Furthermore, chia contains the bioflavonoids like rutin and hesperidin, both of which have been shown to have high inhibitory activity towards lipid peroxidation activity (121). Amaranth and buckwheat are also rich in rutin, which helps to keep capillaries and arteries strong and flexible (53, 67, 122). Polyunsaturated fatty acids (PUFAs), which are considered to be healthy fats and whose consumption is generally healthier, and are an important component in avoiding any fat-related health complications. Despite its health-promoting properties, the human body is unable to produce PUFAs, which must be ingested through the diet to meet the daily requirements (123). Furthermore, omega-3 PUFAs have been shown to be a helpful dietary intervention for the prevention and treatment of cardiovascular disease (124). Chia, amaranth, and quinoa are excellent sources of oil rich in omega-3 fatty acids (33, 45, 125). In addition, ingestion of monounsaturated fatty acids, which include oleic acid, has beneficial effects in lowering the risk of cardiovascular disease. Amaranth, buckwheat, quinoa, and chia have all been reported to have oleic acid in their nutritional profiles (41, 45, 60, 126, 127). Squalene is found in pseudocereals like amaranth, buckwheat, quinoa, and chia; it is an unsaponifiable lipid that acts as a precursor for the biosynthesis of all steroids in plants and animals. In addition, as a lubricant that can resist oxidation, squalene finds use in the cosmetics sector. Squalene is mostly derived from deep-sea dogfish, sharks, and various plants, which include grapes, wheat, rice, etc. Squalene is biologically important due to its potential to reduce cholesterol levels by blocking cholesterol production in the liver (14, 41, 48, 115, 128). Furthermore, amaranth oil contains tocotrienols and squalene, which have been demonstrated to greatly reduce the concentration in blood of low-density lipoprotein (LDL) cholesterol (129). It has been reported in a study conducted on Wistar rats that amaranth protein showed hypotriglyceridemic effects and influenced liver lipid metabolism. The amaranth protein also imparts enhanced antioxidant protection in rats (95, 119, 130). Chia seeds also contain eicosapentaenoic acid and docosahexaenoic acid, both of which have cardioprotective properties (95).

According to studies, type 2 diabetic patients who are at risk of developing cardio-vascular disease and were reported to have consumed 37g of chia seeds per day had lower systolic blood pressure (6.3 mm Hg) than a control group of patients (131). Quinoa contains stigmasterol, which also has cholesterol-lowering activity (101). Thus, the nutritional content of pseudocereals has the ability to reduce the likeli-hood of developing cardiovascular disease as well as any other risk factors that may contribute to cardiovascular disease.

14.3.6 Other Potential Health Benefits of Pseudocereals on Diseases

Food plays a critical role in managing various metabolic diseases and age-related disorders. Pseudocereals, with their remarkable nutritional profile, serve as a cor-nerstone for human health and disease management (132, 133). The nutritional status of pseudocereals has shown improvement in clinical symptoms of various diseases, such as neurological disorders, osteoarthritis, inflammatory disease, Alzheimer's disease, obesity, etc. Neuroprotective compounds found in pseudocere-als, like quinoa, include stigmasterol, β-sitosterol, gallotannins, proanthocyanidins, α-linolenic acid, lignans, and others. Additionally, it has also been suggested that these compounds have antioxidant, antimicrobial, and anti-obesity properties (35, 101). Quinoa further contains saponins, which are a blend of various compounds, including triterpene glycosides derived from oleanolic acid, hederagenin, phytolac-cagenic acid, serjanic acid, and 3b,23,30-trihydroxyolean-12-en-28-oic acid, which have hydroxyl and carboxylate groups at C-3 and C-28. Saponins are generally considered to be anti-nutritional, bitter-tasting compounds. Along with their indi-gestible nature, they are present in the outer coat of the seed. Despite their negative properties, they also have many positive ones as well, including antiviral, antimi-crobial, antioxidant, anti-inflammatory, cytotoxic, analgesic, hypocholesterolemic, antithrombotic, neuroprotective and diuretic activities, modulator of mineral and vitamin absorption, neuroprotective activity, and immune-stimulatory properties (38, 49, 134, 135). Lunasin, which is found in amaranth albumin, globulin, and prolamin, may be a potential source of bioactive peptides with antihypertensive characteristics (112, 119). The compound rutin, which can be found in amaranth, as well as buckwheat, has been exploited medicinally in different countries due to its ability to minimize capillary fragility associated with various hemorrhagic dis-eases, and to reduce hypertension in humans (50, 53, 122). Furthermore, amaranth oil consists of compounds such as tocotrienols and squalene, which have choles-terol-lowering properties in the blood. Amaranth oil has also shown healing prop-erties against duodenal peptic ulcers and chronic gastritis induced by *Helicobacter pylori* (6, 129). The red-seeded varieties of *Amaranthus* grains are rich in various minerals, proteins, vitamins, and beta-carotene. Such nutrients play a major role in managing conditions like eye diseases, coronary artery diseases, etc. Furthermore, amaranth seeds have been shown to inhibit IgE and increase the production of the cytokine Th1, which reduces an individual's allergic reaction. Moreover, amaranth seeds facilitate the growth of helper T cells and therefore help in regulating allergic diseases like asthma and atopic dermatitis (136–138). Quinoa supplies linoleic and linolenic acids which can be converted into arachidonic acid and eicosapentaenoic or docosahexaenoic acids. Eicosapentaenoic and docosahexaenoic acids are essential for fetal development, as well as serving as anticoagulants and playing a role in managing Alzheimer's disease (139, 140). Buckwheat, with its high antioxidant characteristics, also aids in the delay of DNA and lipid oxidation, thereby further avoiding illnesses like inflammation, hypercholesterolemia, and a wide range of neurological disorders from developing (141). Thereby, the importance of pseudo-cereals is enormous in terms of combating diseases, and they may thus be incorporated into a balanced diet regularly.

14.4 Conclusion

Intake of a nutritious and balanced diet is important to prevent or slow the progression of diseases. Pseudocereals as food provide not only the essential elements necessary for an individual to maintain a disease-free life but also ensure food security under adverse environmental conditions. Among pseudocereals, the most popular ones are amaranth, buckwheat, quinoa, and chia. These pseudocereals have not only gained popularity for their nutritive profile but also for their stress tolerance when being cultivated, which enables them to grow under harsh environmental conditions. Furthermore, their gluten-free nature gives them an edge over other gluten-free options when it comes to fulfilling nutritional requirements during diet management among individuals suffering from celiac disease. Moreover, an arsenal of bioactive compounds that includes saponins, phenols, flavonoids, antioxidants, polysaccharides, and various biopeptides are found in pseudocereals to promote health and manage many chronic diseases. It has been observed that pseudocereals have excellent effects on diseases such as cardiovascular diseases, diabetes, hypertension, cancer, etc. In addition the prebiotic effects of pseudocereals have a positive impact on the gastrointestinal microbiota. Pseudocereals have unequivocally been demonstrated to be essential functional foods with several health advantages. However, the need of the hour is to gain a better understanding of the health benefits and mechanisms of action of these bioactive compounds for various chronic disorders.

References

1. Das, S., 2016. Pseudocereals: An efficient food supplement. In *Amaranthus: A Promising Crop of Future* (pp. 5–11). Singapore: Springer.
2. Sharma, A., 2017. Amaranth: A pseudocereal. *Nutrition & Food Science International Journal*, 3, pp. 7–9.

3. Pirzadah, T.B. and Malik, B., 2020. Pseudocereals as super foods of 21st century: Recent technological interventions. *Journal of Agriculture and Food Research*, 2, p. 100052.

4. Morales, D., Miguel, M. and Garcés-Rimón, M., 2021. Pseudocereals: A novel source of biologically active peptides. *Critical Reviews in Food Science and Nutrition*, 61(9), pp. 1537–1544.

5. Weerasekera, A.C., Samarasinghe, K. and Waisundara, V.Y., 2022. Introductory Chapter: Nutritive Value of Pseudocereals. Pseudocerealsin Viduranga Y. Waisundara (Ed.) (p. 1). Sri Lanka: Intechopen.

6. Upasana and Yadav, L., 2022. *Pseudocereals: A Novel Path Towards Healthy Eating.* Sri Lanka: Intechopen.

7. Ňorbová, M., Vollmannová, A., Harangozo, Ľ., Franková, H., Čeryová, N., Jančo, I. and Fandrová, A., 2022. Risk elements, antioxidant activity and polyphenols in pseudocereal grains. *Agrobiodiversity for Improving Nutrition, Health and Life Quality*, 6(1).

8. Martínez-Villaluenga, C., Peñas, E. and Hernández-Ledesma, B., 2020. Pseudocereal grains: Nutritional value, health benefits and current applications for the development of gluten-free foods. *Food and Chemical Toxicology*, 137, p. 111178.

9. Shahbaz, M., Raza, N., Islam, M., Imran, M., Ahmad, I., Meyyazhagan, A., Pushparaj, K., Balasubramanian, B., Park, S., Rengasamy, K.R. and Gondal, T.A., 2022. The nutraceutical properties and health benefits of pseudocereals: A comprehensive treatise. *Critical Reviews in Food Science and Nutrition*, pp. 1–13.

10. Alencar, N.M.M. and de Carvalho Oliveira, L., 2019. Advances in Pseudocereals: Crop cultivation, food application, and consumer perception. In Mérillon, JM., Ramawat, K. (Eds.) *Bioactive Molecules in Food* (pp. 1695–1713). Springer, Cham.

11. Bender, D. and Schönlechner, R., 2021. Recent developments and knowledge in pseudocereals including technological aspects. *Acta Alimentaria*, 50(4), pp. 583–609.

12. Betalleluz-Pallardel, I., Inga, M., Mera, L., Pedreschi, R., Campos, D. and Chirinos, R., 2017. Optimisation of extraction conditions and thermal properties of protein from the Andean pseudocereal cañihua (*Chenopodium pallidicaule* Aellen). *International Journal of Food Science and Technology*, 52(4), pp. 1026–1034.

13. Joshi, D.C., Meena, R.P. and Chandora, R., 2021. Genetic resources: Collection, characterization, conservation, and documentation. In *Millets and Pseudo Cereals* (pp. 19–31). Woodhead Publishing, Shimla.

14. Alvarez-Jubete, L., Arendt, E.K. and Gallagher, E., 2010. Nutritive value of pseudocereals and their increasing use as functional gluten-free ingredients. *Trends in Food Science and Technology*, 21(2), pp. 106–113.

15. Comino, I., de Lourdes Moreno, M., Real, A., Rodríguez-Herrera, A., Barro, F. and Sousa, C., 2013. The gluten-free diet: Testing alternative cereals tolerated by celiac patients. *Nutrients*, 5(10), pp. 4250–4268.

16. Maselli, L. and Hekmat, S., 2016. Microbial vitality of probiotic milks supplemented with cereal or pseudocereal grain flours. *Journal of Food Research*, 5(2), pp. 41–49.

17. Kocková, M., Dilongová, M. and Hybenová, E., 2013. Evaluation of cereals and pseudocereals suitability for the development of new probiotic foods. *Journal of Chemistry*, 2013, Article ID 414303. http://dx.doi.org/10.1155/2013/414303

18. Giménez-Bastida, J.A., Hamdi, S. and Llopis, J.M.L., 2017. Nutritional and health implications of pseudocereal intake. In *Pseudocereals: Chemistry and Technology*, Mainz, (pp. 217–232).

19. Pojić, M. and Tiwari, U., 2020. Processing technologies for healthy grains: Introduction. *Innovative Processing Technologies for Healthy Grains*, pp. 1–7.
20. FAO, IFAD, UNICEF, WFP and WHO, 2022. *The State of Food Security and Nutrition in the World 2022. Repurposing Food and Agricultural Policies to Make Healthy Diets More Affordable*. FAO, Rome. https://doi.org/10.4060/cc0639en.
21. Angioloni, A. and Collar, C., 2011. Nutritional and functional added value of oat, Kamut®, spelt, rye and buckwheat versus common wheat in breadmaking. *Journal of the Science of Food and Agriculture*, 91(7), pp. 1283–1292.
22. Erley, G.S.,Kaul, H.P., Kruse, M. and Aufhammer, W., 2005. Yield and nitrogen utilization efficiency of the pseudocereals amaranth, quinoa, and buckwheat under differing nitrogen fertilization. *European Journal of Agronomy*, 22(1), pp. 95–100.
23. Thakur, P., Kumar, K. and Dhaliwal, H.S., 2021. Nutritional facts, bio-active components and processing aspects of pseudocereals: A comprehensive review. *Food Bioscience*, 42, p. 101170.
24. Mikulajová, A., Takácsová, M., Rapta, P., Brindzová, L., Zalibera, M. and Nemeth, K., 2007. Total phenolic contents and antioxidant capacities of cereal and pseudocereal genotypes. *Journal of Food and Nutrition Research*, 46, pp. 150–157.
25. Ugural, A. and Akyol, A., 2022. Can pseudocereals modulate microbiota by functioning as probiotics or prebiotics? *Critical Reviews in Food Science and Nutrition*, 62(7), pp. 1725–1739.
26. Chlopicka, J., Pasko, P., Gorinstein, S., Jedryas, A. and Zagrodzki, P., 2012. Total phenolic and total flavonoid content, antioxidant activity and sensory evaluation of pseudocereal breads. *LWT – Food Science and Technology*, 46(2), pp. 548–555.
27. Stamatovska, V., Nakov, G., Uzunoska, Z., Kalevska, T. and Menkinoska, M., 2018. Potential use of some pseudocereals in the food industry. *ARTTE*, 6(1), pp. 54–61.
28. Mošovská, S., Mikulášová, M., Brindzova, L., Valík, Ľ. and Mikušova, L., 2010. Genotoxic and antimutagenic activities of extracts from pseudocereals in the Salmonella mutagenicity assay. *Food and Chemical Toxicology*, 48(6), pp. 1483–1487.
29. Patiranage, D.S., Rey, E., Emrani, N., Wellman, G., Schmid, K., Schmöckel, S.M., Tester, M. and Jung, C., 2020. Genome-wide association study in the pseudocereal quinoa reveals selection pattern typical for crops with a short breeding history. bioRxiv.
30. Modgil, R. and Sood, P., 2017. Effect of roasting and germination on carbohydrates and anti-nutritional constituents of indigenous and exotic cultivars of pseudo-cereal (*Chenopodium*). *Journal of Life Sciences*, 9(1), pp. 64–70.
31. Sunil, M., Hariharan, A.K., Nayak, S., Gupta, S., Nambisan, S.R., Gupta, R.P., Panda, B., Choudhary, B. and Srinivasan, S., 2014. The draft genome and transcriptome of *Amaranthus hypochondriacus*: A C4 dicot producing high-lysine edible pseudo-cereal. *DNA Research*, 21(6), pp. 585–602.
32. Fabio, A.D. and Parraga, G., 2017. Origin, production and utilization of pseudocereals. In *Pseudocereals: Chemistry and Technology* (pp. 1–27), John Wiley & Sons, Ltd.
33. Rodríguez, J.P., Rahman, H., Thushar, S. and Singh, R.K., 2020. Healthy and resilient cereals and pseudo-cereals for marginal agriculture: Molecular advances for improving nutrient bioavailability. *Frontiers in Genetics*, 11, p. 49.
34. Škrovánková, S., Válková, D. and Mlček, J., 2020. Polyphenols and antioxidant activity in pseudocereals and their products. *Potravinarstvo Slovak Journal of Food Sciences*, 14, pp. 365–370.
35. Thakur, P. and Kumar, K., 2019. Nutritional importance and processing aspects of Pseudocereals. *Journal of Agricultural Engineering and Food Technology*, 6, pp. 155–160.

36. Topwal, M., 2019. A review on amaranth: Nutraceutical and virtual plant for providing food security and nutrients. *Acta Scientific Agriculture*, 3(1), pp. 9–15.
37. Berghofer, E. and Schoenlechner, R., 2008. *South American Traditional Pseudocereals. Traditional Grains for Low Environmental Impact and Good Health* (pp. 28–31). Gothenburg.
38. De Bock, P., Daelemans, L., Selis, L., Raes, K., Vermeir, P., Eeckhout, M. and Van Bockstaele, F., 2021. Comparison of the chemical and technological characteristics of wholemeal flours obtained from amaranth (*Amaranthus* spp.), quinoa (*Chenopodium quinoa*) and buckwheat (*Fagopyrum* spp.) seeds. *Foods*, 10(3), p. 651.
39. Fasuan, T.O., Asadu, K.C., Anyiam, C.C., Ojokoh, L.O., Olagunju, T.M., Chima, J.U. and Okpara, K.O., 2021. Bioactive and nutritional characterization of modeled and optimized consumer-ready flakes from pseudocereal (*Amaranthus viridis*), high-protein soymeal and modified corn starch. *Food Production, Processing and Nutrition*, 3(1), pp. 1–13.
40. Yeşil, S. and Levent, H., 2022. The influence of fermented buckwheat, quinoa and amaranth flour on gluten-free bread quality. *LWT*, 160, p. 113301. https://doi.org/10.1016/j.lwt.2022.113301
41. Choudhury, S.S. and Singh, H. Psendocereals: Functional Foods or Larder Folderol?. Vigyan Varta an International E-Magazine for Science Enthusiasts.
42. Mendonça, S., Saldiva, P.H., Cruz, R.J. and Arêas, J.A., 2009. Amaranth protein presents cholesterol-lowering effect. *Food Chemistry*, 116(3), pp. 738–742.
43. Mosyakin, S. and Schwartau, V., 2015. Quinoa as a promising pseudocereal crop for Ukraine. *Agricultural Science and Practice*, 2(1), pp. 3–11.
44. Bhathal, S., Grover, K. and Gill, N., 2015. Quinoa-a treasure trove of nutrients. *Journal of Nutrition Research*, 3(1), pp. 45–49.
45. Demir, M.K., 2014. Use of quinoa flour in the production of gluten-free tarhana. *Food Science and Technology Research*, 20(5), pp. 1087–1092.
46. Bilgiçli, N. and İbanoğlu, Ş., 2015. Effect of pseudo cereal flours on some physical, chemical and sensory properties of bread. *Journal of Food Science and Technology*, 52(11), pp. 7525–7529.
47. Banu, I. and Aprodu, I., 2022. Investigations on functional and thermo-mechanical properties of gluten free cereal and pseudocereal flours. *Foods*, 11(13), p. 1857.
48. Mota, C., Nascimento, A.C., Santos, M., Delgado, I., Coelho, I., Rego, A., Matos, A.S., Torres, D. and Castanheira, I., 2016. The effect of cooking methods on the mineral content of quinoa (*Chenopodium quinoa*), amaranth (*Amaranthus* spp.) and buckwheat (*Fagopyrum esculentum*). *Journal of Food Composition and Analysis*, 49, pp. 57–64.
49. Schoenlechner, R., 2016. Properties of pseudocereals, selected specialty cereals and legumes for food processing with special attention to gluten-free products. *Die Bodenkultur: Journal of Land Management, Food and Environment*, 67(4), pp. 239–248.
50. Kwon, S.J., Roy, S.K., Choi, J., Park, J., Cho, S., Sarker, K. and Woo, S.H., 2018. Recent research updates on functional components in Buckwheat. *Journal of Agricultural Science*, 34, pp. 1–8.
51. Lone, A.H., Ganie, S.A., Jan, N. and Lone, M.A., 2018. *In vivo* germination of *Fagopyrum tataricum* (L.) Seeds an economically and medicinally important pseudo-cereal of Kashmir Himalaya. *Journal of Pharmacognosy and Phytochemistry*, 7(6), pp. 2788–2791.
52. Léder, I., 2009. *Buckwheat, Amaranth, and Other Pseudocereal Plants. Encyclopedia of Life Support Systems* (pp. 1–17). 1st ed. EOLSS Publishers Co Ltd, Ramsey.

53. Jancurová, M., Minarovičová, L. and Dandár, A., 2009. Rheological properties of doughs with buckwheat and quinoa additives. *Chemical Papers*, 63(6), pp. 738–741.
54. Gálová, Z., Pálenčárová, E., Chňapek, M. and Balažová, Ž., 2021. 2-DE proteome maps of amaranth and buckwheat seeds. *Journal of Microbiology, Biotechnology and Food Sciences*, 3 pp. 74–76.
55. Grancieri, M., Martino, H.S.D. and Gonzalez de Mejia, E., 2019. Chia seed (*Salvia hispanica* L.) as a source of proteins and bioactive peptides with health benefits: A review. *Comprehensive Reviews in Food Science and Food Safety*, 18(2), pp. 480–499.
56. Din, Z.U., Alam, M., Ullah, H., Shi, D., Xu, B., Li, H. and Xiao, C., 2021. Nutritional, phytochemical and therapeutic potential of chia seed (*Salvia hispanica* L.). A mini-review. *Food Hydrocolloids for Health*, 1, p. 100010.
57. Malik, A.M. and Singh, A., 2021. Pseudocereals Proteins-A comprehensive review on its isolation, composition and quality evaluation techniques. *Food Chemistry Advances*, p. 100001.
58. Niro, S., D'Agostino, A., Fratianni, A., Cinquanta, L. and Panfili, G., 2019. Gluten-free alternative grains: Nutritional evaluation and bioactive compounds. *Foods*, 8(6), p. 208.
59. Prathyusha, P., Kumari, B.A., Suneetha, W.J. and Srujana, M.N.S., 2019. Chia seeds for nutritional security. *Journal of Pharmacognosy and Phytochemistry*, 8(3), pp. 2702–2707.
60. Ayerza, R. and Coates, W., 2005. Ground chia seed and chia oil effects on plasma lipids and fatty acids in the rat. *Nutrition Research*, 25(11), pp. 995–1003.
61. Ayerza, R. and Coates, W., 2011. Protein content, oil content and fatty acid profiles as potential criteria to determine the origin of commercially grown chia (*Salvia hispanica* L.). *Industrial Crops and Products*, 34(2), pp. 1366–1371.
62. Karel, A. Phytochemical profile of chia incorporation in snacks. *Magnesium*, 197, p. 350.
63. Junker, Y., Zeissig, S., Kim, S.J., Barisani, D., Wieser, H., Leffler, D.A., Zevallos, V., Libermann, T.A., Dillon, S., Freitag, T.L., Kelly, C.P. and Schuppan, D., 2012. Wheat amylase trypsin inhibitors drive intestinal inflammation via activation of Toll-like receptor 4. *Journal of Experimental Medicine*, 209(13), pp. 2395–2408.
64. Di Cagno, R., De Angelis, M., Auricchio, S., Greco, L., Clarke, C., De Vincenzi, M., Giovannini, C., D'Archivio, M., Landolfo, F., Parrilli, G. and Minervini, F., 2004. Sourdough bread made from wheat and nontoxic flours and started with selected lactobacilli is tolerated in celiac sprue patients. *Applied and Environmental Microbiology*, 70(2), pp. 1088–1096.
65. Moreno, ML, Comino, I, Sousa, C. Alternative Grains as Potential Raw Material for Gluten-Free Food Development in The Diet of Celiac and Gluten-Sensitive Patients. *Austin J Nutri Food Sci*, 2(3): p. 1016.
66. Zevallos, V.F., Ellis, H.J., Šuligoj, T., Herencia, L.I. and Ciclitira, P.J., 2012. Variable activation of immune response by quinoa (*Chenopodium quinoa* Willd.) prolamins in celiac disease. *The American Journal of Clinical Nutrition*, 96(2), pp. 337–344.
67. Dabija, A., Ciocan, M.E., Chetrariu, A. and Codină, G.G., 2022. Buckwheat and amaranth as raw materials for brewing, a review. *Plants*, 11(6), p. 756.
68. Condés, M.C., Scilingo, A.A. and Añón, M.C., 2009. Characterization of amaranth proteins modified by trypsin proteolysis. Structural and functional changes. *LWT – Food Science and Technology*, 42(5), pp. 963–970.

69. Zannini, E., Jones, J.M., Renzetti, S. and Arendt, E.K., 2012. Functional replacements for gluten. *Annual Review of Food Science and Technology*, 3, pp. 227–245.

70. Montoya-Rodríguez, A., Gómez-Favela, M.A., Reyes-Moreno, C., Milán-Carrillo, J. and González de Mejía, E., 2015. Identification of bioactive peptide sequences from amaranth (Amaranthus hypochondriacus) seed proteins and their potential role in the prevention of chronic diseases. *Comprehensive Reviews in Food Science and Food Safety*, 14(2), pp. 139–158.

71. Paucar-Menacho, L.M., Castillo-Martínez, W.E., Simpalo-Lopez, W.D., Verona-Ruiz, A., Lavado-Cruz, A., Martínez-Villaluenga, C., Peñas, E., Frias, J. and Schmiele, M., 2022. Performance of thermoplastic extrusion, germination, fermentation, and hydrolysis techniques on phenolic compounds in cereals and pseudocereals. *Foods*, 11(13), p. 1957.

72. Singh, R., Kaur, N. and Grover, K., 2022. Development of pseudo cereal quinoa based gluten free product to manage celiac disease among children, 41(14), pp. 14–32.

73. Wang, M., Guo, X., Ma, Y. and Gao, J., 2012. Buckwheat: A novel pseudocereal. In *Cereals and Pulses: Nutraceutical Properties and Health Benefits* (pp. 131–148), John Wiley & Sons.

74. Kiss, A., Takács, K., Nagy, A., Nagy-Gasztonyi, M., Cserhalmi, Z., Naár, Z., Halasi, T., Csáki, J. and Némedi, E., 2019. *In vivo* and *in vitro* model studies on noodles prepared with antioxidant-rich pseudocereals. *Journal of Food Measurement and Characterization*, 13(4), pp. 2696–2704.

75. Costantini, L., Lukšič, L., Molinari, R., Kreft, I., Bonafaccia, G., Manzi, L. and Merendino, N., 2014. Development of gluten-free bread using Tartary buckwheat and chia flour rich in flavonoids and omega-3 fatty acids as ingredients. *Food Chemistry*, 165, pp. 232–240.

76. Markowiak, P. and Śliżewska, K., 2017. Effects of probiotics, prebiotics, and Synbiotics on human health. *Nutrients*, 9(9), p. 1021.

77. Zhang, Y.J., Li, S., Gan, R.Y., Zhou, T., Xu, D.P. and Li, H.B., 2015. Impacts of gut bacteria on human health and diseases. *International Journal of Molecular Sciences*, 16(4), pp. 7493–7519.

78. Gallego, C.G. and Salminen, S., 2016. Novel probiotics and prebiotics: How can they help in human gut microbiota dysbiosis? *Applied Food Biotechnology*, 3(2), pp. 72–81.

79. Sánchez, B., Delgado, S., Blanco-Míguez, A., Lourenço, A., Gueimonde, M. and Margolles, A., 2017. Probiotics, gut microbiota, and their influence on host health and disease. *Molecular Nutrition and Food Research*, 61(1), p. 1600240.

80. Kaur, A.P., Bhardwaj, S., Dhanjal, D.S., Nepovimova, E., Cruz-Martins, N., Kuča, K., Chopra, C., Singh, R., Kumar, H., Şen, F. and Kumar, V., 2021. Plant prebiotics and their role in the amelioration of diseases. *Biomolecules*, 11(3), p. 440.

81. Liptáková, D., Matejčeková, Z. and Valík, L., 2017. Lactic acid bacteria and fermentation of cereals and pseudocereals. *Fermentation Processes*, 10, p. 65459.

82. Medveďová, M.K.J.M.A. and Valík, E.Š.Ľ., 2013. Cereals and pseudocereals as substrates for growth and metabolism of a probiotic strain *Lactobacillus rhamnosus* GG. *Journal of Food and Nutrition Research*, 52(1), pp. 25–36.

83. Soltani, M., Hekmat, S. and Ahmadi, L., 2018. Microbial and sensory evaluation of probiotic yoghurt supplemented with cereal/pseudo-cereal grains and legumes. *International Journal of Dairy Technology*, 71(S1), pp. 141–148.

84. Cruz-Rubio, J.M., Loeppert, R., Viernstein, H. and Praznik, W., 2018. Trends in the use of plant non-starch polysaccharides within food, dietary supplements, and pharmaceuticals: Beneficial effects on regulation and wellbeing of the intestinal tract. *Scientia Pharmaceutica*, 86(4), p. 49.

85. Bianchi, F., Rossi, E.A., Gomes, R.G. and Sivieri, K., 2015. Potentially synbiotic fermented beverage with aqueous extracts of quinoa (*Chenopodium quinoa* Willd) and soy. *Food Science and Technology International*, 21(6), pp. 403–415.

86. Bustamante, M., Oomah, B.D., Rubilar, M. and Shene, C., 2017. Effective *Lactobacillus plantarum* and *Bifidobacterium infantis* encapsulation with chia seed (*Salvia hispanica* L.) and flaxseed (*Linum usitatissimum* L.) mucilage and soluble protein by spray drying. *Food Chemistry*, 216, pp. 97–105.

87. Kwon, H.C., Bae, H., Seo, H.G. and Han, S.G., 2019. Chia seed extract enhances physiochemical and antioxidant properties of yogurt. *Journal of Dairy Science*, 102(6), pp. 4870–4876.

88. Wilcox, G., 2005. Insulin and insulin resistance Gisela. *Insulin and Insulin Resistance Gisela*, 22(2), pp. 61–63.

89. Narula, A., 2021. Probiotics: Origin, products, and regulations in India. In *Microbial Products for Health, Environment and Agriculture* (pp. 59–101). Springer.

90. Khan, M.A.B., Hashim, M.J., King, J.K., Govender, R.D., Mustafa, H. and Al Kaabi, J., 2020. Epidemiology of type 2 diabetes–global burden of disease and forecasted trends. *Journal of Epidemiology and Global Health*, 10(1), p. 107.

91. Qin, L., Knol, M.J., Corpeleijn, E. and Stolk, R.P., 2010. Does physical activity modify the risk of obesity for type 2 diabetes: A review of epidemiological data. *European Journal of Epidemiology*, 25(1), pp. 5–12.

92. https://www.who.int/news-room/fact-sheets/detail/diabetes.

93. Ranilla, L.G., Apostolidis, E., Genovese, M.I., Lajolo, F.M. and Shetty, K., 2009. Evaluation of indigenous grains from the Peruvian Andean region for antidiabetes and antihypertension potential using in vitro methods. *Journal of Medicinal Food*, 12(4), pp. 704–713.

94. Gabrial, S.G., Shakib, M.C.R. and Gabrial, G.N., 2016. Effect of pseudocereal-based breakfast meals on the first and second meal glucose tolerance in healthy and diabetic subjects. *Open Access Macedonian Journal of Medical Sciences*, 4(4), p. 565.

95. Ullah, R., Nadeem, M., Khalique, A., Imran, M., Mehmood, S., Javid, A. and Hussain, J., 2016. Nutritional and therapeutic perspectives of Chia (*Salvia hispanica* L.): A review. *Journal of Food Science and Technology*, 53(4), pp. 1750–1758.

96. Gong, L., Feng, D., Wang, T., Ren, Y., Liu, Y. and Wang, J., 2020. Inhibitors of α-amylase and α-glucosidase: Potential linkage for whole cereal foods on prevention of hyperglycemia. *Food Science and Nutrition*, 8(12), pp. 6320–6337.

97. Scheen, A.J., 2003. Is there a role for α-glucosidase inhibitors in the prevention of type 2 diabetes mellitus? *Drugs*, 63(10), pp. 933–951.

98. Randhir, R., Kwon, Y.I. and Shetty, K., 2008. Effect of thermal processing on phenolics, antioxidant activity and health-relevant functionality of select grain sprouts and seedlings. *Innovative Food Science and Emerging Technologies*, 9(3), pp. 355–364.

99. Taylor, J.R.N., Belton, P.S., Beta, T. and Duodu, K.G., 2014. Increasing the utilisation of sorghum, millets and pseudocereals: Developments in the science of their phenolic phytochemicals, biofortification and protein functionality. *Journal of Cereal Science*, 59(3), pp. 257–275.

100. Conforti, F., Statti, G., Loizzo, M.R., Sacchetti, G., Poli, F. and Menichini, F., 2005. In vitro antioxidant effect and inhibition of α-amylase of two varieties of *Amaranthus caudatus* seeds. *Biological and Pharmaceutical Bulletin*, 28(6), pp. 1098–1102.
101. Gupta, N. and Morya, S., 2022. Bioactive and pharmacological characterization of *Chenopodium quinoa*, *Sorghum bicolor* and *Linum usitassimum*: A review. *Journal of Applied and Natural Science*, 14(3), pp. 1067–1084.
102. Seyfried, T.N. and Shelton, L.M., 2010. Cancer as a metabolic disease. *Nutrition and Metabolism*, 7(1), pp. 1–22.
103. Dhull, S.B., Punia, S., Kidwai, M.K., Kaur, M., Chawla, P., Purewal, S.S., Sangwan, M. and Palthania, S., 2020b. Solid-state fermentation of lentil (*Lens culinaris* L.) with *Aspergillus awamori*: Effect on phenolic compounds, mineral content, and their bioavailability. *Legume Science*, p. e37.
104. Dhull, S.B., Punia, S., Kumar, R., Kumar, M., Nain, K.B., Jangra, K. and Chudamani, C., 2020c. Solid state fermentation of fenugreek (*Trigonella foenum-graecum*): Implications on bioactive compounds, mineral content and in vitro bio-availability. *Journal of Food Science and Technology*, pp. 1–10.
105. Poprac, P., Jomova, K., Simunkova, M., Kollar, V., Rhodes, C.J. and Valko, M., 2017. Targeting free radicals in oxidative stress-related human diseases. *Trends in pharmacological sciences*, 38(7), pp. 592–607.
106. Lee, M.J. and Sim, K.H., 2018. Nutritional value and the kaempferol and quercetin contents of quinoa (*Chenopodium quinoa* Willd.) from different regions. *Korean Journal of Food Science and Technology*, 50(6), pp. 680–687.
107. Alvarez-Jubete, L., Wijngaard, H., Arendt, E.K. and Gallagher, E., 2010. Polyphenol composition and in vitro antioxidant activity of amaranth, quinoa buckwheat and wheat as affected by sprouting and baking. *Food Chemistry*, 119(2), pp. 770–778.
108. Das, S.K., Avasthe, R.K., Ghosh, G.K. and Dutta, S.K., 2019. Pseudocereal buckwheat with potential anticancer activity. *Bulletin of Pure and Applied Sciences Section B-Botany*, 38(2), pp. 94–95.
109. Orona-Tamayo, D., Valverde, M.E. and Paredes-López, O., 2019. Bioactive peptides from selected Latin American food crops–A nutraceutical and molecular approach. *Critical Reviews in Food Science and Nutrition*, 59(12), pp. 1949–1975.
110. Maldonado-Cervantes, E., Jeong, H.J., León-Galván, F., Barrera-Pacheco, A., De León-Rodríguez, A., De Mejia, E.G., Ben, O. and De La Rosa, A.P.B., 2010. Amaranth lunasin-like peptide internalizes into the cell nucleus and inhibits chemical carcinogen-induced transformation of NIH-3T3 cells. *Peptides*, 31(9), pp. 1635–1642.
111. Vilcacundo, R., Miralles, B., Carrillo, W. and Hernández-Ledesma, B., 2018. *In vitro* chemopreventive properties of peptides released from quinoa (*Chenopodium quinoa* Willd.) protein under simulated gastrointestinal digestion. *Food Research International*, 105, pp. 403–411.
112. Silva-Sánchez, C., De La Rosa, A.B., León-Galván, M.F., de Lumen, B.O., de León-Rodríguez, A. and De Mejía, E.G., 2008. Bioactive peptides in amaranth (*Amaranthus hypochondriacus*) seed. *Journal of Agricultural and Food Chemistry*, 56(4), pp. 1233–1240.
113. Jayaprakasam, B., Zhang, Y. and Nair, M.G., 2004. Tumor cell proliferation and cyclooxygenase enzyme inhibitory compounds in *Amaranthus tricolor*. *Journal of Agricultural and Food Chemistry*, 52(23), pp. 6939–6943.

114. Gazem, R.A.A., Puneeth, H.R., Shivmadhu, C. and Madhu, A.C.S., 2017. In vitro anticancer and anti-lipoxygenase activities of chia seed oil and its blends with selected vegetable oils. *In Vitro*, 10(10), p. 124–128.

115. Szabóová, M., Záhorský, M., Gažo, J., Geuens, J., Vermoesen, A., D'Hondt, E. and Hricová, A., 2020. Differences in seed weight, amino acid, fatty acid, oil, and squalene content in γ-irradiation-developed and commercial amaranth varieties (*Amaranthus* spp.). *Plants*, 9(11), p. 1412.

116. https://www.who.int/news-room/fact-sheets/detail/cardiovascular-diseases-(cvds).

117. Stewart, J., Manmathan, G. and Wilkinson, P., 2017. Primary prevention of cardiovascular disease: A review of contemporary guidance and literature. *JRSM Cardiovascular Disease*, 6, p. 2048004016687211.

118. Sabbione, A.C., Scilingo, A. and Añón, M.C., 2015. Potential antithrombotic activity detected in amaranth proteins and its hydrolysates. *LWT – Food Science and Technology*, 60(1), pp. 171–177.

119. Gorinstein, S., Vargas, O.J.M., Jaramillo, N.O., Salas, I.A., Ayala, A.L.M., Arancibia-Avila, P., Toledo, F., Katrich, E. and Trakhtenberg, S., 2007. The total polyphenols and the antioxidant potentials of some selected cereals and pseudocereals. *European Food Research and Technology*, 225(3), pp. 321–328.

120. Henrion, M., Labat, E. and Lamothe, L., 2020. Pseudocereals as healthy grains: An overview. *Innovative Processing Technologies for Healthy Grains*, pp. 37–59.

121. Ding, Y., Lin, H.W., Lin, Y.L., Yang, D.J., Yu, Y.S., Chen, J.W., Wang, S.Y. and Chen, Y.C., 2018. Nutritional composition in the chia seed and its processing properties on restructured ham-like products. *Journal of Food and Drug Analysis*, 26(1), pp. 124–134.

122. Pastor, K. and Acanski, M., 2018. The chemistry behind amaranth grains. *Journal of Nutritional Health & Food Engineering*, 8(5), pp. 358–360.

123. Shen, Y., Zheng, L., Jin, J., Li, X., Fu, J., Wang, M., Guan, Y. and Song, X., 2018. Phytochemical and biological characteristics of Mexican chia seed oil. *Molecules*, 23(12), p. 3219.

124. Ander, B.P., Dupasquier, C.M., Prociuk, M.A. and Pierce, G.N., 2003. Polyunsaturated fatty acids and their effects on cardiovascular disease. *Experimental and Clinical Cardiology*, 8(4), p. 164.

125. Bozorov, S.S., Berdiev, N.S., Ishimov, U.J., Olimjonov, S.S., Ziyavitdinov, J.F., Asrorov, A.M. and Salikhov, S.I., 2018. Chemical composition and biological activity of seed oil of amaranth varieties. *Nova Biotechnologica et Chimica*, 17(1), p. 66–78.

126. Alvarez-Jubete, L., Arendt, E.K. and Gallagher, E., 2009. Nutritive value and chemical composition of pseudocereals as gluten-free ingredients. *International Journal of Food Sciences and Nutrition*, 60(sup4), pp. 240–257.

127. Alvites-Misajel, K., García-Gutiérrez, M., Miranda-Rodríguez, C. and Ramos-Escudero, F., 2019. Organically vs conventionally-grown dark and white chia seeds (*Salvia hispanica* L.): Fatty acid composition, antioxidant activity and techno-functional properties. *Grasas y Aceites*, 70(2), pp.e299–e299.

128. Dąbrowski, G., Konopka, I. and Czaplicki, S., 2018. Variation in oil quality and content of low molecular lipophilic compounds in chia seed oils. *International Journal of Food Properties*, 21(1), pp. 2016–2029.

129. Rollán, G.C., Gerez, C.L. and LeBlanc, J.G., 2019. Lactic fermentation as a strategy to improve the nutritional and functional values of pseudocereals. *Frontiers in Nutrition*, 6, p. 98.

130. Czerwiński, J., Bartnikowska, E., Leontowicz, H., Lange, E., Leontowicz, M., Katrich, E., Trakhtenberg, S. and Gorinstein, S., 2004. Oat (*Avena sativa* L.) and amaranth (*Amaranthus hypochondriacus*) meals positively affect plasma lipid profile in rats fed cholesterol-containing diets. *The Journal of Nutritional Biochemistry*, 15(10), pp. 622–629.
131. Vuksan, V., Whitham, D., Sievenpiper, J.L., Jenkins, A.L., Rogovik, A.L., Bazinet, R.P., Vidgen, E. and Hanna, A., 2007. Supplementation of conventional therapy with the novel grain Salba (*Salvia hispanica* L.) improves major and emerging cardiovascular risk factors in type 2 diabetes: Results of a randomized controlled trial. *Diabetes Care*, 30(11), pp. 2804–2810.
132. Graf, B.L., Rojas-Silva, P., Rojo, L.E., Delatorre-Herrera, J., Baldeón, M.E. and Raskin, I., 2015. Innovations in health value and functional food development of quinoa (*Chenopodium quinoa* Willd.). *Comprehensive Reviews in Food Science and Food Safety*, 14(4), pp. 431–445.
133. Punia, S. and Dhull, S.B., 2019. Chia seed (*Salvia hispanica* L.) mucilage (a heteropolysaccharide): Functional, thermal, rheological behaviour and its utilization. *International Journal of Biological Macromolecules*, 140, pp. 1084–1090.
134. Hernández-Ledesma, B., 2019. Quinoa (*Chenopodium quinoa* Willd.) as source of bioactive compounds: A review. *Bioactive Compounds in Health and Disease*, 2(3), pp. 27–47.
135. Dhull, S.B., Kidwai, M.K., Noor, R., Chawla, P. and Rose, P.K., 2022a. A review of nutritional profile and processing of faba bean (*Vicia faba* L.). *Legume Science*, 58(5), p. e129.
136. Hibi, M., Hachimura, S., Hashizume, S., Obata, T. and Kaminogawa, S., 2003. Amaranth grain inhibits antigen-specific IgE production through augmentation of the IFN-γ response in vivo and in vitro. *Cytotechnology*, 43(1–3), p. 33.
137. Malik, M., Sindhu, R., Dhull, S.B., Bou-Mitri, C., Singh, Y., Panwar, S. and Khatkar, B.S., 2023. Nutritional composition, functionality, and processing technologies for amaranth. *Journal of Food Processing and Preservation*, p. 1753029. https://doi.org/10.1155/2023/1753029.
138. Iftikhar, M. and Khan, M., 2019. Amaranth. In *Bioactive Factors and Processing Technology for Cereal Foods* (pp. 217–232). Springer.
139. Montemurro, M., Pontonio, E. and Rizzello, C.G., 2019. Quinoa flour as an ingredient to enhance the nutritional and functional features of cereal-based foods. In *Flour and Breads and their Fortification in Health and Disease Prevention* (pp. 453–464). Academic Press.
140. Punia, S., Sandhu, K.S., Siroha, A.K. and Dhull, S.B., 2019. *Omega 3-Metabolism, Absorption, Bioavailability and health benefits-A review*. p. 100162, PharmaNutrition.
141. Gimenez-Bastida, J.A. and Zielinski, H., 2015. Buckwheat as a functional food and its effects on health. *Journal of Agricultural and Food Chemistry*, 63(36), pp. 7896–7913.

Chapter 15

Quality Management System for Pseudocereals

Baneeprajnya Nayak, Atul Anand Mishra, Neha Singh

15.1 Introduction

Quality is a result that is created when a need, anticipation, requirement, or demand is met or achieved. Poor quality is defined as something that does not meet the demand, and high quality as something that completely fulfils the need. In all of their endeavours, in all organisations and families, most people aim to achieve this result (high quality). The majority of us desire the outcome of our efforts to fulfil the need, expectation, requirement, or demand, regardless of how it is articulated. So how do we approach it? If all you have is a hammer, everything would appear to be a nail, as Abraham Maslow once said. Quality is similar, which is why over the past 50 years or more, a number of strategies have emerged to attain, maintain, and enhance quality [1].

Food quality and safety refer to each of the factors that affect a food product's capacity to meet all legal, consumer, and regulatory standards [2]. Although there may be some overlap, it is important to highlight that food safety is not the same as food quality. Although safety refers to all actions taken to protect human health, quality comprises all characteristics of a product that affect its worth to consumers. Complete food safety is an impossible ideal. Nonetheless, food is deemed safe if there is a reasonable degree of certainty that, under the usage scenarios that are predicted, no harm will come from consuming it [3, 4].

Quality has become increasingly important in the food industry over the past few decades as a result of rising consumer aspirations, stiffer government restrictions, and escalating market competition. In response, food businesses have recently been pursuing quality management (QM) procedures more and more. According

DOI: 10.1201/9781003325277-15

to ISO 8401, quality management is "all activities of the overall management function that determine the quality policy, objectives and responsibilities, and implement them by means such as quality planning, quality control, quality assurance and quality improvement within the quality system" (ISO 8402, p. 1) [5]. Quality management definition by ISO 9000 again states that it entails coordinated actions to guide and regulate an organization's quality [1]. Food manufacturing businesses who want to create a Quality Management System that covers product development, production, distribution, and customer service would employ ISO 9001 in the food industry. The production and delivery of food products would fall under the scope of ISO 9002, the most popular of the three standards. Being unrelated to production and primarily focussing on the testing and inspection of purchased goods prior to resale, ISO 9003 is seldom ever utilised in any industry. Each version is made up of a number of phases, or elements, that specify what constitutes an appropriate management practice for quality assurance [6].

The use of a quality management system allows a business to specify and illustrate the procedures needed to guarantee that the goods or service meets customer requirements. Planning is a fundamental component of quality management. The planning process outlines the quality, people, financial, and material resource goals that must be attained [7]. When it comes to pseudocereals, quality management is a whole new process because pseudocereals are yet to be explored, unlike cereals. Pseudocereals are a type of grains known as "unexploited foods" which are non-grass plant species that are not part of the cereal family but have traits and roles that are similar to cereals. Although pseudocereals are dicotyledonous, unlike cereals (monocotyledonous), according to botanic characteristics the term "pseudocereals" is applied because of the resemblance of their seeds and texture, starch content, cooking methods, and palatability [8].

Over the past two decades, research on under-utilised crops has attracted more attention. Local Indigenous populations have been consuming pseudocereals for centuries. These non-cereal plants, which have traits and applications resembling those of cereals, stand out as under-utilised foods. The most representative species include buckwheat, amaranth, chia, and quinoa. Although they don't include gluten, they do contain high-quality proteins and peptides as well as other dietary and bioactive elements like phenolic acids, flavonoids, vitamins, and minerals. Pseudocereal protein-derived peptides have been reported to have anticarcinogenic, antioxidant, anti-inflammatory, cholesterol-reducing, and blood pressure-lowering effects. Pseudocereals are becoming more important as a result of their intriguing properties [9].

15.2 What are Quality Management Systems?

The quality management system is the type of management that allows a corporation to accomplish its goals and objectives. Organisations only have one rule:

therefore, formalising a portion of a system that prioritises quality does not offer any advantages. There are no advantages to focussing solely on one aspect of performance when it is a combination of factors that deliver organisational performance, so how well the system helps the organisation accomplish its goals, to function effectively and satisfy its intended purpose and goal is how thoroughly the system is judged for its competency, appropriateness, and effectiveness [1].

15.3 General Characteristics of Pseudocereals

Pseudocereals are dicotyledonous crops of different families like buckwheat of the family Polygonacae, quinoa of the family Amaranthaceae, and amaranth of the family Amaranthaceae. They were the staple food of civilizations in South and Central America like the Incas and Aztecs from the tenth to eleventh centuries AD. The organic cultivation of pseudocereals has recently become very popular due to the high nutritional value of the grains.

These crops have a wide variety of genotypes since they were used and acclimated to different habitats; as a result, their growing introductions into new ecosystems differ in how their genetic potential for production is expressed and how long it takes for crops to attain full maturity. The principal constraints on production in Europe are the absence of ice during the growing season and the required temperatures for active photosynthesis, with the exception of buckwheat, in which the genotype-specific requirements for short days means that temperatures above 24 °C stop pollination and stunt growth. The key advantages include the ability of amaranth and quinoa to withstand arid circumstances, soil pH, and salinity, as well as the C_4 photosynthesis route of amaranth [10].

Buckwheat is very tolerant of weeds, possibly as a result of allelopathy and its rapid early growth [11]. With the exception of emerging plants being attacked by nematodes, buckwheat does not suffer from disease or insect attack that could result in significant economic losses. Quinoa and amaranths shouldn't need any significant pest and disease management methods either. However, a number of polyphagous pest species, such as *Agrotis ipsilon* Hufnagel of the nuctuid family of Lepidoptera, may result in financial losses. *Eursacca melanocampta* Meyrick and *Eursacca quinoae* Povolyny are among the common pests unique to the Andes that can reduce quinoa yield by up to 50% [12] (Table 15.1).

15.4 Why are Quality Management Systems Needed for Pseudocereals?

Despite the prospective benefits of the adequate use of these neglected grain crops, a number of barriers, including a dearth of research into these crops, prevent their widespread integration into food systems and breeding programmes. Regardless of

Table 15.1 Environmental Requirements and Cultivation Characteristics for Pseudocereals

Requirements	Buckwheat	Quinoa	Amaranth
Growth season (days)	110	125–240	105–160
Minimum germination temperature (°C)	7	5–7	12
Frost tolerance (°C)	−1.3 to −2.9	−3 to −15	0
Assimilation stop temperature (°C)	10	9	16
Water-use efficiency (kg water kg^{-1} of dry plant matter)	510–600	400	200–335

the neglected grain in issue, these elements, which range from agricultural (growing area, yield capacity), technological (trait development), societal (knowledge transmission), and economic issues, share striking commonalities. These underused grain crops' agronomic potential is largely unexplored. Quinoa, teff, chia, and amaranth are examples of plants grown in countries that do not profit from the high-capital agriculture typical of the production of major cereals. As a result, we are only able to estimate these under-utilised crops' potential in comparison with major staple cereals because much of the information we have about their yield and quality comes from low-input systems. For example, canary seed, buckwheat, and broomcorn millet are under-utilised grains that would profit from high-input agriculture yet are frequently produced as a supplementary crop in place of damaged crops of other species or as a quick replacement crop for summer fallow. While being a logical choice for a producer, the short (and substandard) growing season allocated to these grains makes it difficult to correlate their output and compare them with those of their counterparts in the staple cereals family. This method of use may be reflected in the small acreage of planting space devoted to under-utilised grains, which is significantly less than that of the major cereal crops. Yet, during the past three decades, quinoa production has gradually increased [13].

For several under-utilised (pseudo)cereal crops, genetic constraints exist. For example, buckwheat shows self-incompatibility and is naturally cross-pollinated. As a consequence, creating buckwheat lines which are self-compatible is essential for production and trait enhancement. Furthermore, because pipelines for mutation and transformation have not been built or need to be optimised, breeding for these grain crops is mostly dependent on exploiting natural variability. For these under-utilised grains, there is currently little or no overlap between genomics and breeding. Overall, the lack of systematic breeding efforts aimed at increasing the use of under-utilised grain crops in high-input agriculture systems is now a constraint.

Most of the under-utilised grains mentioned here are largely bred and cultivated by local communities, but this knowledge is frequently prevented from spreading to the larger global community of farmers. Amaranth, quinoa, chia, and even buckwheat are examples of foods where both conventional wisdom and scientific research are progressively making their way over linguistic barriers into popularly spoken languages in international science and agriculture. Except for quinoa, the aforementioned under-used grains haven't received much media or commercial attention to date. Sponsorship in a farm-to-fork system for under-used grains is fruitless without a significant marketing push. In this sense, quinoa is an illustration of an advertising triumph. Quinoa's global increase in production was facilitated by the integration of producers into the world market, which was a result of high-profile public figures endorsing it as a health-promoting grain. To emerge from obscurity and gain a firmer footing in the global market, other under-valued grains would need to find a way to get beyond such marketing obstacles [14].

15.5 Quality Management Systems for Pseudocereals

Pseudocereal quality management systems are shown in Figure 15.1.

15.5.1 Position in Rotation

Every pseudocereal should not be planted repeatedly in the same field, with rotations of no less than three years; 5–6 years is recommended in organic cultivation

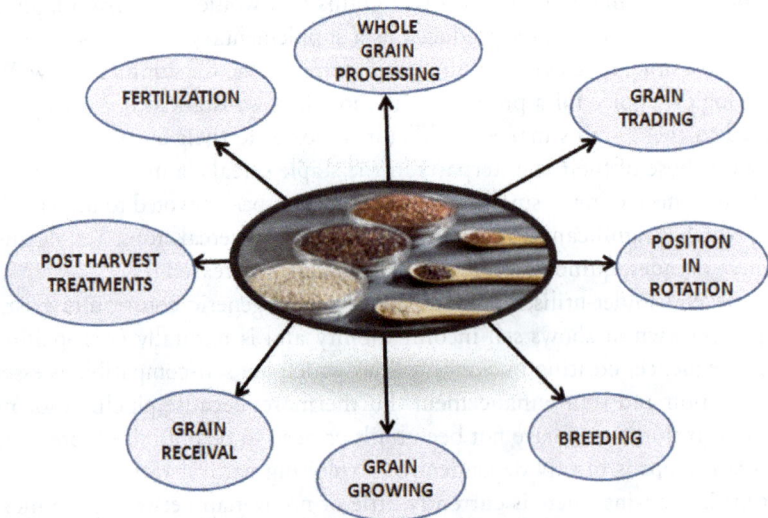

Figure 15.1   Quality management systems in pseudocereals.

due to the aforementioned plant diseases, potential polyphagous pathogens (and pests) that could arise, and an increase in weeds. Increased plant mortality and the emergence of various disorders, particularly root infections by *Gaeumannomyces* sp., are other consequences of inappropriate buckwheat crop rotation. Unlike amaranth, quinoa seeds exhibit exhibit dormancy and will sprout when the conditions are right; in the undomesticated form, seeds may spend up to three years in the soil before sprouting.

The residual accessible nitrogen required by amaranth (after harvesting and/or extra fertilisation with bio-fertiliser) is higher than for buckwheat because of lodging [15]. Cereals, perennial and grain legumes are suitable crops prior to pseudocereals in the rotation. Due to its strong competitive ability against dicotyledonous and monocotyledonous weeds, including the problematic couch grass, buckwheat seems to be an ideal crop prior to fibre flax. Given its short growing season, buckwheat can be utilised as a fallow/green manure/cover crop after earlier harvesting of previous crops (such as early potatoes, early harvested barley, and early combinations of grains and legumes for green feeding). For such prior crops to successfully produce organic cash grains and pseudocereals, competing weeds must be eliminated. Buckwheat can create allelophatic chemicals that could restrict weed growth whether it is planted as a crop, a cover crop, or as a green manure crop and can be used as a useful rotation crop for controlling specialist pests and diseases as it belongs to a family (Polygonaceae) to which no major crop belongs [11].

15.5.2 Breeding

The plant breeder is the backbone of the process that ultimately leads right to the customer and must be influenced by the consumer in order to set the agenda for grain quality. The same consumer can be involved in a market in a foreign land, so the breeder needs to get to know the quality parameters of farmers in that market. The agronomist is also in charge of creating novel varieties that meet the requisites of the grain grower, who has particular agricultural needs to maximise grain yields and needs for grain quality to guarantee good returns for the harvested grain.

Goals of genetic improvement of grain genotypes:

1. Increasing grain output;
2. Removing obstacles posed by biotic and abiotic stressors;
3. Enhancing the grain's processing and end-use quality.

In the early phases of the breeding procedure, as opposed to the ultimate stage, the breeder uses different management strategies to achieve the third and final goal (grain quality). The initial goal of management is to determine and eliminate strains that are unlikely to fulfil the grade standards when there are a lot of lines to be checked. By deleting them at an early stage, resources can be conserved for the superior lines because they won't be spread needlessly. Tests that help in recognizing DNA markers and specific proteins are particularly beneficial at this point because they focus on gene-related problems [16].

15.5.3 Grain Growing

The roles of the seed and grain farmers come in after the breeder.
    Factors affecting grain growing procedure:

1. The seed being sown is of the right variety;
2. The seed being sown is free of disease and weed seeds.

Due to the requirement for maximization of financial returns, which are expressed as a combination of grain volume and market value, controlling the grain quality for the farmer interacts with grain output. Thus, the association between growth environment, management of grain cultivation by the farmer and variety adds up to become an interaction. A crucial supervisory duty is placed on the pure-seed supplier to deliver seed of the right variety that is devoid of contaminated kernels and seed-borne diseases. At the time of planting, when management duty shifts to the farmer, thorough evaluations of purity and identification must be done by the farmer as a guarantee of the features of the grade. The grade of the grain in the harvested grain will be based on seed quality, farm management, and environmental factors. There are now numerous online systems available to help with real-time farm controls in the direction of specified quality targets, but none of these methods can fully eliminate the uncertainties of climate influences. However, the "precision agriculture" method of farm management presents opportunities for achieving quality as well as quantity of grain [17, 18].

15.5.4 Grain Receipt

1. The crucial step of evaluating the numerous factors of grain quality after it leaves the farm is what will decide the amount to be paid to the farmer. In order to ensure that it is blended with grain of a comparable grade type, this quality assessment also defines how the grain supplied is classified.
2. Harvested grain is often transported directly from a local storage facility. Beyond this point, it is necessary to maintain the quality of the collected grain that has been determined to be of a particular quality grade by preserving its quality grade during storage and transportation.
3. When grain is being harvested, there is a significant potential to control grain quality—providing prompt access to sufficient analytical data. The precedence is usually to deliver the grains into silos in advance so that rains do not damage quality, rather than spending valuable time making sure the grain standard is correctly measured to allow the independent binning of kernels of varying types. This is because harvest time is busy for everyone involved.

Despite the time restrictions for evaluating grain quality prior to deliveries being combined in a single storage silo, it is still possible to keep small subsamples of each batch that goes into each silo so that it can later be thoroughly analysed and provide details about the overall grain quality [19].

15.5.5 Fertilization

The nutritional intake of pseudocereals ranges from minimal to very high. In the organic farming method, residual organic fertilisers of the previous plant or bio-manure applied in the fall season power appropriate yields. Low yields are caused by the little amount of nutrients that are available, with the exception of certain heavy metals in amaranth (a plant used for heavy metal remediation of soil) and phosphorus in buckwheat, whose productivity may be affected by the acidification of the rhizosphere in comparison with spring wheat [20, 21]. Due to potential lodging issues, nitrogen manure fertilisation of buckwheat shouuld be applied to the previous crop in the crop rotation cycle [15].

15.5.6 Harvesting and Post-harvest Requisites

In contrast to cereals, harvesting, postharvest technology, and food manufacturing of the various pseudocereals differ according to botanical differences, with the exception of buckwheat's sowing, harvesting, and milling machinery [20]. These are a few important parameters for the harvesting of pseudocereals:

1. The optimal harvesting season for amaranth can be identified by the yellow colour of the leaves and the dry kernels [15].
2. Buckwheat can be harvested when around 75% of the seeds have reached maturation.
3. Quinoa is harvested when the crops start thinning, becoming more yellow, and the leaves wither off the plants.

Following harvesting, which can be accomplished using specialised cereal or clover threshers, the seed is immediately dried to a moisture content of 12% and put into channelled ground storage containers, which need to be free of moisture and well ventilated.

15.5.7 Whole Grain Processing

A fundamental element in deciding quality-based classification and management is the suitability of grain for processing. When the grain is delivered for milling, malting, or any other type of manufacturing, quality assessment may once again take precedence unless individual preservation can be assured. The information so acquired will dictate the way a particular grain batch may be used, managed, and preserved in the series of processes resulting in the final product that is presented for selling to the consumer. It is still possible to modify the processing procedures to make up for inadequate grain quality when it is received [19].

15.5.8 Grain Trading

The farmer must give particular consideration to the grain trade stage. The grower has to decide whether to aim for maximum grain output, regardless of poor quality and lower price, or whether to aim for a greater cost for the plant (but reduced output), based on an amalgamation of suitable variety, superior protein percentage, and freedom from flaws and contaminants. This decision is made before the crop is even sown.

If direct delivery to the processing factory is intended, these latter parts of grain quality assessment must be made at the time of delivery. The installation of storage facilities on the farm could be a substitute management strategy. This tactic enables the grower to delay trading the grain in the hopes of improved sale circumstances [22].

15.6 Conclusion

Although it is unlikely that the under-utilised grains will become a significant part of the food supply in the time ahead, increasing their area cultivated and their production may create a nutrient symbiosis with the primary cereals. If money is invested in the genetics and marketing of these under-utilised crops, climates that aren't suitable for growing staple crops, such as parched and dry, semi-desert lands, might be used to increase food growth, using pseudocereals. This potential for wider cultivation is supported by the range of photorespirating traits in the underexplored grains, with the presence of C_4 species eliminating the requirement for intensive efforts in gene changes to improve net photosynthesis like those carried out in rice.

Superior varieties that are well adapted to different climates can be created for the underexploited grain crops by using the accessible germplasm collections and developing genetic resources. The evolution of novel methods, such as mechanical harvesting, food technology, and postharvest preservation will focus on clear approaches for exploitation of under-utilised grains that can be traced through tactical crop production. Target traits include lodging resistance and stress tolerance, seed size, seed shattering, and grain breaking. The execution of these breeding techniques also depends on making cutting-edge breeding techniques and unified crop breeding policies available to domestic breeders and local growers. Overall, in addition to biofortifying the main cereal grains, better dietary utilisation of under-utilised grains will have a significant impact on reducing the hunger crisis.

15.7 Future Recommendations

Before we can achieve the intended results regarding pseudocereals, there is still much work to be done because the economic potential of these foods is not fully understood. Pseudocereals can help improve the quality of life by adding antioxidants that reduce the effects of oxidative stress on the body, which in turn improves life quality and increases life expectancy. Pseudocereals can be added to our regular diets either on their own or in conjunction with true cereals. There is enormous untapped potential in the industrial approach to pseudocereals, which demands attention. Even in advanced nations with predominantly caloric-dense diets, value-added products can be created on an industrial scale, and the market may be established to address nutrient-related malnutrition.

Coordination of efforts at the local, regional, and global levels, as well as via a multi-stakeholder strategy, is necessary to realize the full potential of pseudocereals. This endeavour must be conducted in accordance with the global United Nations sustainable development goals (SDGs) and must take into account the sustainable building of supply chains, especially for pseudocereals and their products, as well as the enhancement of the local population's sustainable standard of living. For these new super foods, there are successful approaches for introducing pseudocereals into national, regional, and international markets. Additionally, there are successful examples of recently released food products in Europe. For the continued effective production and dissemination of more food products, germplasm preservation has to be taken into account to sustain biodiversity.

References

1. Hoyle, David. *Quality Management Essentials*. Routledge, 2007.
2. Will, M., Food, D. Guenther. Quality and safety standards as required by EU Law and the private industry with special reference to MEDA countries' exports of fresh and processed fruits and vegetables, herbs and spices. *A Practitioners' Reference* 2nd edition. GTZ – Division 45, 2007.
3. WHO. Safety aspects of genetically modified foods of plant origin, A report of joint FAO/WHO experts consultation on foods derived from biotechnology. Geneva, 29 May–2 June 2000.
4. Oloo, J.E.O. Food safety and quality management in Kenya: An overview of the roles played by various stakeholders. *African Journal of Food, Agriculture, Nutrition and Development*, 2010, 10(11), 4379–4397.
5. Dora, Manoj, Kumar, Maneesh, Van Goubergen, Dirk, Molnar, Adrienn, Gellynck, Xavier. Food quality management system: Reviewing assessment strategies and a feasibility study for European food small and medium-sized enterprises. *Food Control*, 2013;31(2), 607–616, ISSN 0956-7135, doi: 10.1016/j.foodcont.2012.12.006.
6. Early, Ralph. *Guide to Quality Management Systems for the Food Industry*. Springer Science & Business Media, New york, 2012.

7. Constantinescu, Pop Cristina Gabriela, Huțanu, G. Study on the impact of quality management systems on food safety case study: Expanded cereals. In *Modern Technologies in the Food Industry*, Institutional Repository of the Technical University of Moldova, Chișinău, 2016, (pp. 170–175).

8. Ciudad-Mulero, M., Fernandez-Ruiz, V., Matallana-Gonzalez, M.C., Morales, P. Dietary fiber sources and human benefits: The case study of cereal and pseudocereals. *Advances in Food and Nutrition Research*, Academic Press, Cambridge, MA, 2019, 83–134.

9. Morales, Diego, Miguel, Marta, Garcés-Rimón, Marta. Pseudocereals: A novel source of biologically active peptides. *Critical Reviews in Food Science and Nutrition*, 2021;61(9), 1537–1544, doi: 10.1080/10408398.2020.1761774.

10. Kaul, H.P. et al. The suitability of amaranth genotypes for grain and fodder use in Cen- tral Europe. *Die Bodenkultur*, 1996;47, 173–181.

11. Iqbal, Z. et al. Allelopathic activity of buckwheat: Isolation and characterization of phenolics. *Weed Science*, 2003;51(5), 657–662.

12. Rasmussen, C., Lagnaoui, A., Esbjerg, P. Advances in the knowledge of quinoa pests. *Food Reviews International*, 2003;19(1–2), 61–75.

13. Ueno, M., Yasui, Y., Aii, J., Matsui, K., Sato, S., Ota, T. 2016. Genetic analyses of the heteromorphic self-incompatibility (S) locus in buckwheat. In Zhou, M., Kreft, I., Woo, S.-H., Chrungoo, N., Wieslander, G. (Eds.), *Molecular Breeding and Nutritional Aspects of Buckwheat*. Academic Press, Cambridge, MA, 411–421.

14. Bazile, D., Jacobsen, S.-E., Verniau, A. The global expansion of quinoa: Trends and limits. *Frontiers in Plant Science*, 2016;7, 622, doi: 10.3389/fpls.2016.00622.

15. Bavec, F., Bavec, M. *Organic Production and Use of Alternative Crops*. Taylor and Francis, Boca Raton, New York, London, 250, 2006.

16. Mauria, S. DUS testing of crop varieties—A synthesis on the subject of new PVP-opting countries'. *Plant Varieties and Seeds*, 2000;13, 69–90.

17. Wrigley, C.W. Precision agriculture—A means of improving grain quality? *Cereal Foods World*, 2005;50, 143–144.

18. Gelinas, P., David, C. Organic grain production and food processing. In Wrigley, C., Corke, H., Seetharaman, K., Faubion, J. (Eds.), *Encyclopedia of Food Grains*, vol. 4. Elsevier Ltd, Oxford, 154–161, 2016.

19. Godber, J.S. *Oil from Rice and Maize*, 453–457, 2016.

20. Zhu, Y.G. et al. Buckwheat (*Fagopyrum esculentum* Moench) has high capacity to take up phosphorus (P) from a calcium (Ca)-bound source. *Plant and Soil*, 2002;239(1), 1–8.

21. Bekkering Cody, S., Li, Tian. Thinking outside of the cereal box: Breeding underutilized (pseudo) cereals for improved human nutrition. *Frontiers in Genetics*, 2019;10, doi: 10.3389/fgene.2019.01289, ISSN=1664-8021https://www.frontiersin.org/articles/10.3389/fgene.2019.01289.

22. USDA, F.S.A. Warehouses licensed under the U.S. Warehouse Act: As of December 31, 2014. USDA-FSA, Washington, DC, 2008. http://www.fsa.usda.gov/Internet/FSA_ File/whselst2014.pdf.

Index

329

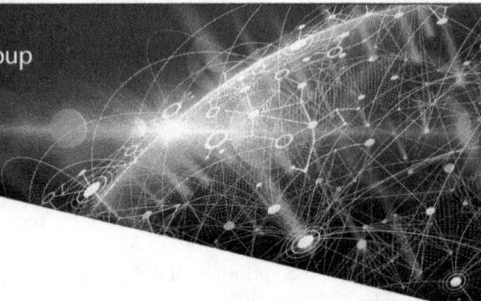

For Product Safety Concerns and Information please contact our EU
representative GPSR@taylorandfrancis.com
Taylor & Francis Verlag GmbH, Kaufingerstraße 24, 80331 München, Germany

www.ingramcontent.com/pod-product-compliance
Lightning Source LLC
Chambersburg PA
CBHW060804220326
41598CB00022B/2538